Advances in Intelligent Systems and Computing

Volume 853

The series "Advances in Intelligent Systems and Computing" contains publications on theory, applications, and design methods of Intelligent Systems and Intelligent Computing. Virtually all disciplines such as engineering, natural sciences, computer and information science, ICT, economics, business, e-commerce, environment, healthcare, life science are covered. The list of topics spans all the areas of modern intelligent systems and computing such as: computational intelligence, soft computing including neural networks, fuzzy systems, evolutionary computing and the fusion of these paradigms, social intelligence, ambient intelligence, computational neuroscience, artificial life, virtual worlds and society, cognitive science and systems, Perception and Vision, DNA and immune based systems, self-organizing and adaptive systems, e-Learning and teaching, human-centered and human-centric computing, recommender systems, intelligent control, robotics and mechatronics including human-machine teaming, knowledge-based paradigms, learning paradigms, machine ethics, intelligent data analysis, knowledge management, intelligent agents, intelligent decision making and support, intelligent network security, trust management, interactive entertainment, Web intelligence and multimedia.

The publications within "Advances in Intelligent Systems and Computing" are primarily proceedings of important conferences, symposia and congresses. They cover significant recent developments in the field, both of a foundational and applicable character. An important characteristic feature of the series is the short publication time and world-wide distribution. This permits a rapid and broad dissemination of research results.

More information about this series at http://www.springer.com/series/11156

Jerzy Świątek · Leszek Borzemski
Zofia Wilimowska
Editors

Information Systems Architecture and Technology: Proceedings of 39th International Conference on Information Systems Architecture and Technology – ISAT 2018

Part II

Editors
Jerzy Świątek
Faculty of Computer Science
and Management
Wrocław University of Science
and Technology
Wrocław, Poland

Leszek Borzemski
Faculty of Computer Science
and Management
Wrocław University of Science
and Technology
Wrocław, Poland

Zofia Wilimowska
University of Applied Sciences in Nysa
Nysa, Poland

ISSN 2194-5357 ISSN 2194-5365 (electronic)
Advances in Intelligent Systems and Computing
ISBN 978-3-319-99995-1 ISBN 978-3-319-99996-8 (eBook)
https://doi.org/10.1007/978-3-319-99996-8

Library of Congress Control Number: 2018952643

This Springer imprint is published by the registered company Springer Nature Switzerland AG
The registered company address is: Gewerbestrasse 11, 6330 Cham, Switzerland

Preface

Variability of the environment increases the risk of the business activity. Dynamic development of the IT technologies creates the possibility of using them in the dynamic management process modeling and decision-making processes supporting. In today's information-driven economy, companies uncover the most opportunities. Contemporary organizations seem to be knowledge-based organizations, and in connection with that information becomes the most critical resource. Knowledge management is the process through which organizations generate value from their intellectual and knowledge-based assets. It consists of the scope of strategies and practices used in corporations to explore, represent, and distribute knowledge. It is a management *philosophy,* which combines good practice in purposeful information management with a culture of organizational learning, to improve business performance. An improvement of the decision-making process is possible to be assured by the analytical process supporting. Applying some analytical techniques such as computer simulation, expert systems, genetic algorithms can improve the quality of managerial information. Combining analytical techniques and building computer hybrids give synergic effects—additional functionality—which makes managerial decision process better. Different technologies can help in accomplishing the managerial decision process, but no one is in favor of information technologies, which offer differentiable advantages. Information technologies take place a significant role in this area. A computer is a useful machine in making managers' work more comfortable. However, we have to remember that the computer can become a tool only, but it cannot make the decisions. You can not build computers that replace the human mind. Computers can collect, select information, process it and create statistics, but decisions must be made by managers based on their experience and taking into account computer use. Different technologies can help in accomplishing the managerial decision process, but no one like information technologies, which offer differentiable advantages.

Computer science and computer systems, on the one hand, develop in advance of current applications, and on the other hand, keep up with new areas of application. In today's all-encompassing cyber world, nobody knows who motivates. Hence, there is a need to deal with the world of computers from both points of view.

In our conference, we try to maintain a balance between both ways of development. In particular, we are trying to get a new added value that can flow from the connection of the problems of two worlds: the world of computers and the world of management. Hence, there are two paths in the conference, namely computer science and management science.

This three-volume set of books includes the proceedings of the 2018 39th International Conference Information Systems Architecture and Technology (ISAT), or ISAT 2018 for short, held on September 16–18, 2018, in Nysa, Poland. The conference was organized by the Department of Computer Science and Department of Management Systems, Faculty of Computer Science and Management, Wrocław University of Science and Technology, Poland, and University of Applied Sciences in Nysa, Poland.

The International Conference Information Systems Architecture has been organized by the Wrocław University of Science and Technology from the seventies of the last century. The purpose of the ISAT is to discuss a state-of-the-art of information systems concepts and applications as well as architectures and technologies supporting contemporary information systems. The aim is also to consider an impact of knowledge, information, computing, and communication technologies on managing of the organization scope of functionality as well as on enterprise information systems design, implementation, and maintenance processes taking into account various methodological, technological, and technical aspects. It is also devoted to information systems concepts and applications supporting the exchange of goods and services by using different business models and exploiting opportunities offered by Internet-based electronic business and commerce solutions.

ISAT is a forum for specific disciplinary research, as well as on multi-disciplinary studies to present original contributions and to discuss different subjects of today's information systems planning, designing, development, and implementation.

The event is addressed to the scientific community, people involved in a variety of topics related to information, management, computer and communication systems, and people involved in the development of business information systems and business computer applications. ISAT is also devoted as a forum for the presentation of scientific contributions prepared by MSc. and Ph.D. students. Business, Commercial, and Industry participants are welcome.

This year, we received 213 papers from 34 countries. The papers included in the three proceedings volumes have been subject to a thoroughgoing review process by highly qualified peer reviewers. The final acceptance rate was 49%. Program Chairs selected 105 best papers for oral presentation and publication in the 39th International Conference Information Systems Architecture and Technology 2018 proceedings.

The papers have been grouped into three volumes:

Part I—discoursing about essential topics of information technology including, but not limited to, computer systems security, computer network architectures, distributed computer systems, quality of service, cloud computing and high-performance computing, human–computer interface, multimedia systems, big data, knowledge discovery and data mining, software engineering, e-business systems, web design, optimization and performance, Internet of things, mobile systems and applications.

Part II—addressing topics including, but not limited to, model-based project and decision support, pattern recognition and image processing algorithms, production planning and management systems, big data analysis, knowledge discovery and knowledge-based decision support and artificial intelligence methods and algorithms.

Part III—is gain to address very hot topics in the field of today's various computer-based applications—is devoted to information systems concepts and applications supporting the managerial decisions by using different business models and exploiting opportunities offered by IT systems. It is dealing with topics including, but not limited to, knowledge-based management, modeling of financial and investment decisions, modeling of managerial decisions, organization and management, project management, risk management, small business management, software tools for production, theories, and models of innovation.

We would like to thank the program committee and external reviewers, essential for reviewing the papers to ensure a high standard of the ISAT 2018 conference and the proceedings. We thank the authors, presenters, and participants of ISAT 2018, without them the conference could not have taken place. Finally, we thank the organizing team for the efforts this and previous years in bringing the conference to a successful conclusion.

September 2018

Leszek Borzemski
Jerzy Świątek
Zofia Wilimowska

ISAT 2018 Conference Organization

General Chair

Zofia Wilimowska, Poland

Program Co-chairs

Leszek Borzemski, Poland
Jerzy Świątek, Poland
Zofia Wilimowska, Poland

Local Organizing Committee

Zofia Wilimowska (Chair)
Leszek Borzemski (Co-chair)
Jerzy Świątek (Co-chair)
Mariusz Fraś (Conference Secretary, Website Support)
Arkadiusz Górski (Technical Editor)
Anna Kamińska (Technical Secretary)
Ziemowit Nowak (Technical Support)
Kamil Nowak (Website Coordinator)
Danuta Seretna-Sałamaj (Technical Secretary)

International Program Committee

Zofia Wilimowska (Chair), Poland
Jerzy Świątek (Co-chair), Poland
Leszek Borzemski (Co-chair), Poland
Witold Abramowicz, Poland
Dhiya Al-Jumeily, UK
Iosif Androulidakis, Greece

Patricia Anthony, New Zealand
Zbigniew Banaszak, Poland
Elena N. Benderskaya, Russia
Janos Botzheim, Japan
Djallel E. Boubiche, Algeria
Patrice Boursier, France
Anna Burduk, Poland
Andrii Buriachenko, Ukraine
Udo Buscher, Germany
Wojciech Cellary, Poland
Haruna Chiroma, Malaysia
Edward Chlebus, Poland
Gloria Cerasela Crisan, Romania
Marilia Curado, Portugal
Czesław Daniłowicz, Poland
Zhaohong Deng, China
Małgorzata Dolińska, Poland
Ewa Dudek-Dyduch, Poland
Milan Edl, Czech Republic
El-Sayed M. El-Alfy, Saudi Arabia
Peter Frankovsky, Slovakia
Mariusz Fraś, Poland
Naoki Fukuta, Japan
Bogdan Gabryś, UK
Piotr Gawkowski, Poland
Arkadiusz Górski, Poland
Manuel Graña, Spain
Wiesław M. Grudewski, Poland
Katsuhiro Honda, Japan
Marian Hopej, Poland
Zbigniew Huzar, Poland
Natthakan Iam-On, Thailand
Biju Issac, UK
Arun Iyengar, USA
Jürgen Jasperneite, Germany
Janusz Kacprzyk, Poland
Henryk Kaproń, Poland
Yury Y. Korolev, Belarus
Yannis L. Karnavas, Greece
Ryszard Knosala, Poland
Zdzisław Kowalczuk, Poland
Lumír Kulhanek, Czech Republic
Binod Kumar, India
Jan Kwiatkowski, Poland
Antonio Latorre, Spain
Radim Lenort, Czech Republic
Gang Li, Australia
José M. Merigó Lindahl, Chile
Jose M. Luna, Spain
Emilio Luque, Spain
Sofian Maabout, France
Lech Madeyski, Poland
Zbigniew Malara, Poland
Zygmunt Mazur, Poland
Elżbieta Mączyńska, Poland
Pedro Medeiros, Portugal
Toshiro Minami, Japan
Marian Molasy, Poland
Zbigniew Nahorski, Poland
Kazumi Nakamatsu, Japan
Peter Nielsen, Denmark
Tadashi Nomoto, Japan
Cezary Orłowski, Poland
Sandeep Pachpande, India
Michele Pagano, Italy
George A. Papakostas, Greece
Zdzisław Papir, Poland
Marek Pawlak, Poland
Jan Platoš, Czech Republic
Tomasz Popławski, Poland
Edward Radosinski, Poland
Wolfgang Renz, Germany
Dolores I. Rexachs, Spain
José S. Reyes, Spain
Małgorzata Rutkowska, Poland
Leszek Rutkowski, Poland
Abdel-Badeeh M. Salem, Egypt
Sebastian Saniuk, Poland
Joanna Santiago, Portugal
Habib Shah, Malaysia
J. N. Shah, India
Jeng Shyang, Taiwan
Anna Sikora, Spain
Marcin Sikorski, Poland
Małgorzata Sterna, Poland
Janusz Stokłosa, Poland
Remo Suppi, Spain
Edward Szczerbicki, Australia

Eugeniusz Toczyłowski, Poland
Elpida Tzafestas, Greece
José R. Villar, Spain
Bay Vo, Vietnam
Hongzhi Wang, China
Leon S. I. Wang, Taiwan
Junzo Watada, Japan
Eduardo A. Durazo Watanabe, India
Jan Werewka, Poland
Thomas Wielicki, USA
Bernd Wolfinger, Germany
Józef Woźniak, Poland
Roman Wyrzykowski, Poland
Yue Xiao-Guang, Hong Kong
Jaroslav Zendulka, Czech Republic
Bernard Ženko, Slovenia

ISAT 2018 Reviewers

Hamid Al-Asadi, Iraq
Patricia Anthony, New Zealand
S. Balakrishnan, India
Zbigniew Antoni Banaszak, Poland
Piotr Bernat, Poland
Agnieszka Bieńkowska, Poland
Krzysztof Billewicz, Poland
Grzegorz Bocewicz, Poland
Leszek Borzemski, Poland
Janos Botzheim, Hungary
Piotr Bródka, Poland
Krzysztof Brzostkowski, Poland
Anna Burduk, Poland
Udo Buscher, Germany
Wojciech Cellary, Poland
Haruna Chiroma, Malaysia
Witold Chmielarz, Poland
Grzegorz Chodak, Poland
Andrzej Chuchmała, Poland
Piotr Chwastyk, Poland
Anela Čolak, Bosnia and Herzegovina
Gloria Cerasela Crisan, Romania
Anna Czarnecka, Poland
Mariusz Czekała, Poland
Y. Daradkeh, Saudi Arabia
Grzegorz Debita, Poland
Anna Dobrowolska, Poland
Maciej Drwal, Poland
Ewa Dudek-Dyduch, Poland
Jarosław Drapała, Poland
Tadeusz Dudycz, Poland
Grzegorz Filcek, Poland
Mariusz Fraś, Poland
Naoki Fukuta, Japan
Piotr Gawkowski, Poland
Dariusz Gąsior, Poland
Arkadiusz Górski, Poland
Jerzy Grobelny, Poland
Krzysztof Grochla, Poland
Bogumila Hnatkowska, Poland
Katsuhiro Honda, Japan
Zbigniew Huzar, Poland
Biju Issac, UK
Jerzy Józefczyk, Poland
Ireneusz Jóźwiak, Poland
Krzysztof Juszczyszyn, Poland
Tetiana Viktorivna Kalashnikova, Ukraine
Anna Kamińska-Chuchmała, Poland
Yannis Karnavas, Greece
Adam Kasperski, Poland
Jerzy Klamka, Poland
Agata Klaus-Rosińska, Poland
Piotr Kosiuczenko, Poland
Zdzisław Kowalczyk, Poland
Grzegorz Kołaczek, Poland
Mariusz Kołosowski, Poland
Kamil Krot, Poland
Dorota Kuchta, Poland
Binod Kumar, India
Jan Kwiatkowski, Poland
Antonio LaTorre, Spain
Arkadiusz Liber, Poland
Marek Lubicz, Poland

Emilio Luque, Spain
Sofian Maabout, France
Lech Madeyski, Poland
Jan Magott, Poland
Zbigniew Malara, Poland
Pedro Medeiros, Portugal
Vojtěch Merunka, Czech Republic
Rafał Michalski, Poland
Bożena Mielczarek, Poland
Vishnu N. Mishra, India
Jolanta Mizera-Pietraszko, Poland
Zbigniew Nahorski, Poland
Binh P. Nguyen, Singapore
Peter Nielsen, Denmark
Cezary Orłowski, Poland
Donat Orski, Poland
Michele Pagano, Italy
Zdzisław Papir, Poland
B. D. Parameshachari, India
Agnieszka Parkitna, Poland
Marek Pawlak, Poland
Jan Platoš, Czech Republic
Dolores Rexachs, Spain
Paweł Rola, Poland
Stefano Rovetta, Italy
Jacek, Piotr Rudnicki, Poland
Małgorzata Rutkowska, Poland
Joanna Santiago, Portugal
José Santos, Spain
Danuta Seretna-Sałamaj, Poland
Anna Sikora, Spain
Marcin Sikorski, Poland
Małgorzata Sterna, Poland
Janusz Stokłosa, Poland
Grażyna Suchacka, Poland
Remo Suppi, Spain
Edward Szczerbicki, Australia
Joanna Szczepańska, Poland
Jerzy Świątek, Poland
Paweł Świątek, Poland
Sebastian Tomczak, Poland
Wojciech Turek, Poland
Elpida Tzafestas, Greece
Kamila Urbańska, Poland
José R. Villar, Spain
Bay Vo, Vietnam
Hongzhi Wang, China
Shyue-Liang Wang, Taiwan, China
Krzysztof Waśko, Poland
Jan Werewka, Poland
Łukasz Wiechetek, Poland
Zofia Wilimowska, Poland
Marek Wilimowski, Poland
Bernd Wolfinger, Germany
Józef Woźniak, Poland
Maciej Artur Zaręba, Poland
Krzysztof Zatwarnicki, Poland
Jaroslav Zendulka, Czech Republic
Bernard Ženko, Slovenia
Andrzej Żołnierek, Poland

ISAT 2018 Keynote Speaker

Professor Dr. Abdel-Badeh Mohamed Salem, Faculty of Science, Ain Shams University, Cairo, Egypt
Topic: Artificial Intelligence Technology in Intelligent Health Informatics

Contents

Model Based Project and Decision Support

Model Order Reduction Adapted to Steel Beams Filled with a Composite Material

Paweł Dunaj(✉), Michał Dolata, and Stefan Berczyński

West Pomeranian University of Technology Szczecin, Szczecin, Poland
{pawel.dunaj,michal.dolata,
stefan.berczynski}@zut.edu.pl

Abstract. In presented paper, an analysis of model order reduction (MOR) techniques applied to steel beams filled with a composite material is presented. This research concerns specific construction solutions used in technological machines. The analyzes concern three reduction methods: Guyan reduction also referred as static condensation, Craig-Bampton reduction and Kammer reduction. These techniques are applied to matrix equations describing steel beams filled with a composite material model, established by the finite element method (FEM). The article contains information about preparation of the full model and model parameters identification process. To verify FEM model quality its results are compared to experimental modal analysis results. The analysis compares and contrasts the MOR techniques by considering the nature of the individual algorithms and analyzing results of numerical example. The comparison of reduced models computational time at subsequent stages have also been made.

Keywords: Model order reduction · Guyan reduction
Craig-Bampton reduction · Kammer reduction · Composite beams

1 Introduction

Despite the significant development of high-performance computing technologies, high-dimensional and multiparametric problems remain difficult to tackle even by advanced simulation methods. Therefore, the methods allowing to reduce the dimensionality of the problem are becoming more and more popular. One of such is Model Order Reduction (MOR), its purpose is to find a low order model (reduced model) to approximate the original large-scale model with high accuracy. Reduced model causes storage memory saving and shortening computation time. It can be used to replace the original model as a component in a larger simulation (e.g. substructuring method) or it might be used as a simplified and hence faster to compute model suitable in real time applications.

MOR has been used in many fields e.g. computational electromagnetics [3, 16], computational fluid dynamics [4, 12], thermal analysis [8], vibrostability analysis [13] and structural dynamics [2, 5, 7, 11, 14]. In structural dynamic, MOR based on Guyan reduction or Craig-Bampton reduction has been extensively use to speed up dynamic simulations, especially solving eigenvalue problem. These two methods can be found

J. Świątek et al. (Eds.): ISAT 2018, AISC 853, pp. 3–13, 2019.
https://doi.org/10.1007/978-3-319-99996-8_1

nowadays in almost any commercial finite element software packages. However, neither Guyan reduction nor Craig-Bampton reduction produces an optimal reduced model [1]. The analysis compares and contrasts the abovementioned techniques with reduction method proposed by Kammer [10], based on modal coordinates.

2 Research Object

The research object was an unconstrained finite element model of a steel beam with square cross-section of 70 × 70 mm, a wall thickness of 3 mm and a length of 1000 mm, filled with a composite material. Such beams are the basic components of a welded machine tool body shown in Fig. 1. Material properties of a composite beam were determined on the basis of static test results which are shown in Table 1.

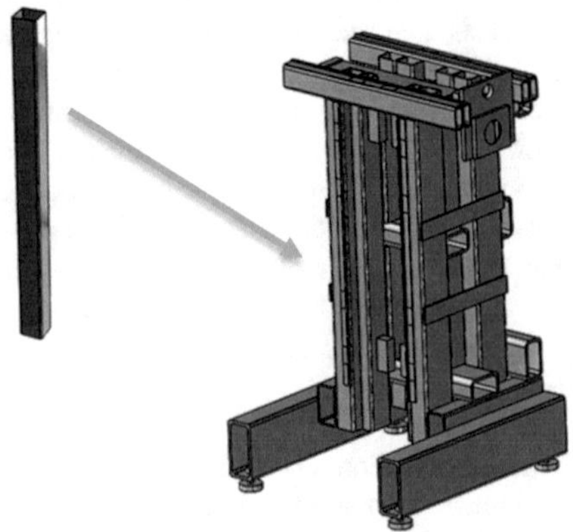

Fig. 1. Steel beam filled with a composite material as a component of the welded machine body

The discretized model shown in Fig. 2 was developed using Midas NFX software. The steel coating and composite filing were discretized using CHEXA, which are 3-D six-sided isoparametric solid element with eight nodes. Contact between steel coating and composite filling was modelled as nodes coincidence. Structured meshing technique was taken to improve the efficiency of the FEM. The uneven division of the grid is dictated by the nature of the connection between the element and the rest of the structure, thus it is possible to use the model in substructuring method. Summarizing, the developed model consists of 2019 degrees of freedom (DOFs), which amounts to the mass [M] and stiffness [K] matrix of the dimensions 2019 × 2019.

For the determined model, the following eigenproblem can be formulated:

$$([\mathrm{K}]-[\lambda_{\mathrm{f}}][\mathrm{M}])[\phi_{\mathrm{f}}]=0 \tag{1}$$

where: $[\lambda_f]$ – full system eigenvalues matrix, $[\phi_f]$ – full system eigenvectors matrix. As a result of solving the formulated eigenproblem a set of eigenvectors (mode shapes) and eigenvalues (natural frequencies) was obtained. In order to obtain reliable results,

Table 1. Material properties

Parameter	Steel	Composite material
Young's modulus	212 ± 5 GPa	16,8 ± 0,2 GPa
Poisson's ratio	0,28 ± 0,03	0,20 ± 0,05
Density	2118 kg/m^3	2118 kg/m^3

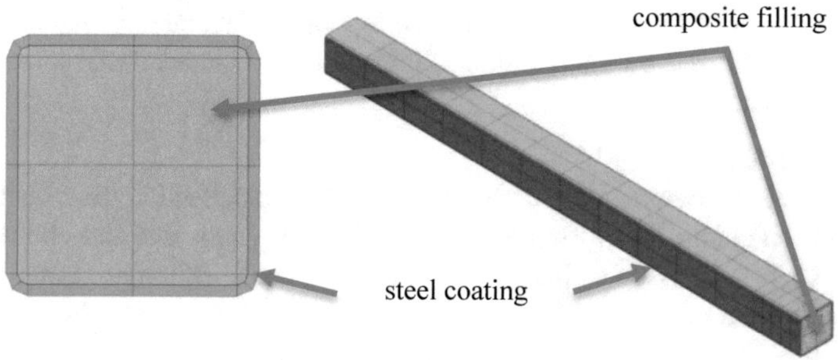

Fig. 2. Discretized model

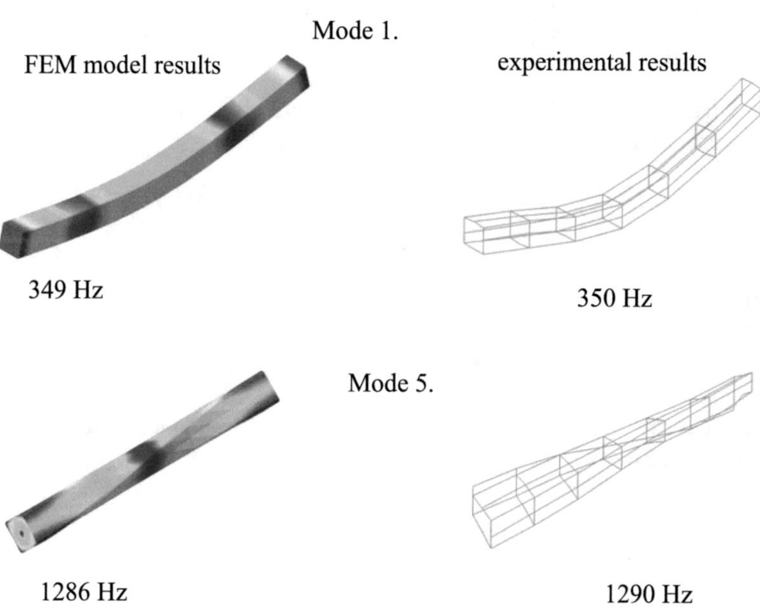

Fig. 3. A comparison of chosen mode shapes

the model was subjected to the experimental identification of mode shapes determined on the basis of impact test. Comparison of exemplary mode shapes was shown in Fig. 3.

3 Model Order Reduction Techniques

In this section three abovementioned reduction methods are presented. The aim is to reduce the number of DOFs in the model while retaining its quality. All of these three methods were analyzed by many authors, following condensed description was made on the basis of [15].

Guyan Reduction

The first reduction method is the static condensation also called Guyan reduction [9]. In this method remaining DOFs (master) are denoted by $\{u_m\}$, and the eliminated ones (slave) by $\{u_s\}$. We assume that forces acting on slave DOFs are equal to 0. The equation of motion is:

$$[M]\{\ddot{u}\} + K\{u\} = \{F\} \tag{2}$$

We can divide the mass and stiffness matrices as follows:

$$\begin{bmatrix} M_{mm} & M_{ms} \\ M_{sm} & M_{ss} \end{bmatrix} \begin{Bmatrix} \ddot{u}_m \\ \ddot{u}_s \end{Bmatrix} + \begin{bmatrix} K_{mm} & K_{ms} \\ K_{sm} & K_{ss} \end{bmatrix} \begin{Bmatrix} u_m \\ u_s \end{Bmatrix} = \begin{Bmatrix} F_m \\ F_s \end{Bmatrix} = \begin{Bmatrix} F_m \\ 0 \end{Bmatrix} \tag{3}$$

Since the inertia loads $[M_{mm}]\ \{\ddot{u}_m\}$ are significantly larger than remaining loads the parts of the mass matrix other than $[M_{mm}]$ can be zeroed.

$$\begin{bmatrix} M_{mm} & 0 \\ 0 & 0 \end{bmatrix} \begin{Bmatrix} \ddot{u}_m \\ \ddot{u}_s \end{Bmatrix} + \begin{bmatrix} K_{mm} & K_{ms} \\ K_{sm} & K_{ss} \end{bmatrix} \begin{Bmatrix} u_m \\ u_s \end{Bmatrix} = \begin{Bmatrix} F_m \\ F_s \end{Bmatrix} = \begin{Bmatrix} F_m \\ 0 \end{Bmatrix} \tag{4}$$

Using equations above we can express $\{u_s\}$ by $\{u_m\}$:

$$\{u_s\} = -[K_{ss}]^{-1}[K_{sm}]\ \{u_m\} = [G_{sm}]\ \{u_m\} \tag{5}$$

Since only stiffness is used we can consider this as a static condensation. The displacement vector $\{u\}$ can be described as:

$$\{u\} = \begin{Bmatrix} u_m \\ u_s \end{Bmatrix} = \begin{bmatrix} I \\ G_{sm} \end{bmatrix} \{u_m\} = [T_{sm}]\{u_m\} \tag{6}$$

where $[T_{sm}]$ is the Guyan transformation matrix. If we write the total kinetic energy equation:

$$T = \frac{1}{2}\{\dot{u}\}^T[M]\{\dot{u}\} = \frac{1}{2}\{\dot{u}_m\}^T[T_{sm}]^T[M][T_{sm}]\{\dot{u}_m\} \tag{7}$$

and the master mass matrix $[\overline{M}_{mm}]$ is:

$$[\overline{M}_{mm}] = [T_{sm}]^T[M][T_{sm}] \tag{8}$$

From the potential energy equation in analogy to the kinetic energy equation the reduced stiffness matrix $[\overline{K}_{mm}]$ is:

$$[\overline{K}_{mm}] = [T_{sm}]^T[K][T_{sm}] \tag{9}$$

The biggest challenge in this method is the process of selecting master nodes. Generally speaking DOFs of large masses should be considered as master ones. There is a guideline for selecting master DOFs:

$$\frac{1}{2\pi}\sqrt{\frac{k_{ii}}{m_{ii}}} \leq 1.5f_{max} \tag{10}$$

where k_{ii} and m_{ii} are diagonal terms (translational and rotational) of stiffness and mass matrices and f_{max} is maximum frequency of interest. At least these DOFs should be selected which does not mean that the results will be similar to the full model. This method is a static one so it gives acceptable results only for rather low frequencies of the system. At higher frequencies neglecting moments of inertia have a strong influence.

Craig-Bampton Reduction

The second method is Craig-Bampton reduction [6]. Craig-Bampton reduction unlike Guyan reduction accounts for both inertia and stiffness making it more accurate. In this method the displacement vector is written on a basis of static modes $[\Phi_s]$ with $\{u_s\} = [I]$ and elastic mode shapes $[\Phi_p]$ with fixed external degrees of freedom $\{u_s\} = \{0\}$ and the eigenvalue problem:

$$([K_{ss}] - \lambda[M_{ss}])\phi = 0 \tag{11}$$

Than $\{u\}$ can be expressed as:

$$\{u\}=[\phi_s]\{u_s\} + [\phi_p]\{\eta_p\} = [\phi_s, \phi_p]\begin{Bmatrix} u_s \\ \eta_p \end{Bmatrix} = [\Psi]\{U\} \tag{12}$$

If inertia effects are assumed to be zero $\{F_m\} = \{0\}$ and boundary DOFs $\{u_s\} = [I]$, static modes can be obtained. The static transformation:

$$\{u\} = \begin{Bmatrix} u_m \\ u_s \end{Bmatrix} = \begin{bmatrix} \phi_{ms} \\ I \end{bmatrix}\{u_s\} = [\phi_s]\{u_s\} \tag{13}$$

If external degrees of freedom are assumed to be fixed $\{x_j\} = 0$ The eigenvalue problem can be expressed as:

$$([K_{mm}] - \langle \lambda_m \rangle [M_{mm}]) [\Phi_{mp}] = \{0\} \tag{14}$$

The internal degrees of freedom are expressed on modal matrix:

$$\{u_m\} = [\Phi_{mp}] \{\eta_p\} \tag{15}$$

The modal transformation:

$$\{u\} = \begin{Bmatrix} u_m \\ u_s \end{Bmatrix} = \begin{bmatrix} \phi_{mp} \\ 0 \end{bmatrix} \{\eta_p\} = [\phi_p] \{\eta_p\} \tag{16}$$

In this method static displacements are enclosed in static modes but dynamic properties are connected with elastic modes. With equal potential and kinetic energies:

$$[\Psi]^T [M][\Psi]\{\ddot{U}\} + [\Psi]^T [K][\Psi]\{U\} = [\Psi]^T \{F(t)\} \tag{17}$$

The following expression can be delivered:

$$[M_{CB}]\{\ddot{U}\} + [K_{CB}]\{U\} = [\Psi]^T \{F\} \tag{18}$$

Craig–Bampton, reduction compensates for the neglected inertia terms by including a set of generalized coordinates. These coordinates represent the amplitudes ratios of mode shapes calculated for the slave structure, with the master DOFs being fixed. Assuming a harmonic solution and that there are no loads acting on the slave DOFs.

As with Guyan reduction, the accuracy of Craig-Bampton reduction depends on the selection of the master DOFs, which affects both the static modes and the eigenmodes of the slave structure. In addition, it should be noted that the accuracy of Craig-Bampton reduction also depends on the choice of eigenmodes, some eigenmodes have a greater impact on the result than others. Generally the more modes specified the better the accuracy at the cost of increased computation time.

Kammer Reduction

In this method [10] the displacement vector $\{u(t)\}$ is projected on the modal matrix $[\Phi]$. The number of remaining mode shapes is much less than the total number of DOFs:

$$\{u(t)\} = [\Phi]\{\eta(t)\} \tag{19}$$

$\{\eta(t)\}$ is a vector of generalized coordinates. The displacement vector $\{u(t)\}$ can be expressed by master DOFs (*m*) and slave (removed) DOFs (s):

$$\begin{Bmatrix} u_m \\ u_s \end{Bmatrix} = \begin{bmatrix} \Phi_m \\ \Phi_s \end{bmatrix} \{\eta\} \tag{20}$$

expressing in $\{u_m\}$:

$$\begin{Bmatrix} u_m \\ u_s \end{Bmatrix} = \begin{bmatrix} I \\ T_{sm} \end{bmatrix} \{u_m\} = [T_{Kammer}]\{u_m\} \tag{21}$$

$$\{u_m\} = [\Phi_m]\{\eta\} \tag{22}$$

Expressing the vector of generalized coordinates $\{\eta\}$ in $\{u_m\}$ is not so obvious because the inverse of $[\Phi_m]$ matrix does not exist. However due to further transformations we get the following expression:

$$[\Phi_m]^{-1} = \left([\Phi_m]^T[\Phi_m]\right)^{-1}[\Phi_m]^T \tag{23}$$

the matrix $\left([\Phi_m]^T[\Phi_m]\right)^{-1}[\Phi_m]^T$ is called the pseudo-inverse matrix of the modal matrix $[\Phi_m]$. By analogy we can obtain the vector of eliminated displacements:

$$\{u_s\} = [\Phi_s]\left([\Phi_m]^T[\Phi_m]\right)^{-1}[\Phi_m]^T\{u_m\} = [T_{sm}]\{u_m\} \tag{24}$$

The displacement vector can be expressed as:

$$\begin{Bmatrix} u_m \\ u_s \end{Bmatrix} = \begin{bmatrix} I \\ [\Phi_s]\left([\Phi_m]^T[\Phi_m]\right)^{-1}[\Phi_m]^T \end{bmatrix} \{u_m\} = \begin{bmatrix} I \\ T_{sm} \end{bmatrix} \{u_m\} = [T_{Kammer}]\{u_m\} \tag{25}$$

The reduced mass and stiffness matrices are:

$$[M_{Kammer}] = [T_{Kammer}]^T[M][T_{Kammer}] \tag{26}$$

$$[K_{Kammer}] = [T_{Kammer}]^T[K][T_{Kammer}] \tag{27}$$

Reduction Process

Due to the fact that in the commercial FEM software the Kammer method is not implemented, the calculations were carried out using the Matlab environment. First, using the Midas NFX preprocessor, geometric model was meshed. Next, on the basis of the defined grid using NeiNastran solver, matrices describing the mass and stiffness properties of the structure were built. The final step was to export the matrices to a.bdf file, using the EXTSEOUT (DMIGBDF) command, then, using a specially prepared script, Matlab matrices were imported, and calculations were carried out i.e. sorting mass and stiffness matrices, reducing the structure and finally solving the eigenproblem. Figure 4 shows model order reduction workflow.

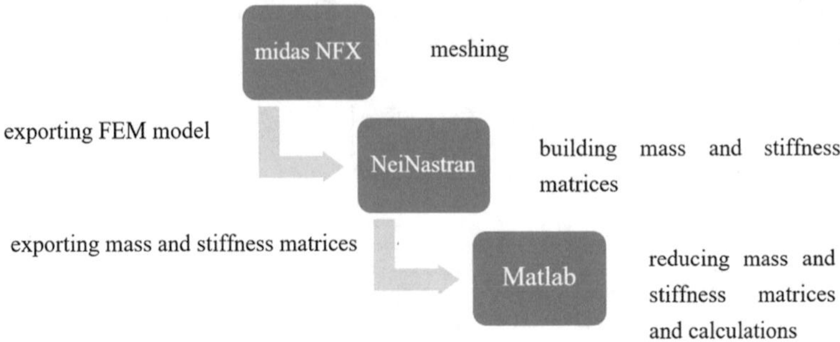

Fig. 4. Model order reduction workflow

4 Results

The established model was subjected to reduction procedure, resulting in reducing matrix dimensions form 2019 × 2019 to 30 × 30. Figure 5 shows nodes related to master DOFs selected in reduction process. Table 2 compares the impact of reduction methods on the accuracy of the computed natural frequencies.

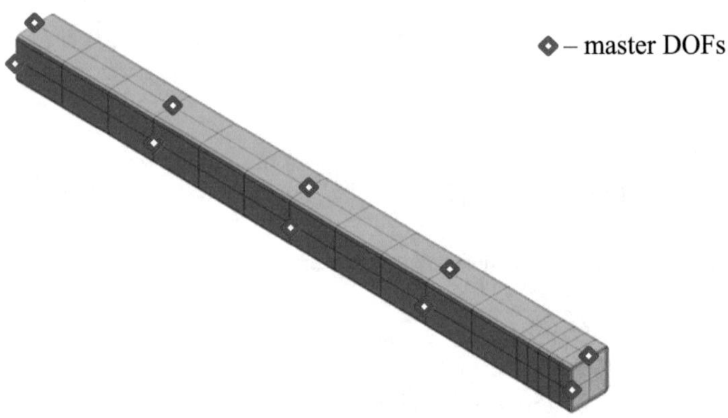

Fig. 5. Master DOFs selected in the reduction process

Analyzing the results contained in Table 3 it can be seen that the biggest differences in eigenfrequencies values appear in the case of Guyan reduction. This method gives acceptable results for first two frequencies, due to its static nature, at higher frequencies larger errors related to the neglected inertia terms appears. Use of Craig–Bampton method gives a better accuracy at higher frequencies, the neglected inertia terms in Guyan reduction are compensated by the use of an additional set of generalized coordinates. The use of Kammer reduction, gives an exact solution due to the fact that the method is based on generalized coordinates.

Table 2. Effect of applied reduction method on flexible mode shapes eigenfrequencies values

Mode number	Full model	Guyan reduction	Craig-Bampton reduction	Kammer reduction
1.	349,58 Hz	351,83 Hz	350,18 Hz	349,58 Hz
2.	349,58 Hz	353,39 Hz	351,03 Hz	349,58 Hz
3.	945,42 Hz	991,75 Hz	956,64 Hz	945,42 Hz
4.	945,42 Hz	1023,58 Hz	964,51 Hz	945,42 Hz
5.	1286,36 Hz	1469,68 Hz	1377,14 Hz	1286,36 Hz
6.	1811,00 Hz	2113,82 Hz	1921,09 Hz	1811,00 Hz
7.	1811,00 Hz	2230,27 Hz	1996,67 Hz	1811,00 Hz
8.	2017,15 Hz	2352,99 Hz	2322,33 Hz	2017,15 Hz

Table 3 presents the calculation times for each reduction method related to the calculations performed on the full model, taking into account the individual stages of the analysis.

Table 3. Comparison of calculation times for individual reduction methods

Algorithm step	Full model	Guyan reduction	Craig-Bampton reduction	Kammer reduction
Sorting DOFs	–	44%	44%	44%
Reducing matrices	–	31%	176%	1%
Solving eigenproblem	100%	0,17%	0,15%	0,2%
Total	100%	75,17%	220,15%	45,2%

Summarizing, it can be seen that in general the accuracy of analyzed reduction methods is satisfactory, and the time needed for solving an eigenproblem for reduced model is noticeably shorter. However, taking into account all stages of reduction process a significant part of the time is intended for sorting the mass and stiffness matrices due to the adopted degrees of freedom division strategy. It should also be noted that the time needed to sort the mass and stiffness matrices, largely depends on the matrices size and sorting algorithm.

In the case of Guyan and Craig-Bampton reduction, the significant computation cost is associated with the matrices reduction stage. It is related to the need to invert the matrices containing slave DOFs. In addition, in the case of Craig-Bampton reduction, to determine transformation matrix, it is necessary to solve the eigenproblem for slave DOFs.

While in the case of Kammer reduction, a pseudo-inverse matrix for the master DOFs is determined, the step of determining the reduced matrix is much shorter than in the case of abovementioned reduction methods.

5 Findings

This paper presents the comparison of reduction process carried out on FEM model of steel beams filled with a composite material, conducted with three methods: Guyan, Craig-Bampton and Kammer. In order to obtain a reliable results, a special script using Matlab environment was developed to compare the calculation times at subsequent stages of reduction process.

Comparing computation times of presented methods, one can conclude that Kammer method is the fastest one. This method gives the best results also in terms of accuracy and it is related to the definition of transformation matrix. Despite these facts the Kammer method is omitted in commercial implementations giving way to Guyan and Craig-Bampton approaches.

The other important thing that came out after considering all pros and cons of reduction methods is that choosing master and slave DOFs is crucial in case of Guyan and Craig-Bampton methods and have a big impact on the results, when analyzing structure composed of two different materials. In the case of Kammer method, the choice master DOFs has no effect on the result due to fact that transformation matrix is based on modal coordinates.

Acknowledgements. This work was funded by EU grant: “Light construction vertical lathe” POIR.04.01.02-00-0078/16.

References

1. Antoulas, A.C.: Approximation of large-scale dynamical systems. Society for Industrial and Applied Mathematics, Siam (2005)
2. Besselink, B., Tabak, U., Lutowska, A., Van De Wouw, N., Nijmeijer, H., Rixen, D.J., Schilders, W.H.A.: A comparison of model reduction techniques from structural dynamics, numerical mathematics and systems and control. J. Sound Vib. **332**(19), 4403–4422 (2013)
3. Bonotto, M., Cenedese, A., Bettini, P.: Krylov subspace methods for model order reduction in computational electromagnetics. IFAC-PapersOnLine **50**(1), 6355–6360 (2017)
4. Chen, G., Li, D., Zhou, Q., Da Ronch, A., Li, Y.: Efficient aeroelastic reduced order model with global structural modifications. Aerosp. Sci. Technol. **76**, 1–13 (2018)
5. Craig, R.R.: Coupling of substructures for dynamic analysis: an overview. In: Proceedings of the 41st AIAA/ASME/ASCE/AHS/ASC Structures, Structural Dynamics, and Materials Conference, Atlanta, USA (2000)
6. Craig, R., Bampton, M.: Coupling of substructures for dynamic analyses. AIAA J. **6**(7), 1313–1319 (1968)
7. Flodén, O., Sandberg, G., Persson, K.: Reduced order modelling of elastomeric vibration isolators in dynamic substructuring. Eng. Struct. **155**, 102–114 (2018)
8. Gouda, M.M., Danaher, S., Underwood, C.P.: Building thermal model reduction using nonlinear constrained optimization. Build. Environ. **37**(12), 1255–1265 (2002)
9. Guyan, R.J.: Reduction of stiffness and mass matrices. AIAA J. **3**(2), 380 (1965)
10. Kammer, D.C.: Test-analysis model development using an exact modal reduction. Int. J. Anal. Exp. Modal Anal. **2**(4), 174–179 (1987)

11. Klerk, D.D., Rixen, D.J., Voormeeren, S.N.: General framework for dynamic substructuring: history, review and classification of techniques. AIAA J. **46**(5), 1169–1181 (2008)
12. Pagliuca, G., Timme, S.: Model reduction for flight dynamics simulations using computational fluid dynamics. Aerosp. Sci. Technol. **69**, 15–26 (2017)
13. Pajor, M., Marchelek, K., Powałka, B.: Method of reducing the number of DOF in the machine tool-cutting process system from the point of view of vibrostability analysis. Modal Anal. **8**(4), 481–492 (2002)
14. Rösner, M., Lammering, R., Friedrich, R.: Dynamic modeling and model order reduction of compliant mechanisms. Precis. Eng. **42**, 85–92 (2015)
15. Wijker, J.J.: Spacecraft structures, pp. 265–280. Springer, Heidelberg (2008)
16. Wittig, T., Schuhmann, R., Weiland, T.: Model order reduction for large systems in computational electromagnetics. Linear Algebra Appl. **415**(2–3), 499–530 (2006)

Case-Based Parametric Analysis: A Method for Design of Tailored Forming Hybrid Material Component

Renan Siqueira(✉), Mehdi Bibani, Iryna Mozgova, and Roland Lachmayer

Leibniz Universität, Hannover, Germany
siqueira@ipeg.uni-hannover.de

Abstract. Between the recent advances in manufacturing engineering stands Tailored Forming, a process chain that produces massive hybrid material components through the use of different forming techniques. The motivation behind such a process is the achievement of higher performance parts, such as lightweight or local integrated functions. Thereby, new restrictions take place in the design of these parts, requiring the implementation of a suitable multi-material design methodology to attend user requirements. One of these new challenges is the design of the joining zone between the two metals, which presents limited controllability during the manufacturing process. With this objective, here is proposed the use of a Case Based Reasoning (CBR) system as design method. For that, a parametric model is created and, through an interface between CAD and finite element systems, a solution space is generated and analyzed, forming the first case-base. A comparison analysis of these results is executed, bringing valuable information for the current research. At the end, a similarity method is implemented in order to propose the most suitable solution among all variations based on specified requirements. With that, this tool will support the user on the creation of new cases and the machine learning process on storing the knowledge.

Keywords: Tailored Forming · Multi-material design · Case Based Reasoning

1 Introduction

The fast development of technologies with new application possibilities requires components with properties more and more specific. That makes the search for mechanical components with higher performance a constant goal of the industry [17]. These properties can be, for example, lighter weight, longer life-circle, stronger materials or higher stiffness. To do so, new technologies in the field of manufacturing process are being developed, so that the common limitations found in the design phase are reduced.

One of these new technologies is Tailored Forming, which presents a process chain for the manufacture of hybrid components [5]. This multi-material design brings, however, a big amount of degrees of freedom, requiring a systematic methodology for its construction, as seen in the works of Kleemann et al. [13] and Brockmoeller et al. [9].

J. Świątek et al. (Eds.): ISAT 2018, AISC 853, pp. 14–28, 2019.
https://doi.org/10.1007/978-3-319-99996-8_2

Not only in the methodological point of view, the design parameters must also be carefully studied and analyzed, in order to explore all the advantages that it brings in the most efficient way.

The objectives of the present study are two: creation of a parametric analysis framework that generates a solution space, using the example of the joining zone geometry of a hybrid shaft; and the coupling of a case-based system in this framework by using the data generated as initial case-base, in order to support future application of Tailored Forming. Since Tailored Forming is a technology still in research, the results here presented deliver valuable information for future design though this new technology.

2 Background Research

2.1 Tailored Forming

Tailored Forming is a new manufacturing technique that allows the construction of hybrid high performance components, being the aim of the Collaborative Research Project (CRC) 1153 established at the Leibniz University of Hannover. This technology consists in a process chain that combines different manufacturing techniques in order to produce functional final parts. The process includes the following sequence: generation of the separated mono-material parts; joining process performed by friction welding, laser welding or compound profile extrusion to generate semi-finished hybrid workpieces; metal forming process through high temperatures, such as cross wedge rolling, forging or impact extrusion; and finally machining and heat treatments are executed to obtain the final piece [3]. Figure 1 shows a diagram with this sequence.

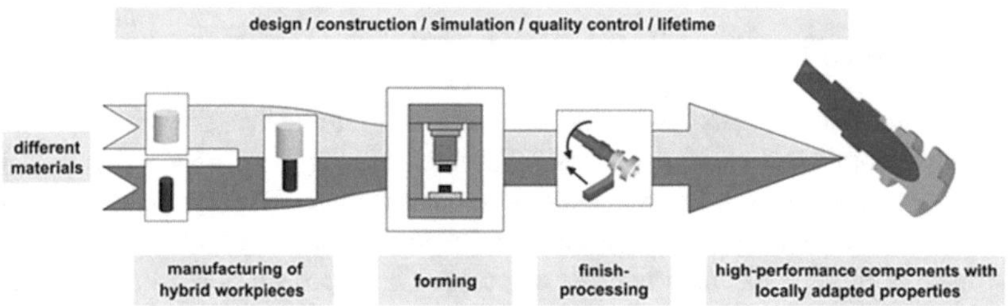

Fig. 1. Process chain of Tailored Forming technology [16].

Motivated by the new possibilities that this technology brings, new design methodologies must be implemented in order to investigate the effects of having two materials in a single part. The use of multiple material increases considerably the degree of complexity in design, due to the large number of distribution possibilities and interfacial geometries. However, this large solution space has many restrictions provided by the manufacturing process described. These geometric constraints must be adjusted according to the technique used.

This creates an iterative process that starts with the concept generation for a component based on the available forming technology, followed by a design process that must generate a solution space. Within this solution space, the best fit design can be chosen based on its ability to deliver user's specifications [14]. As example, we take the design of a shaft, which is one of the demonstrators from the CRC 1153, seen in Fig. 2. This shaft is manufactured through a laser welding of the mono-material workpieces, followed by a process of impact extrusion or cross wedge rolling, finishing with a machining technology step [4].

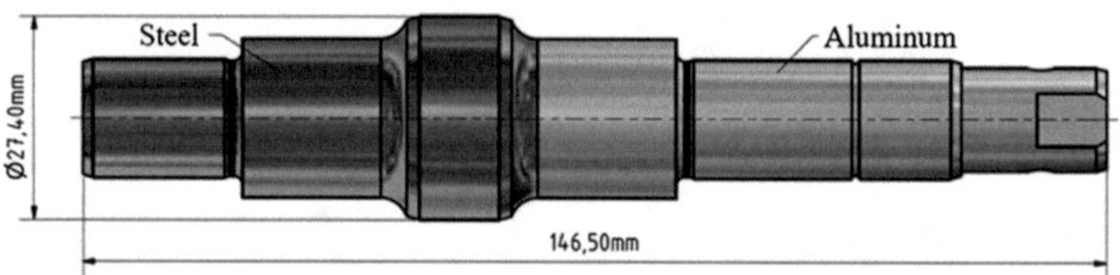

Fig. 2. Hybrid shaft example manufactured by Tailored Forming.

The shaft presented in Fig. 2 is made of steel and aluminum alloy, with the intent of having steel at the region where the mechanical load is more intense (section with higher diameter). This is a concept solution created from the forming techniques used. The geometry of the interface between the two materials must be, however, analyzed and optimized in a way that the manufacturing constraints are obeyed. One example of constraints here is the fact the joining region must contain steel inside the aluminum, and never the opposite. For this task, a design configuration process will be implemented in order to create the base information for our case-based tool.

2.2 Case-Based Reasoning

Case-based reasoning (CBR) is a method of artificial intelligence for solving problems by learning from precious experiences. This methodology can be applied in a large range of areas, such as exact sciences or mundane tasks. A case-base is the main component of a CBR-system, consisting of a collection of problem-solution pairs, where each of these pairs are defined as one case. The approach of reasoning consists in searching similar problems in the case-base and adapting the solution of these problems for the creation of a new one [1, 15].

The CBR system can be built in three configurations: Textual CBR, Conversational Based CBR and Structural CBR (Table 1). The variants differ on how the case-base is used, the way in which the cases are described, and the process of finding similar cases for the described problem [7].

A common CBR process can be described by the following cycle: Retrieve, Reuse, Revise and Retain (Fig. 3) [1, 7].

The process starts with the search for the solution of an actual problem, defining a new case. In the Retrieve process, the description of this new case is used to find one or more similar cases in the case-base. Then, with a success of the previous step, the

Table 1. Configuration types for CBR-System [8].

CBR Type	Case Representation	Case Search	Ex. Application
Textual CBR	Information entities	Text comparison	Facts, repair reports
Conversational CBR	Question-answer pairs	Conversation CBR-users	Call Center Agent
Structural CBR	Attribute-value pairs	Similarity	Design, Configuration

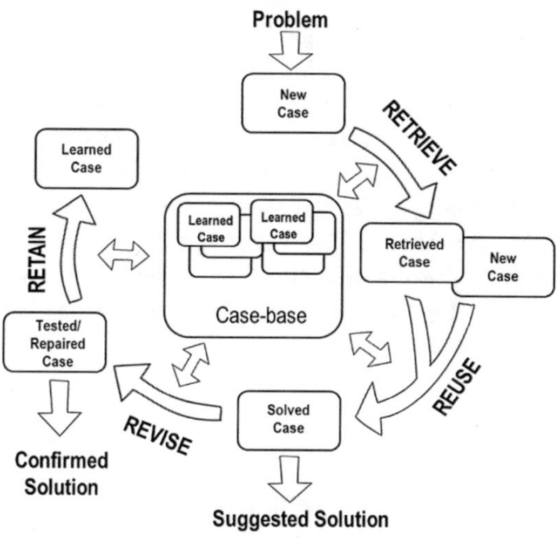

Fig. 3. CBR Cycle [7].

solution for the selected cases will be adapted during the Reuse phase. A proposed solution for the initial problem will be tested in the Revise process. After the testing and possible repair of the solved case, the validated case is retained in the Retain phase and saved back in the case-base for future use. With the conclusion of this cycle, the CBR system is able to learn by itself [1].

In order to minimize the search effort and response time of the system, the individual cases are described by descriptors. Descriptors can refer to technical aspects as well as to the tasks of the system [6, 8]. In the current work, a Structural CBR is used, so that the case is presented as attributes-values pairs. Then, the search for the similar case in the case-base is made by the use of similarity models, such as Hamming-similarity [10]. This is a model based on Hamming-distance, as seen in Eq. 1.

$$h_s'(\mathrm{x},\mathrm{y}) = 1 - \frac{h_d(x,y)}{\sum_{i=1}^{n} w_i} \tag{1}$$

Where h_s' is the Hamming-similarity; x and y are the properties of two different cases; h_d is the Hamming-distance; and w_i is the weight of an attribute i; and n is the number of attributes. With this formula, the new case is compared to all cases in the case-base and the similarity between them is measured. At the end, the designer will be able to see the most similar cases and, based on their parameters, design a new model that suits the requirements.

3 CBR Implementation for a Hybrid Shaft

3.1 System Architecture

In the current work, a generative design system is implemented using the software Autodesk Inventor (2017) [12] and Abaqus CAE (2014) [2]. This step is used here for a parametrical analysis of the design and it will serve as the first cases for the case-base system. For the parametric model generation, Inventor presents all the tools needed to perform the task, as well as a Visual Basic for Applications (VBA) interface [11]. This interface allows full automatization for the variation of parameters. For the Finite Element Analysis automation, Abaqus presents the advantage of having a Python development interface, the Abaqus PDE. Through this language, all the inputs and outputs can be accessed, allowing also our iterative run. The framework between the two software is shown in Fig. 4, where the whole design configuration process and end-user interface are represented.

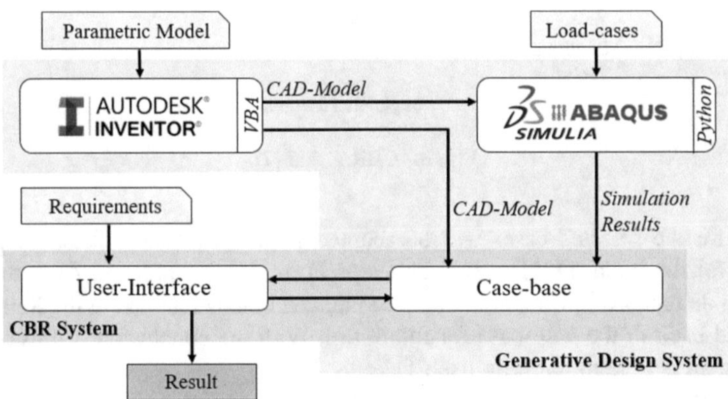

Fig. 4. Interface process showing the whole process.

As seen, the purpose here is the creation of an interface between our CAD generator and our expert Finite Element system, so that all the information generated is saved in a databank. Later on, this database will serve as the case-base for the CBR system, proposing a solution according to user's specific requirements. In next sections, we introduce in more details how each interface was built and how the design engineer's inputs were performed for the example of the hybrid shaft.

3.2 Parametric CAD Model

For the construction of our CAD Model, the first step is to have a full described model, which is here the pre-determined concept for the hybrid shaft presented before. Then, the joining zone must be represented and this is a crucial moment where the user must define his input wisely. This joining zone could be constructed in a vast number of possibilities, but it is important to keep in mind that the more parameters needed for its description, the more complex the system becomes. This complexity can bring difficulty to the analysis of the isolate influence of every parameter. So, it is important to keep this description as simple as possible, using the minimal number of parameters that can describe the allowable geometry in an efficient way.

For our shaft, we have also the manufacturing restrictions earlier mentioned. These restrictions must be also obeyed in this description. Since we have a symmetric component, our joining zone should also be symmetric. This reduces our problem to the description of a 2D-function profile $f(x)$ that will be fully revolved to create the joining zone surface, as can be seen in Fig. 5.

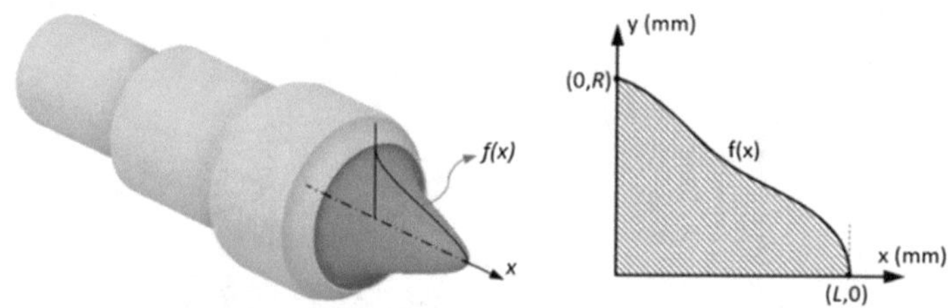

Fig. 5. Revolution of the profile to create the joining surface and its function.

As one of the manufacturing restrictions, we have the fact that the inner part of the joining zone must be Steel and the outer part Aluminum. This surface should also be as smooth as possible, since it will be manufactured in a forming process where very little control over small details is present. The position of the surface, as seen before, was fixed in the concept proposal. So, in order to create this function in a practical way, two physical parameters were chosen to describe it: length and volume. The physical meaning makes the interpretation more expressive and intuitive. We looked then for a function in which: my initial point is $(0, R)$, where R is the radius of the shaft at the pre-determined interfacial point; my final point is at $(L, 0)$, where L is the length of the joining zone; and the definite integral between these two points after a revolution is equal to V, where V is the volume of steel added by the surface and our second control parameter. A last constraint was implemented to avoid a pointed shape, which is performed by making the derivative at $(L, 0)$ to be $-\infty$. This definition of the problem is showed in Eq. 2.

$$
\begin{array}{c}
\pi \int_0^L f(x)^2 dx = V \\
f(0) = R \\
f(L) = 0 \\
\frac{df}{dx_{x=L}} = -\infty
\end{array} \tag{2}
$$

A solution for this problem is presented in Eq. 3.

$$
f(x) = R\sqrt{1 - \left(\frac{x}{L}\right)^{\left(\frac{V}{\pi R^2 L - V}\right)}}; \quad L > \frac{V}{\pi R^2} \tag{3}
$$

The function found makes a simple and effective description of the joining zone with only two parameters. The restriction of the L parameter has a physical meaning that, when L tends to the value of $V/\pi R^2$, this function tends to assume a "square" shape. So, values close to this limit must not be considered, since it gives a solution similar to a plane in a different location. Figure 6 presents some of the possible geometries according to the choice of parameters, where the dotted line represents the cited restriction.

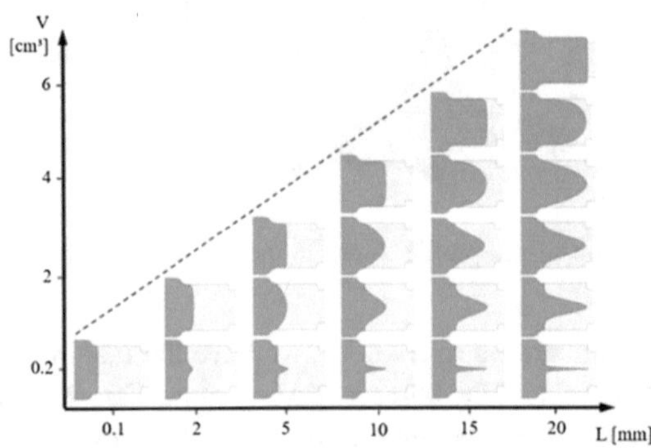

Fig. 6. Possible design configurations of the Joining Zone according to the parametric description provided (R =10,75 mm).

As seen, this function allows a big variety of possibilities, proving a suitability for the solution space generation. Other functions could also be used here, but since no other specific information can be given about the manufacturing restrictions, the function presented is a suitable alternative.

The step of choosing how to make the parametric description cannot always be solved with simplicity. In some cases, many parameters may appear, increasing considerably the computational cost and bringing more difficulty to the interpretation of the result. For that reason, this method is a suitable choice for highly constrained systems, as the one seen in our shaft.

Finally, after this description is determined and fixed, a macro-based script exports all possible configurations as CAD models. For that, we determine the range of values for each parameter that must be executed, considering the restrictions that exists between the parameters itself. At the end, a library of CAD models is generated, which will serve as input for the next phase of the process with the expert system.

This library presents our solution space. It is known that, although the parametric description tries to take into consideration all manufacturing constraints, there will be a range of parameter combinations in our proposal formula that will still violate some of the manufacturing constraints. For example, a very small value for *V* with a large value for *L* generates a long and thin shape that cannot be manufactured. The purpose here is, however, a behavior analysis and a theoretical solution generation. Also, the Tailored Forming manufacturing restrictions are still not clearly formulated. So, in future, after experimental validations, the not suitable designs generated here must be removed from the solution space.

3.3 Finite Element Model and Load Cases

After the construction of the CAD data library, with variated parameters, all the files must be imported iteratively into Abaqus, where they will be simulated. For that, the first step is the construction by the user of a primary simulation model that will work as standard case, where the material properties, load cases, boundary conditions, mesh specifications, among others, must be specified. This is nothing more than a full model that can run successfully for any of the cases generated in the previous step.

Here, one of the biggest advantages is that this file must not be reconfigured during the interactive running, since my parametric model presents no change in the topology, keeping the same number of faces, edges and corners. So, a simple replacement of the CAD drawing is enough to make the model ready for a new run, requiring only the definition of materials for each part of the component and a new meshing. In all simulations it was used quadratic mesh elements, in order to have a better approximation of maximum stress results.

With that, our base model is constructed here for our two cases of interest: bending and torsion. Figure 7 presents the scheme for both load cases studied here, where *F* is the bending force and *M* the torsion moment.

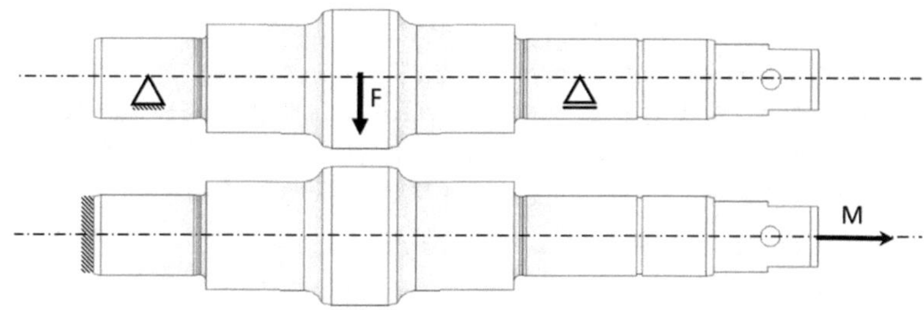

Fig. 7. Load cases of the shaft for torsion and flexion.

After the definition of the base model, a script must be executed, so that all CAD files can be imported iteratively. This script was written in the language Python and, in this step, the output data desired must be defined. For our particular case, we don't search just for generic results of our model, but mainly specific data about the joining zone, so that we can reach a better understanding about it. So, firstly our script executes an isolation of the joining zone, selecting just the set of elements that have a connection with it, as seen in Fig. 8.

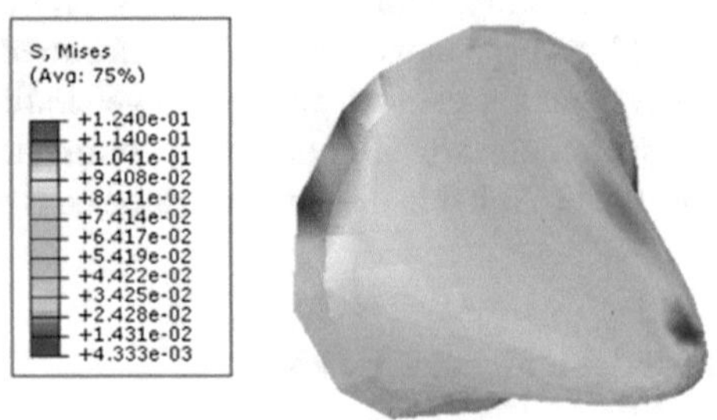

Fig. 8. Stress field result example for an arbitrary joining zone of the shaft.

Secondly, the results at these elements are extracted. Finally, in order to make an analysis from the joining zone behavior, we calculate with this data three parameters:

- Maximum Stress at the joining zone (σ_{JZmax})
- Average Stress at the joining zone ($\overline{\sigma}_{JZ}$)
- Standard Deviation of the stress at the joining zone (s_{JZ})

All of these calculations are made using Von Mises Stress failure theory, so that we deal with just absolute values of stress instead of separated components. The calculation of Average Stress and Standard Deviation are made using weights. These weights are proportional to the area that the element occupies at the joining zone. The equations used in these calculations are presented in Eq. 4.

$$\overline{\sigma}_{JZ} = \frac{\sum_{i=1}^{N} w_i \sigma_i}{\sum_{i=1}^{N} w_i}; \quad s_{JZ} = \sqrt{\frac{N \sum_{i=1}^{N} w_i (\sigma_i - \overline{\sigma})^2}{(N-1) \sum_{i=1}^{N} w_i}} \tag{4}$$

Where N is the number of elements in the joining zone set; σ_i is the Von Mises stress of the element; and w_i is the weight, which is here the area of the element. In addition, some other results of the simulation were also computed, so that the end-user can also use as criteria in a later stage, according to other specifications. They are:

- Maximum Von Mises Stress in each material ($\sigma_{1max}, \sigma_{2max}$)
- Global Safety Factor (n)
- Maximum Bending (d)

With this data we are able to make a characterization of all the designs and use this information as basis for our case-based system.

3.4 Case-Base Generation and Results Analysis

In this section, we performed the configuration process for our shaft, in order to build the database. According to prior discussions, the process was executed for a certain range of the parameters *L* and *V*. This range includes here all the geometric allowable minimum and maximum values. For a better sensibility, a smaller discretization was used for low values of *V*. This resulted in a number of 2200 CAD model variations of our joining zone.

Here, we focus on the results that are more relevant to Tailored Forming, which are the stress behavior at the joining zone, in order to make the qualitative analysis of it. As said before, two load cases were analyzed: pure bending and pure torsion. In addition, a run with both loads combined was also executed.

Pure Bending. For the case of pure bending, using a force *F* of 100 N, we obtained the following results shown in Fig. 9. Since only here we check only elastic behavior, the absolute value of the force doesn't interfere the relation between the results. Every point in these graphs represents one simulation, or iteration, from the process.

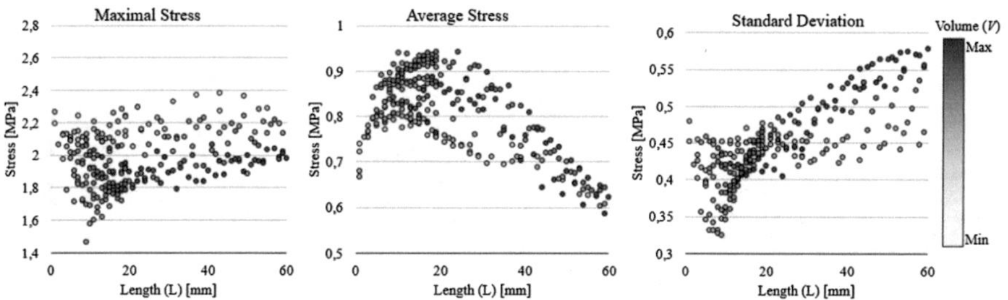

Fig. 9. Simulation results at the Joining Zone for pure bending.

In these results, the first parameter, the Length *L*, is represented in the *x-axis* of the graphs. The second parameter, the Volume *V*, is represented here as a gray-color gradient, where the darker points are for designs with higher steel volume and consequently higher mass.

After the analysis of these results, it is noticeable the existence of a point where the maximum stress reaches a minimum, as well as the standard deviation. These points come from the same simulation, where $V = 1090$ mm^3 and $L = 9$ mm.

Pure Torsion. The following Fig. 10 presents the results for the case of pure torsion, using a torsion moment *M* of 10 Nmm. The same representation method from before is utilized.

For this load case, the result presented doesn't have any strong minimum peaks as seen in the case of bending. However, it has a strong direct dependence with the parameter *V*. The lowest values for maximum stress and standard deviation are found when *V* and *L* are minimal. This result suggests that for torsion, a flat surface connection would be provided a homogeneous stress distribution.

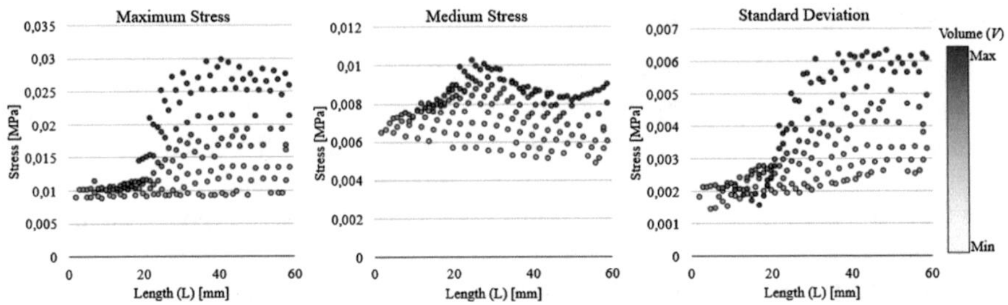

Fig. 10. Simulation results at the Joining Zone for pure torsion.

Combined Bending and Torsion. Finally, we want to investigate a generic case where bending and torsion are combined. This describes better a real application of a shaft. We applied then a bending force strong enough to show its influence in a torsion model. These values were 10 N for F and 100 Nmm for M. Although they are different from the previous values, the point of interest here is the relation between them, which in this case is 1:10. The results are shown in Fig. 11.

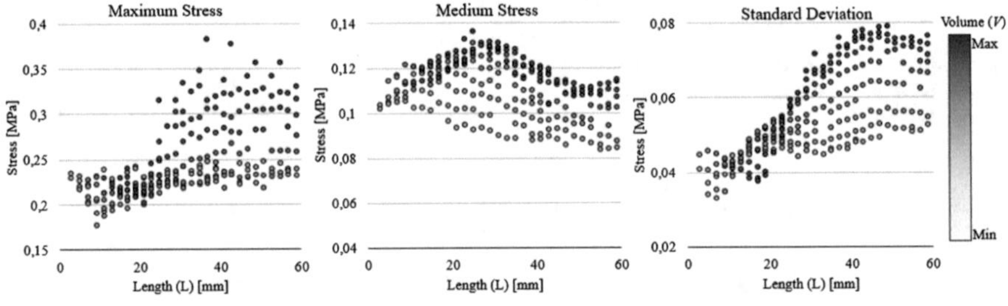

Fig. 11. Simulations result at the Joining Zone for combined Bending and Torsion.

By resemblance, the graphs shown in Fig. 8 are more similar to the case of pure torque, showing a predominance of torsion stress. However, it is seen that the presence of a bending moment already changed the bottom line of the maximum stress, having again the presence of a minimum point.

3.5 Retrieve Step

As commented earlier, the implementation of the CBR system was made using Hamming-similarity approach as search tool in the retrieve step. Figure 12 shows the layout of the user-interface constructed with values inputted as example.

As seen, the parameters of design are not used as attributes in this stage, the user must only specify the desirable properties. As next step, the similarity of this case is calculated for all items in the case-base created. In the second layout presented below, the most similar case is presented (Fig. 13).

NEW CASE					
	Atributes	Weight	Value	Unit
General	Force Applied	1	100	N
	Torsion Applied	1	0,1	N.m
	Maximum Bending	0	0	mm
	Maximum Stress	0	0	MPa
	Safety Factor	0	0	-
Joining Zone	Maximum Stress	2	0,2	MPa
	Medium Stress	1	0,1	MPa
	Standard Deviation	2	0,05	MPa

Fig. 12. Input interface for a new case to be searched as example.

MOST SIMILAR CASE			
	Atributes	Unit	Value
	Similarity	%	87,1%
Model Parameters	Case Index	-	**203**
	L - Length of Joining Zone	mm	21
	V - Volume of Joining Zone	mm³	1089,2
	P - Position of Joining Zone	mm	61,5
General	Force Applied	N	10,000
	Torsion Applied	N.m	0,100
	Maximum Bending	mm	3,24E-04
	Maximum Stress	MPa	0,9397
	Safety Factor	-	329,9
Joining Zone	Maximum Stress	MPa	0,200
	Medium Stress	MPa	0,110
	Standard Deviation	MPa	0,048

Fig. 13. Most similar case, or solved case, found in the case-base for the new case.

For the example presented above, the most similar case had 87.1% of similarity. The FE analysis for this case can be easily found, as shown in Fig. 14. Based on that and other similar cases, the user can modify the parameters to create a new case in have the most suitable design.

The CBR-System gives the possibility to a quick evaluation of the model that presents the most similar result. This visualization, as seen in Fig. 14, allows the user to see, for example, the behavior of the joining zone, serving as an important tool for the Tailored Forming research. With that, the designer may use this case as solution for his problem, or take it as inspiration for the creation of a new case that will be revised, tested and added to the case-base, closing the CBR cycle and supporting the machine learning process.

Fig. 14. Best result of the parametric shaft for the example given: the first is a plane cut from the CAD model in Autodesk Inventor; the second is a plane cut from Abaqus showing the stress field.

4 Conclusions

The design with multi-material presents a challenge by its own complexity and this issue brings even more concern when the manufacturing constraints must be taken into consideration. Tailored Forming provides a new possibility for the creation of hybrid components, but it brings with it many geometry restrictions and new design challenges. In this paper, a parametric model was implemented in order to define these constraints. Although the characterization was successful, it still doesn't guarantee that the geometry can be manufactured. This is still the most sensible step in the process that requires high attention of the designer.

As one of the objective in the research, a first case-base for our CBR was generated for a hybrid shaft, which also permitted the analyze of the joining zone under stress conditions. The results seen in this analysis suggest that the presence of a best and unique solution for our problem depends directly on the load case that the component is subjected to. The analysis was conclusive to determine that a best solution for a case with bending exists, which leads to how the connection zone in hybrid components should be designed. These are valuable information for the creation of first Tailored Forming guideline rules that will guide engineers through this new way of construction.

Following, the retrieve step of CBR system was implemented through a similarity method. This last step provides another essential tool that can help the designer to find the best solution for a problem and store the knowledge. Although the example presented in this study is limited in terms of design requirements, it is showed the potential of this method. In this sense, the case-base can still be enlarged, in order to provide a robust solution space with a vast number possibilities. Within this cycle, this CBR will be able to contain the manufacturing restrictions integrated in the case-base, acting as a self-learning tool of these constraints and bringing essential contribution for the research in the field.

Acknowledgment. The results presented in this paper were obtained under the umbrella of Collaborative Research Centre 1153 "Process Chain for Manufacturing Hybrid High Performance Components by Tailored Forming", preliminary inspection project C2. The authors would like to thank the German Research Foundation (DFG) and the CRC 1153 for its financial and organizational support.

References

1. Aamodt, A.: Plaza. E.: Case-based reasoning: foundational issues, methodological variations, and system approaches. AI Commun. **7**(6), 39–59 (1994)
2. Abaqus CAE Version 6.14-2. Dassault Systemes © (2014)
3. Behrens, B.-A., Bouguecha, A., Frischkorn, C., Huskic, A., Stakhieva A., Duran, D.: Tailored forming technology for three dimensional components: approaches to heating and forming. In: 5th International Conference on Thermomechanical Processing, Milan, Italy, 26–28 October 2016. Associazione Italiana Di Metallurgia, Milan (2016)
4. Behrens, B.-A., Bouguecha, A., Moritz, J., Bonk, C., Stonis, M., Klose, C., Blohm, T., Chugreeva, A., Duran, D., Matthias, T., Golovko, O., Thürer, S. E., Uhe, J.: Aktuelle Forschungsschwerpunkte in der Massivumformung. 22. Umformtechnisches Kolloquium, Hannover, Germany, 15–16 March 2017, Hannoversche Forschungsinstitut für Fertigungsfragen, Hannover (2017)
5. Behrens, B.-A., Overmeyer, L., Barroi, A., Frischkorn, C., Hermsdorf, J., Kaierle, S., Stonis, M., Huskic, A.: Basic study on the process combination of deposition welding and subsequent hot bulk forming. Prod. Eng. **7**(6), 585–591 (2013). https://doi.org/10.1007/s11740-013-0478-y
6. Beierle, C., Kern-Isberner, G.: Methoden wissensbasierter Systeme: Grundlagen. Algorithmen, Anwendungen. Computational Intelligence, German Edition (2008)
7. Bergmann, R., Althoff, K.D., Breen, S., Göker, M., Manago, M., Traphöner, R., Wess, S.: Developing industrial case-based reasoning applications: The INRECA methodology. Springer Science & Business Media (2003)
8. Bibani, M., Gembarski, P.C., Lachmayer, R.: Ein wissensbasiertes System zur Konstruktion von Staubabscheidern. In: DFX 2017: Proceedings of the 28th Symposium Design for X, 4–5 October 2017, Bamburg, Germany (2017)
9. Brockmoeller, T., Gembarski, P.C., Mozgova, I., Lachmayer, R.: Design catalogue in a CAE environment for the illustration of tailored forming. In: 59th Ilmenau Scientific Colloquium, Ilmenau, Germany, 11–15 September 2017. Technische Universität Ilmenau (2017)
10. Freudenthaler, B.: Case-based Reasoning (CBR): Grundlagen und ausgewählte Anwendungsgebiete des fallbasierten Schließens. VDM Verlag, Saarbrücken (2008)
11. Gembarski, P.C., Li, H., Lachmayer, R.: KBE-Modeling techniques in standard CAD-systems: case study – autodesk inventor professional. In: Managing Complexity – Proceedings of the 8th Mass Customization, Personalization and Co-creation Conference, 20–22 October 2015, Montreal, Canada (2015)
12. Inventor Professional, Version 2017 RTM, Autodesk Inc. © (2016)
13. Kleemann, S., Fröhlich, T., Türck, E., Vietor, T.: A methodological approach towards multi-material design of automotive components. Procedia CIRP **60**, 68–73 (2017). https://doi.org/10.1016/j.procir.2017.01.010
14. Maher, M.L., Pu, P.: Issues and Applications of Case-Based Reasoning to Design. Taylor & Francis. https://doi.org/10.4324/9781315805894

15. Richter, M.M., Weber, R.O.: Case-Based Reasoning. Springer, Berlin (2016). https://doi.org/10.1007/978-3-642-40167-1
16. SFB 1153: Prozesskette zur Herstellung hybrider Hochleistungsbauteile durch Tailored Forming, https://www.sfb1153.uni-hannover.de/sfb1153.html. Accessed 08 Apr 2018
17. Ullman, D.G.: The Mechanical Design Process, vol. 2. McGraw-Hill, New York (1992)

Optimal Design of Colpitts Oscillator Using Bat Algorithm and Artificial Neural Network (BA-ANN)

E. N. Onwuka(✉), S. Aliyu, M. Okwori, B. A. Salihu, A. J. Onumanyi, and H. Bello-Salau

Department of Telecommunication Engineering, Federal University of Technology, Minna, Niger State, Nigeria
{onwukaliz, salihu.aliyu, michaelokwori, salbala, adeizal, habeeb.salau}@futminna.edu.ng

Abstract. Oscillators form a very important part of RF circuitry. Several oscillator designs exist among which the Colpitts oscillator have gained widespread application. In designing Colpitts oscillator, different methods have been suggested in the literature. These ranges from intuitive reasoning, mathematical analysis, and algorithmic techniques. In this paper, a new meta-heuristic Bat Algorithm (BA) is proposed for designing Colpitts oscillator. It involves a combination of BA and Artificial Neural Network (ANN). BA was used for selecting the optimum pair of resistors that will give the maximum Thevenin voltage while ANN was used to determine the transient time of the optimized pairs of resistors. The goal is to select, among the several optimized pairs of resistors, the pair that gives the minimum transient response. The results obtained showed that BA-ANN gave a better transient response when compared to a Genetic Algorithm based (GA-ANN) technique and it also consumed less computational time.

Keywords: Artificial Neural Network · Bat algorithm · Colpitts oscillator Genetic Algorithm · RF circuit · Transient response

1 Introduction

An electronic oscillator can be seen either as a circuit capable of converting dc signal to ac signal operating at a very high frequency or a device that generates ac signals of a given waveform such as sine, square, saw tooth, or pulse shape. It provides an AC output signal without necessarily requiring any externally applied input signal. It can also be described as an unstable amplifier.

There are different categories of oscillator depending on the output waveform, operating frequency range and the circuit components used. Based on circuit components used, the Colpitts oscillator falls under the LC type among others such as Clapp, and Hartley oscillators. Conventional methods of designing Colpitts oscillator involves either the use of the following: intuitive techniques, and analytical techniques for the determination of the values of the circuit components used. However, emerging trends in electronic circuit optimization involve the use of artificial intelligent techniques such

J. Świątek et al. (Eds.): ISAT 2018, AISC 853, pp. 29–38, 2019.
https://doi.org/10.1007/978-3-319-99996-8_3

as ANN, PSO, and GA among others [1–6]. In the next section, a brief review of related work where artificial intelligence have been applied in the design of Colpitts oscillator is presented.

The rest of this paper is organized as follows: Sect. 2 presents a review of related work, while the oscillator design is presented in Sect. 3. Section 4 presents resistor selection using artificial intelligence technique followed by results and discussion which are presented in Sect. 5. Conclusion and future recommendations are given in Sect. 6.

2 Related Work

Optimization techniques are fast gaining applications in the area of electronic circuit design. This is particularly due to the ability of most of these algorithm to mimic natural intelligence in animal. For example, in the design of dc-dc converter, three intelligent optimization techniques (GA, Scatter Search (SS), and Simulated Annealing (SA)) have been evaluated for optimality [7]. The converter efficiency in forward mode operation was derived and used as the optimization objective function. The optimal parameters of the converter obtained from Genetic Algorithm method was compared with those obtained using SA and SS intelligent techniques. The waveform resulting from the three approaches both in forward and backup modes were close to the ideal waveform of the converter. However, SS outperformed GA and SA in terms of execution time.

Similarly, radio frequency varactor circuit design has also been improved using optimization techniques. For example, an optimization method for design of RF varactors was proposed by [8]. Generally, varactor behavior is characterized by some set of supporting equations based on technical parameters. Consequently, this makes the accuracy of the results obtained from RF varactor design adaptable to any technology. GA optimization methodology was used to particularly achieve the varactor circuit design. An interesting feature of GA is that it is able to handle continuous as well as discrete variables, thus providing the possibility of adapting it to both technological and layout constraints. A set of working examples for UMC130 technology were used to justify the validity of the proposed model. The results obtained, identified the likelihood of analytical method of varactor design, enhanced with a GA optimization technique [8]. The accuracy of the obtained results was evaluated in comparison with a HSPICE simulator. Similarly, an optimal LC-VCO design using evolutionary algorithm (GA) was proposed by [9]. Considering the challenge in designing the on-chip LC tank, an optimization technique was used. To overcome phase-noise limitation, the approach sought to minimize both VCO phase noise and power consumption. The validity of the results obtained was also verified using HSPICE/RF simulation thus showing GA as a potential algorithm for designing an accurate and efficient oscillator. The same authors went further to compare the performance of three popularly known meta-heuristic algorithms (GA, PSO, and SA) for LC-VCO design [10]. The results obtained showed that GA, despite being the fastest algorithm, gave the worst deviation from the final solution. However, PSO showed a trade-off between convergence and computational time. In addition, PSO also requires less parameter adjustment than GA and SA, while SA gave the best solution.

A neuro-genetic framework for centering of millimeter wave oscillators have been proposed in the literature [11]. Neural Network was used for circuit modeling while GA for parameter optimization. The authors focused on yield enhancement using Monte Carlo based method. The proposed method was used for a design centering on 30 GHz cross-coupled VCO as well as a fixed frequency 60 GHz oscillator. The results obtained showed significant yield improvement from 8% to 91% for 30 GHz and 7% to 70% for the 60 GHz oscillator.

Various intelligent techniques for analogue electronic circuit design were presented by [6, 12], with PSO being the best followed by GA in terms of frequency response and power consumption reduction. A new hybrid artificial intelligence technique for the design of Colpitts oscillator was proposed. The approach involved optimization of the Thevenin resistors of a common based Colpitts oscillator using a combination of Genetic Algorithm (GA) and Artificial Neural Network (ANN) [4, 5]. GA was used for selecting the pair of resistors that gives the maximum Thevenin voltage while ANN was used to determine the transient time of the optimized couple of resistors. From the results obtained, it was reported that the selected resistor pair for the Colpitts oscillator has shortest transient time and stable dc during long-term operation. From the foregoing it could be seen that state-of-the-art researches have shown the benefits in using artificial intelligence methods for circuit design optimization. Thus, in this paper, an approach involving the use of Bat Algorithm (BA) is introduced in combination with ANN. A performance analysis of this approach was also conducted and results obtained were compared with a previous approach to the same problem. In the rest of this paper, we present brief discussion on the Colpitts oscillator design, followed by the proposed resistor selection algorithm for the Colpitts oscillator. Finally we present our results, and compare with previous similar work, and then conclude the paper.

3 Oscillation Design Methodology

Transient time is the time taken for a circuit to move from one steady-state to another steady-state. It is the time taken for the circuit to settle down when turned ON/OFF. It is of utmost for a circuit to have small transient time as such delay in time determines how soon the final output level is reached. In this section, the design of Colpitts oscillator is presented with the goal of achieving minimum transient time using optimization technique. Figure 1, shows the circuit diagram of the Colpitts oscillator whose design is to be optimized using the combination of optimization technique and artificial intelligence (AI) approach. It is a common base Colpitts oscillator consisting of a voltage divider network using R_1 and R_2, an emitter bypass resistor R_3, two coupling capacitors C_3 and C_4, and an LC tank comprising C_1, C_2 and L_2.

The oscillator was designed around a BJT transistor whose base is connected to the LC tank as feedback via a coupling capacitor C_3. The oscillating frequency of the oscillator can be obtained as in Eq. (1).

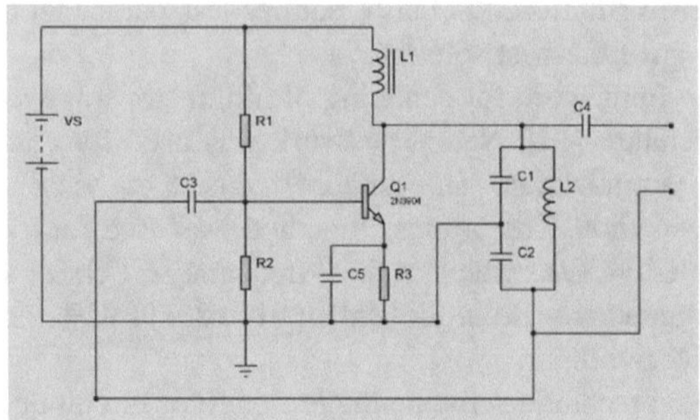

Fig. 1. Circuit diagram of common base Colpitts oscillator

$$f = \frac{1}{2\pi\sqrt{LC_{eqv}}} \tag{1}$$

where C_{eqv} is the parallel combination of C_1 and C_2 and given as

$$C_{eqv} = \frac{C_1C_2}{C_1 + C_2} \tag{2}$$

As obtained [4, 5], using large signal analysis, the equivalent Thevenin resistance R_{th} and voltage V_{th} are given respectively as

$$R_{th} = \frac{R_1R_2}{R_1 + R_2} \tag{3}$$

$$V_{th} = \frac{R_2}{R_1 + R_2}V_{cc} \tag{4}$$

Similarly, from large signal analysis the dc operating point as well as the collector current I_c of the oscillator can be obtained. Thus, the collector current I_c is given as in Eq. (5).

$$I_c = \frac{V_{th} - V_{BE}}{R_E + \frac{R_{th}}{h_{FE}}} \tag{5}$$

Thus, it can be seen from Eq. (5) that I_c is directly proportional to the difference between the Thevenin equivalent voltage V_{th} and V_{BE}. V_{BE} is 0.69 at room temperature and can vary with change in temperature. A change in V_{BE} changes the difference between the V_{th} and V_{BE}, consequently affecting the collector current I_c. Therefore, a slight change in V_{BE} if V_{th} is small can affect the transistor operating point. Consequently, V_{th} should be selected relatively large with respect to V_{BE}. Therefore, the goal of the oscillator design is to maximize V_{th} in order to maintain stable dc operating point of the transistor while minimizing the transient response time. From Eq. (4), it can be seen that V_{th} depends on R_1, R_2, and the supply voltage V_{cc}. Since the V_{cc} is constant, maximizing V_{th} simply requires selection of the best combination of resistors R_1 and R_2 which maximizes the V_{th}. Resistors R_1 and R_2 do not only determine the Thevenin equivalent voltage, they also affect the quality of sine wave obtained from the oscillator. Thus, simulation using LTspice software was conducted to determine the range of resistor values that can give the required oscillator waveform [4]. Different combinations of R_1 and R_2 were utilized and the time taken for the waveform to achieve steady amplitude was recorded. Results obtained for this simulation [4, 5] was also used in this work to train the neural network model. Based on the results, resistance value range of 100 kΩ to 1 MΩ was identified as a suitable range for resistor selection. Thus, an artificial intelligent technique was employed to select the combination of resistors (R_1 and R_2) that gives the maximum Thevenin voltage and minimum transient time using the obtained range of resistance values as constraint.

4 Resistor Selection Using AI Techniques

GA is an evolutionary theory based algorithm that combines crossover, mutation and selection approach in searching for an optimal solution. It has found widespread applications in different fields of engineering and science. Recently, it was introduced in the design of Colpitts oscillator. However, due to its computational complexity, in this work, a new bio-inspired Bat Algorithm (BA) is proposed for the same purpose.

4.1 The Bat Algorithm

Bat Algorithm (BA) is a new bio-inspired algorithm introduced by Yang (2010) and has been established to be a very efficient algorithm for optimization [13]. BA is a recently introduced meta-heuristic algorithm, which imitates the echolocation behavior of bats to carry out global optimization. The excellent performance of this algorithm has been demonstrated among other very well-known algorithms such as GA and PSO [14]. Micro Bats use a type of sonar called echolocation to detect prey, avoid obstacles, and locate their roosting crevices in the dark. These bats emit a very loud sound pulse and listen for reflection from the surrounding objects. The loudness of the released pulse varies from the loudest when searching for prey and to a quieter base when homing towards the prey.

Some important features of BA includes its ability to increase the assortment of the results in the population using frequency-tuning technique, automatic zooming such that it balances between exploration and mistreatment during the search process thus mimicking the changes of pulse emission rates and loudness of bats when looking for prey. BA is based on three idealized rules [13, 15]:

(1) Bats use the concept of echolocation to sense distance, as well as to differentiate between food/prey and background obstacles in some magical way
(2) They fly randomly with a velocity v_i at position x_i using a constant frequency $f_{\min}$, a variable wavelength λ and loudness A_0 while searching for their prey. The

wavelength (or frequency) of their released pulses can automatically be tuned in addition to tuning the rate of pulse emission $r \in [0, 1]$, in accordance to their closeness to their target.
(3) Though the loudness can fluctuate in many ways, it is assumed that the loudness changes from a large (positive) A_0 to a minimum fixed value $A_{\min}$.

Every individual Bat is associated with a velocity v_i^t at position x_i^t at iteration t in a search space or solution space of dimension d. At any given iteration t the current best Bat position (solution) at that iteration is denoted as x_*. The frequency f_i, velocity v_i and solution x_i are updated using Eqs. (6)–(8).

$$f_i = f_{\min} + (f_{\max} - f_{\min})\beta \tag{6}$$

$$v_i^t = v_i^{t-1} + (x_i^{t-1} - x_*)f_i \tag{7}$$

$$x_i^t = x_i^{t-1} + v_i^t \tag{8}$$

where $\beta \in [0\ 1]$ is a random vector drawn from a uniform distribution. After a solution is chosen from the current best solution, a new solution for individual Bat is obtained from Eq. (9) [16].

$$x_{new} = x_{old} + \in A^t \tag{9}$$

where $\in$ is a random number which can be drawn from a uniform distribution. The algorithm starts with initializing the individual Bat with a random frequency or wavelength within the maximum and minimum allowed value. Thus the BA is considered a frequency-tuned algorithm [14]. Bat algorithm was used for determining the optimal resistance values of the two resistors, while ANN was used to predict the transient response of the generated optimized pair of resistors.

4.2 Artificial Neural Network for Oscillator Transient Time Predicting

The use of Artificial Neural Networks (ANNs) for modeling non-linear and complex problems has been largely motivated by the ability of systems to mimic natural intelligence in learning from experience. ANNs learn from training data by creating an input-output mapping without the need to explicitly derive the underlying equations. It has found broad areas of applications including but not limited to areas such as: pattern classification, function approximation, optimization, prediction and automatic control, among others.

Individual link to a neuron has an adaptable weight factor allied with it. Each of the neuron in the network sums up its weighted inputs to give an internal activity level as:

$$a_i = \sum_{j=1}^{n} w_{ij}x_{ij} - w_{io} \tag{10}$$

where w_{ij} is the weight of the link from input j to neuron i, x_{ij} is the input vector (R_1 and R_2 in our case) number j to neuron i, and w_{io} is the threshold associated with unit i.

The internal activity a_i is passed through a nonlinear activation function ϕ to give the output of the neuron y_i

$$y_i = \phi(a_i) \tag{11}$$

The weights of the connections are adjusted during the training process to achieve the desired input/output relation of the network.

4.3 The Proposed BA-ANN Model

The flowchart of the proposed model is as shown in Fig. 2. It consists of two parts, the Bat optimization and the ANN part. BA was used to generate several optimum resistance combination for resistors R_1 and R_2, while ANN was used on the other hand to predict the transient response of the generated combinations. The combination with the minimum transient time will be selected as will be seen in the results section.

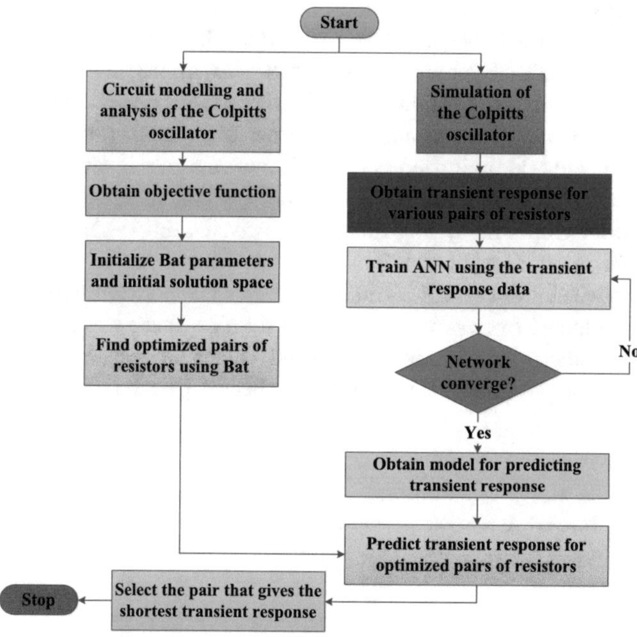

Fig. 2. Flowchart for selection of R_1 and R_2

BA requires an objective function f containing the parameters (R_1, R_2) to be optimized. The goal of the BA is to maximize the Thevenin equivalent voltage (V_{th}). Thus Eq. (4) is rewritten as:

$$f(R_1, R_2) = \frac{R_2}{R_1 + R_2} V_{cc} \tag{12}$$

Vcc is the supply voltage which was set to 12 V, R_1, R_2 were constrained to lower bound of 100 kΩ and upper bound of 1 MΩ which serve as the range that produced pure sine wave. BA was set to a population size and generation of 500, amplitude (A) of 0.6, and pulse rate (r) of 0.5. The minimum ($f_{\min}$) and maximum ($f_{\max}$) frequencies were set to 0 and 3 respectively. While BA searches for the optimum combination of resistance values, there is need to find the combination that gives the minimum transient time. Considering the time to compute transient response for the 500 generated combinations, ANN was used to study and understand the underlying relationship that exists between the resistance combinations and the transient response obtained from the circuit simulation. Consequently, ANN is able to forecast the transient time for the 500 pairs of resistors in just a single step.

5 Results and Discussion

This section presents the results obtained from the BA-ANN algorithm in comparison to previously obtained results using GA-ANN. Both the BA algorithm and ANN were implemented in MATLAB environment. ANN was implemented using the Neural Network toolbox. Back propagation was used with 2 input nodes, a single hidden layer containing three (3) neurons and a single neuron in the outer (output) layer. BA was first used to generate 500 optimized combination of resistors R_1 and R_2. The following parameters were used to tune the BA: A = 0.6, r = 0.5, and $f_{\max} = 3$, etc. The generated population from the Bat optimization was fed into ANN to determine the combination of resistors with the minimum transient time. ANN first learned the relationship between the resistor combination and the corresponding transient time using the data obtained from simulation. Initial solution space was generated randomly between the minimum and maximum allowed values of resistors. Results obtained showed that BA converged in a very few number of iterations as can be seen from Fig. 3.

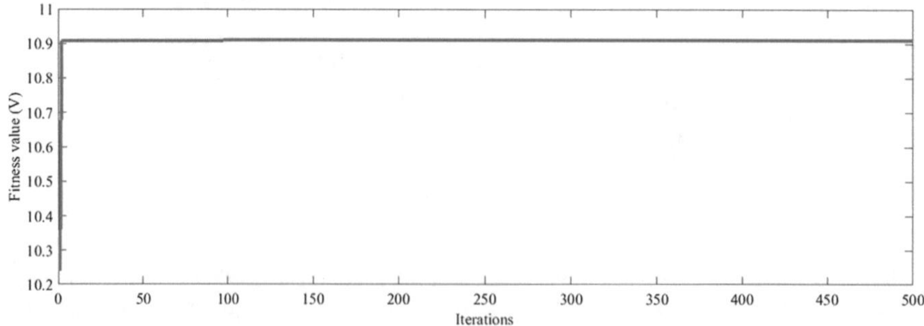

Fig. 3. Plot of fitness value against iteration obtained from Bat algorithm for the 500 optimized pairs of resistor value

The results also showed that irrespective of the initial solution space, BA always converge to the optimum solution. Some of the optimized values obtained are shown in Table 1.

Table 1. Some of the optimized resistor combination using BA

R_1(kΩ)	100.0	100.0	99.999	99.998	99.998	99.997	99.997	99.996	99.994
R_2(kΩ)	999.99	999.9	1000.01	999.9	1000.0	999.9	999.8	999/9	1000

On applying the 500 optimized resistor pair as input to the ANN model, a minimum transient time of 0.952 ms was obtained which occurred at $R_1 = 99.994$ kΩ and $R_2 = 1$ MΩ. The ANN model was also applied on the GA optimized values obtained by [5]. Table 2 shows the transient response obtained in comparison to the GA-ANN approach.

Table 2. Transient response predicted using AI for both GA and BA

R_1(kΩ)	R_2(kΩ)	T (ms), GA-ANN	T (ms) BA-ANN
100	958	1.27	0.939
100	979	1.29	0.946
100	986	1.30	0.948
100	986	1.30	0.948
100	910	1.22	0.925
100	1000	1.32	0.953
100	965	1.28	0.942

From Table 2, it can be seen that the minimum transient time of 0.925 ms occurred at $R_1 = 100$ kΩ and $R_2 = 910$ kΩ. Thus, it can be concluded that both GA and BA gave close range of optimized values, however, BA has a 31.89% reduction in computation time when compared to GA.

6 Conclusion

In this paper, a combination of Bat algorithm and Artificial Neural Network have been introduced for the design of Colpitts oscillator. The objective was to select the best combination of resistors that gives the Colpitts oscillator maximum Thevenin voltage with minimum transient response. BA was used to select the best combination of resistor values, while ANN predicts the transient time of the optimized resistance values. Obtained results was compared with similar work done using GA. Both GA and BA converge to an approximate solution, however, result from the proposed approach yielded 31.89% lesser transient response with less computation time. Future work includes application of the developed Colpitts oscillator for development of a GSM signal booster.

Acknowledgements. The research group, on behalf of Federal University of Technology, Minna, Niger State, appreciates the support of Nigeria Communication Commission (NCC) for this project in which a number of students were trained. This project was funded from grant number NCC/CS/007/15/C/038.

References

1. Aggarwal, V.: Evolving sinusoidal oscillators using genetic algorithms. In: Proceedings of the NASA/DoD Conference on Evolvable Hardware, pp. 67–76 (2003)
2. Fozdar, M., Arora, C.M., Gottipati, V.R.: Recent trends in intelligent techniques to power systems. In: 42nd International Universities Power Engineering Conference, UPEC 2007, pp. 580–591
3. Zhang, J., Shi, Y., Zhan, Z.-H.: Power electronic circuits design: a particle swarm optimization approach. In: Asia-Pacific Conference on Simulated Evolution and Learning, pp. 605–614 (2008)
4. Amsa, M.G.B.A., Aibinu, A.M., Salami, M.J.E.: Application of intelligent technique for development of Colpitts oscillator. In: 2013 IEEE Business Engineering and Industrial Applications Colloquium (BEIAC), pp. 617–622 (2013)
5. Amsa, M.G.B.A., Aibinu, A.M., Salami, M.J.E.: A novel hybrid artificial intelligence technique for colpitts oscillator design. J. Control. Autom. Electr. Syst. **25**(1), 10–21 (2014)
6. Ushie, O.J., Abbod, M.: Intelligent optimization methods for analogue electronic circuits: GA and PSO case study. In: The International Conference on Machine Learning, Electrical and Mechanical Engineering, Dubai, pp. 8–9 (2014)
7. Rao, K.S.R., Chew, C.-K.: Simulation and design of A DC-DC synchronous converter by intelligent optimization techniques. In: 2010 International Conference on Intelligent and Advanced Systems (ICIAS), pp. 1–6 (2010)
8. Pereira, P., Fino, H., Ventim-Neves, M.: RF varactor design based on evolutionary algorithms. In: 2012 Proceedings of the 19th International Conference Mixed Design of Integrated Circuits and Systems (MIXDES), pp. 277–282 (2012)
9. Pereira, P., Fino, M.H., Ventim-Neves, M.: Optimal LC-VCO design through evolutionary algorithms. Analog Integr. Circuits Signal Process. **78**(1), 99–109 (2014)
10. Pereira, P., Kotti, M., Fino, H., Fakhfakh, M.: Metaheuristic algorithms comparison for the LC-Voltage controlled oscillators optimal design. In: 2013 5th International Conference on Modeling, Simulation and Applied Optimization (ICMSAO), pp. 1–6 (2013)
11. Sen, P., et al.: Neuro-genetic design centering of millimeter wave oscillators. In: Digest of Papers. 2006 Topical Meeting on Silicon Monolithic Integrated Circuits in RF Systems, pp. 4–5 (2006)
12. Ushie, O.J., Abbod, M., Ashigwuike, E.: Naturally based optimisation algorithm for analogue electronic circuits: GA, PSO, ABC, BFO, and firefly a case study. J. Autom. Syst. Eng. **9**(3), 173–184 (2015)
13. Yang, X.: A new metaheuristic bat-inspired algorithm. In: Nature Inspired Cooperative Strategies for Optimization (NICSO 2010), pp. 65–74 (2010)
14. Mirjalili, S., Mirjalili, S.M., Yang, X.-S.: Binary bat algorithm. Neural Comput. Appl. **25**(3–4), 663–681 (2014)
15. Yang, X.-S., He, X.: Bat algorithm: literature review and applications. Int. J. Bio-Inspired Comput. **5**(3), 141–149 (2013)
16. Yang, X.-S.: Bat algorithm and cuckoo search: a tutorial. In: Artificial Intelligence, Evolutionary Computing and Metaheuristics, pp. 421–434 (2013)

An Adaptive Observer State-of-Charge Estimator of Hybrid Electric Vehicle Li-Ion Battery - A Case Study

Roxana-Elena Tudoroiu[1], Mohammed Zaheeruddin[2], and Nicolae Tudoroiu[3(✉)]

[1] University of Petrosani, Petrosani, Romania
tudelena@mail.com
[2] Concordia University, Montreal, Canada
zaheer@encs.concordia.ca
[3] John Abbott College, Sainte-Anne-de-Bellevue, Canada
ntudoroiu@gmail.com

Abstract. In this research paper we investigate the procedure design and the implementation in a real time MATLAB SIMULINK R2017a simulation environment of an accurate adaptive observer state estimator. The effectiveness of the observer state estimator design is proved through intensive simulations performed to estimate the state-of-charge of a lithium-ion rechargeable battery integrated in a hybrid electric vehicle Battery Management System structure for a particular Honda Insight Japanese car. The state-of-charge is an essential internal parameter of the lithium-ion battery, but not directly measurable, thus an accurate estimation of battery state-of-charge becomes a vital operation for the Battery Management System. This is the main reason that motivates us to find the most suitable state-of-charge estimator in terms of estimation accuracy, fast convergence and robustness to the possible changes in the state-of-charge initial value, to the temperature effects on the battery, changes in the battery internal resistance and nominal capacity.

Keywords: Hybrid electric vehicle · Adaptive observer · State estimation Battery state-of-charge · Riccati equation · Lithium-ion battery Battery Management System

1 Introduction

1.1 Brief Presentation of Lithium-Ion Batteries

Currently, the lithium-ion (Li-ion) and nickel metal hydride (NiMH) rechargeable batteries are the most two promising technologies widely seen in the automotive industry applications [1–4]. They have a great potential to reduce greenhouse and other exhaust gas emissions, and require extensive research efforts and huge investments, since the environmental impact is a key issue on the enhancing the battery technologies, as is mentioned also in [1, 2]. A great amount of research and development is done for battery size, improved performance and cost, as the main concerns regarding

J. Świątek et al. (Eds.): ISAT 2018, AISC 853, pp. 39–48, 2019.
https://doi.org/10.1007/978-3-319-99996-8_4

the hybrid electric vehicles (HEVs) automotive industry growth. The NiMH and Li-ion batteries have a great potential for a higher efficiency HEVs. Even though more expensive, the Li-ion batteries seem to have become the best choice for HEVs, due to a high storage capacity, their small size, light weight, a tiny "memory effect", and a great capability to hold and distribute large power [1–4]. Moreover, the upcoming improvements of Li-ion batteries technologies is lithium-air batteries with a higher energy density and much lighter due to the oxygen cathode [2]. A big advancement in HEVs requires also to create new designs capable to integrate the Li-ion battery technologies with vehicles engines of high efficiency, as is stated in [1]. Also, the Li-ion batteries design should be aligned in conformity with the international standards specs for "vibration, shocks, temperature effects, acceleration, crush impact, heat, overcharge and over-discharge cycles, and short circuit", as is mentioned in [1, 2]. A Li-ion battery cell has a short term life due to the inside presence of the unwanted irreversible chemical or physical changes that affect significant its electrical performance [2]. Consequently, the Li-Ion battery performance deteriorates over time whether the battery is used or not, process known as "cycle fade" or "calendar fade" [2, 4]. A mature and comprehensive battery management system (BMS) in HEVs is an essential component that performs several functions, ones of them mentioned also in the abstract section [2]. The BMS consists of measurement sensors, controllers, safety circuitry incorporated inside the battery packs, serial communications and specialized hardware equipment, computation software components to monitor, compute and show constantly the state of health of the battery (SOH), the battery state-of-charge (SOC), the temperature inside the battery, the battery performance and its longevity [2]. However, the Li-Ion battery SOC remains one of the most important operational condition battery parameter tightly monitored by BMS, but it cannot be measured directly. Thus, the SOC accurate estimation becomes one of the BMS responsibility task to avoid possible overcharging/over-discharging battery dangerous operation conditions, and to improve the battery life cycle [2–5]. Basically, the battery SOC is defined as the available capacity of a battery, therefore as a percentage of its rated capacity [2]. In the majority of the cases the SOC estimation is based on the available and measurable battery parameters values, such as the current flow within the battery, the battery terminal voltage, as well as the temperature inside the battery. The remainder of this chapter is organized as follows. In Sect. 2 is introduced the proposed selection model criterion, a linear RC series cells electric circuit as a generic third order RC Li-Ion battery equivalent model (3RC EMC), and also the state space dynamic model equations are derived. In Sect. 3, is proposed for design and real time implementation an adaptive observer state estimator (AOSE). The simulation results in MATLAB R201a and SIMULINK followed by a performance analysis are presented in Sect. 4. Section 5 concludes the research paper contributions.

1.2 The Li-Ion Battery Model Selection Criterion

The selection criterion of any battery type depends on several characteristics, among them the weight, power density, cost, size, life cycle, battery state-of-charge, and maintenance [1, 2]. Related to this selection, as was pointed out also in the previous Subsect. 1.1 the Li-Ion battery is the most suitable choice for HEVs. For simulation purpose we choose from literature a linear equivalent electric circuit model consisting of an open circuit voltage connected in series with the internal battery resistance and three series consecutive parallel Resistor-Capacitor polarization cells, easy to be implemented in real time [2, 3]. Thus, it is an OCV-R-RC-RC-RC model called also the third order RC EMC Li-ion battery model, as shown in Figs. 1 [2, 3]. In simulations a specific setup with constant parameters is chosen, whose values are the same as those given in [2], Table 2, p. 29. This model is under investigation to prove the effectiveness of the proposed battery SOC real time estimator developed in Sect. 3. The reason for this model selection is to benefit of its simplicity and its ability to capture accurately the entire dynamics of Li-Ion battery, and to be implemented easily in real time with acceptable range of performance. Also, we are more interested in the "*proof concep*t" algorithmic considerations as motivated by the requirements imposed by the environment and the vehicle. Furthermore, the proposed model choice gives us more flexibility to prove the effectiveness of the adaptive observer SOC estimator in terms of SOC estimation accuracy, speed convergence, robustness to different changes in battery model parameters (i.e. internal resistance, battery capacity affected by aging degradation and repeated charging and discharging cycles) and to the current sensor measurements level noise.

Extensive simulations carried out in MATLAB R2017a simulation environment proved that this electrical circuit model is relatively accurate to capture the main dynamic circuit characteristics of a Li-Ion battery cell, such as the open-circuit voltage, terminal voltage, and transient response. The main drawback still remains since in "real life" the dynamics of the battery cell is seriously affected by the temperature effects and changes in battery SOC.

The role of the resistor-capacitor (RC) series cells integrated in EMCs is to improve models' accuracy and to increase also its structural complexity [2, 3, 5].

1.3 EMC Li-Ion Battery in NREL ADVISOR MATLAB Platform - Case Study

The proposed ECM Li-ion battery validation is done by comparison of the test results using an Advanced Vehicle Simulator (ADVISOR) MATLAB platform, developed by US National Renewable Energy Laboratory (NREL). The NREL Li-Ion battery model

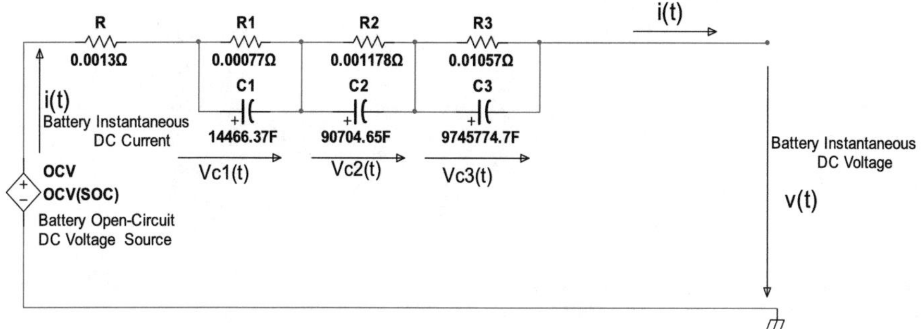

Fig. 1. EMC Li-ion battery electric circuit model (National Instruments 14.1 Editor)

integrated in ADVISOR MATLAB platform is a Li-Ion battery model 6Ah and nominal voltage of 3.6 V produced by the company SAFT America, as is mentioned in [2, 3].

For simulation purpose and comparison of the tests results, the NREL Li-ion battery model is incorporated in a BMS' HEV of a particular Japanese Honda Insight HEV car, as an input vehicle under standard initial conditions (e.g. 70% SOC initial condition) that has the setup shown in Fig. 2. As a driving cycle test for the case study HEV car speed provided by the ADVISOR US Environmental Protection Agency (EPA) is selected an Urban Dynamometer Driving Schedule (UDDS), as is shown in Fig. 3 [2, 3]. The driving UDDS cycle car speed profile and its corresponding Li-Ion battery input current UDDS cycle profile, of the 1370 s time window length, are represented separately in the same ADVISOR MATLAB platform in the top side and the bottom side corresponding graphs of Fig. 4. Also, the gear ratio speed and SOC curves as a results of to the same UDDS cycle tests are shown in the middle two graphs of the same Fig. 4 [2]. The Honda Insight HEV car model is loaded on test data from NREL and Argon National Laboratory (ARL) and data from publishing sources. This model is scalable, and can be used by the user, to define his own control strategy. The ADVISOR MATLAB platform can also be online free downloaded from the website: https://sourceforge.net/projects/adv-vehicle-sim/. The data from NREL and ANL are analyzed to determine the hybrid powertrain characteristics.

The control strategy block receives the value of the torque required into the clutch, and based on this value and the car speed, the electric motor torque contribution is calculated. The remaining torque is demanded from IC engine. The electric motor torque is decided based on torque and rate acceleration, such that for car speed above 10 mph the electric motor assists the IC engine, producing around 10 Nm of torque.

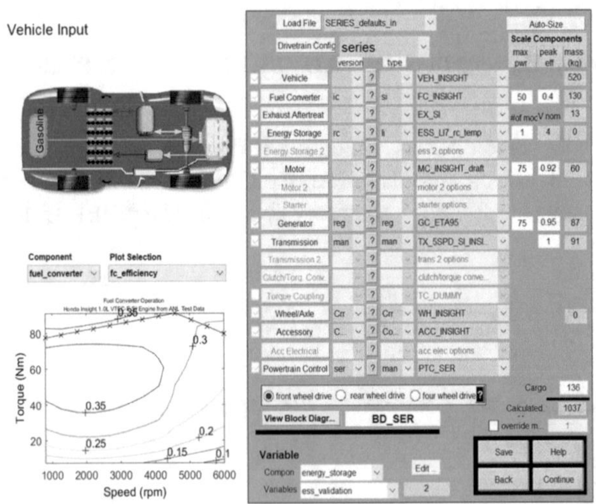

Fig. 2. The setup of the Japanese honda insight HEV car in ADVISOR MATLAB platform under standard initial conditions (initial value of SOC of 70%)

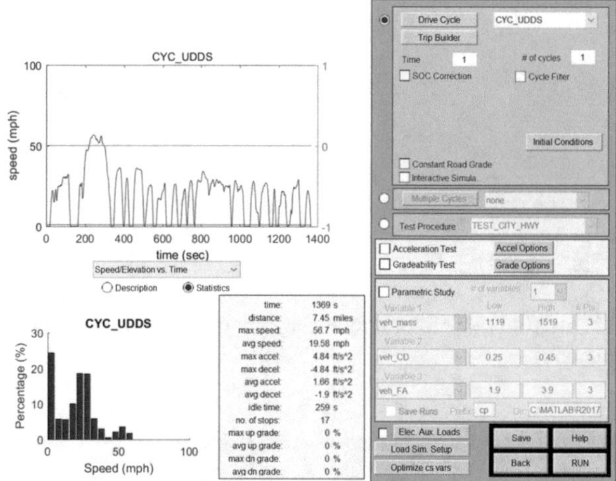

Fig. 3. The UDDS cycle profile speed test for honda insight HEV car in ADVISOR MATLAB platform

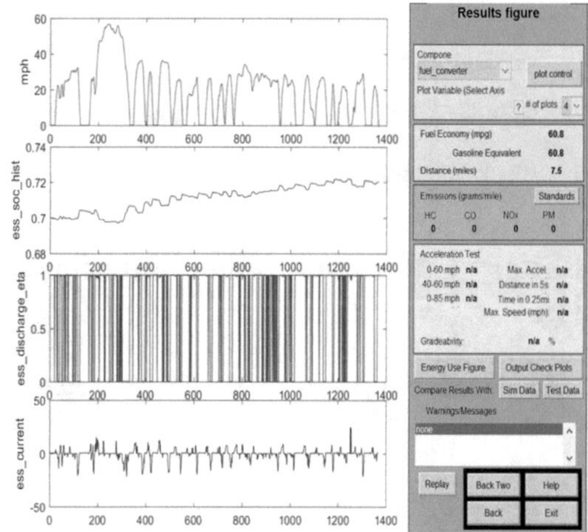

Fig. 4. The corresponding Honda Insight car speed profile (top side), SOC estimate (the second near top side graph), gear ration speed (the third near bottom side graph), and Li-ion battery input current profile (bottom side) on ADVISOR MATLAB platform

During regeneration (i.e. brake is depressed) the electrical motor regens a portion of the negative torque available to the driveline, as is shown in Fig. 5 and SIMULINK Block diagram from Fig. 6. At low car speeds, usually below 10 mph, the braking is primarily only the friction brakes, and there is no electric assist in the first region.

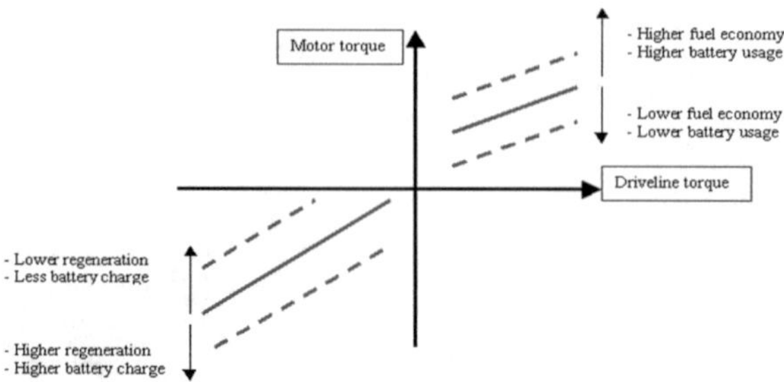

Fig. 5. The Honda Insight HEV motor torque diagram (reproduced from ADVISOR MATLAB platform documentation)

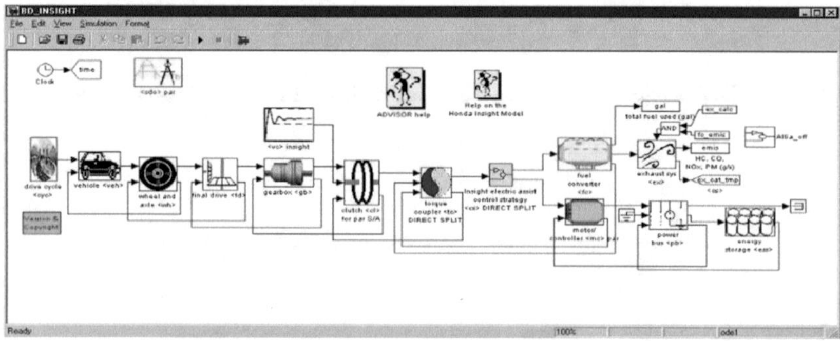

Fig. 6. The simulink block diagram of honda insight in ADVISOR MATLAB platform

2 The Li-Ion Battery Continuous Time State Space Representation

According to the electric circuit setup of the EMC Li-ion battery shown in Fig. 1 the following state-space equations can be written for the battery dynamics [2]:

$$\begin{aligned} \frac{dx_1}{dt} &= -\frac{1}{T_1}x_1 + \frac{1}{C_1}u, \quad T_1 = R_1C_1 \\ \frac{dx_2}{dt} &= -\frac{1}{T_2}x_2 + \frac{1}{C_2}u, \quad T_2 = R_2C_2 \\ \frac{dx_3}{dt} &= -\frac{1}{T_3}x_3 + \frac{1}{C_3}u, \quad T_3 = R_3C_3 \\ \frac{dx_4}{dt} &= -\frac{\eta}{C_{nom}}u, \quad x_4 = SOC, T_1, T_2, T_3 \text{ -- time constants} \end{aligned} \tag{1}$$

$$y = \text{OCV}(x_4) - x_1 - x_2 - x_3 - \text{R}u, \quad \text{and } x_4 = SOC \tag{2}$$

$$\begin{aligned} &OCV(x_4(t)) = K_0 - K_1\frac{1}{x_4(t)} - K_2x_4(t) + K_3\ln(x_4(t)) + K_4\ln(|1 - x_4(t)|) \\ &K_0 = 4.23, K_1 = 0.000036, K_2 = 0.24, K_3 = 0.22, K_4 = -0.04 \end{aligned} \tag{3}$$

where the components of the battery state vector x1(t), x2(t), x3(t) represents the voltages of the RC polarization cells, and x4(t) denotes the battery SOC. The OCV shown in Fig. 7 is a nonlinear function of SOC that combines three additional well-known models, namely Shepherd, Unnewehr universal and Nernst models, defined in [4] with the coefficients set at same values as in [2, 3]. The variables u and y designate the input Li-ion battery charging or discharging current, and the terminal Li-ion battery output voltage respectively. The values of the battery parameters are given at the room temperature (i.e. 25 °C) and are set, for simulation purpose and "proof-concept" considerations, to the same values used in [3], Table 5.4, p. 100, and assumed time constant and independent on the battery SOC changes and temperature effects.

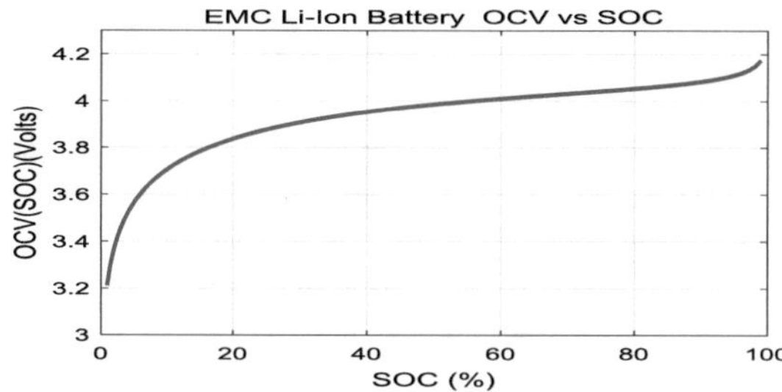

Fig. 7. The Li-ion battery OCV nonlinear function of SOC for charging cycle

In addition, the values of these parameters differ for charging and discharging cycles, as well as the columbic efficiency η, thus the cell's voltage behavior will be described by two sets of parameters, one for charging and other one for discharging cycles, as is shown in [2, 3, 5]. In a realistic environment of operating conditions the battery's parameters are variable with respect to the temperature, the SOC and the current direction, making the overall Li-Ion battery to behave as a nonlinear model. As is stated in [2–5], experimental data and curve fitting techniques are used to find empirical equations relating the parameters with the operating conditions. In similar way, the nominal values of OCV coefficients are chosen to fit the model to the manufacture's data by using a least square curve fitting estimation method, as is suggested in [2–5], where the OCV curve shown in Fig. 7 is assumed to be the average of the charge and discharge curves taken at low direct currents (dc) rates [2–4] (e.g. 1C rate, equivalent to 6A) from fully charged to fully discharged battery.

2.1 ECM Li-Ion Battery Validation

The EMC Li-ion battery SIMULINK model that will be integrated in the Adaptive Observer State Estimator structure, useful also for model validation, is represented in Fig. 8.

The EMC Li-Ion battery model is validated by comparison of EMC Li-ion battery SOC with NREL Li-Ion SOC estimated in ADVISOR MATLAB platform for an UDDS driving cycle test performed on the Japanese Honda Insight HEV. The MATLAB SIMULINK simulation results related to EMC Li-ion battery SOC and ADVISOR SOC estimates are shown in Fig. 9. These simulations reveal a very good match between the both SOC curves, and thus it proves with certainty the ability of the proposed EMC Li-ion battery model to capture with high accuracy the entire dynamics of the battery.

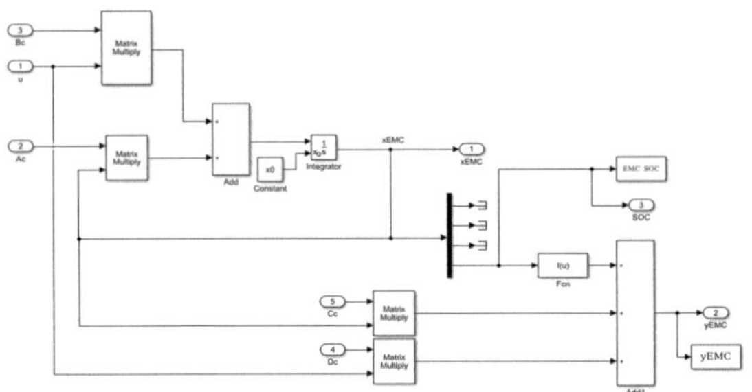

Fig. 8. ECM Li-Ion battery SIMULINK model

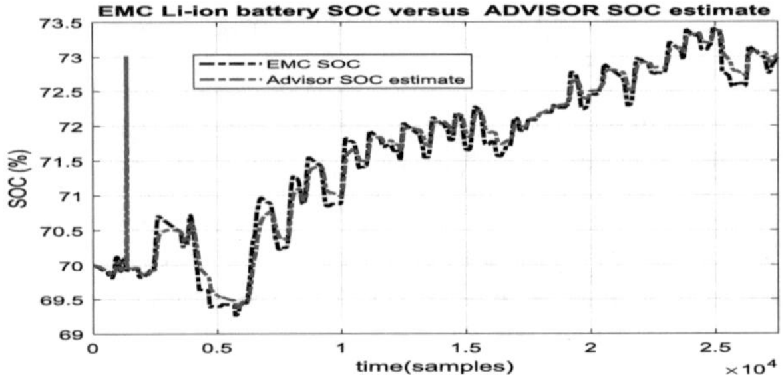

Fig. 9. EMC Li-ion battery SOC versus ADVISOR MATLAB platform SOC estimate

3 The Adaptive Observer State Estimator Design and Real Time Implementation

The Adaptive Observer state estimator (AOSE) design follows the same design procedure developed in [5] adapted to the proposed EMC model described by the Eqs. (1)–(3). The architecture structure of AOSE in SIMULINK representation is shown in Fig. 10.

The dynamics of the AOSE is described by the EMC Li-ion battery model Eqs. (1)–(3) in matrix form in conjunction with the state estimator and Riccati equations and similar as in [5], but adapted to our case study:

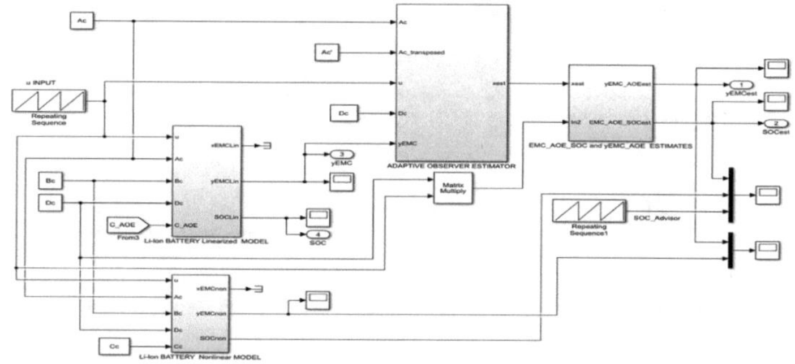

Fig. 10. SIMULINK model of AOSE of Li-ion battery SOC

$$
\begin{array}{l}
\frac{dw(t)}{dt} = [A - K(t)C]w(t) \\
\frac{d\hat{x}(t)}{dt} = A\hat{x}(t) + Bu + [K(t) + \Gamma w(t)Cw(t)](y(t) - C\hat{x}(t) - Du(t)) \\
\frac{dP(t)}{dt} = AP(t) + P(t)A^T - P(t)C^T V^{-1} CP(t) + W, \quad \text{Riccati equation} \\
K(t) = P(t)C^T V^{-1}, \quad \text{gain matrix}
\end{array}
\tag{4}
$$

The state vector $\hat{x}(t)$ is the estimate value of the state vector attached to the Eqs. (1), A, B, C, and D denote the matrices attached to a matrix representation of Eqs. (1)–(3), as is shown in [2] p. 28, [3] p. 88, and the Eq. (3) is linearized around an operating point, set to SOC (0) = 0.7. The matrix W, and the scalars Γ and V denote the tuning parameters of the adaptive observer state estimator, set to the following values: W = diag ([0.1 0.1 0.1 150]), Γ = 150, V = 1.

4 The MATLAB SIMULINK Simulation Results and Performance Analysis

The MATLAB SIMULINK simulation results cover the "real life" behavior of the Li-ion battery, whose dynamics is seriously affected by temperature, reflected in the AOSE robustness for two scenarios including changes in SOC initial value from 70% to 30%, as is shown in the Fig. 11 at 25 °C, and simultaneous changes in SOC initial value, a decrease by 50% in the nominal battery capacity due to aging effect, and a change in temperature from 25 °C to 5 °C, as is shown in Fig. 12. The simulation results reveal a very good performance for AOSE for both cases in terms of SOC estimation accuracy, fast convergence and robustness with respect to 3RC ECM Li-Ion battery model.

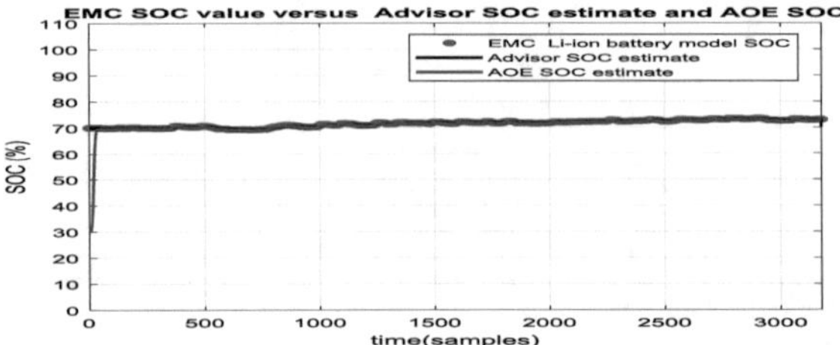

Fig. 11. The AOSE performance such as accuracy, convergence speed, and robustness at 25 °C

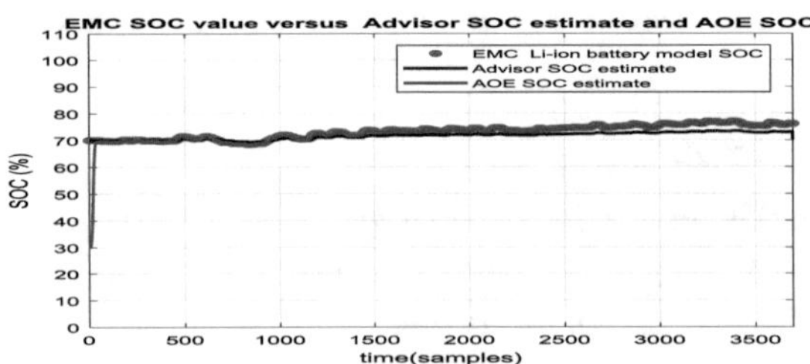

Fig. 12. The AOSE performance such as accuracy, convergence speed, and robustness at 5 °C

5 Conclusions

The MATLAB SIMULINK simulation results of the proposed AOSE real time implementation in terms of SOC estimation accuracy, convergence speed and robustness, are promising. For future work we are focusing to extend AOSE application to a wide range of similar applications in automotive industry field, for different types of batteries.

References

1. Jayam, A.P., Ferdowsi, M.: Comparison of NiMH and Li-Ion batteries in automotive applications. In: Proceedings of the IEEE Vehicle Power and Propulsion Conference, pp. 1–6. IEEE Xplore Digital Library (2008)
2. Tudoroiu, R.-E., Zaheeruddin, M., Radu, M.-S., Tudoroiu, N.: Real-time implementation of an extended kalman filter and a pi observer for state estimation of rechargeable li-ion batteries in hybrid electric vehicle applications—a case study. J. Batteries **4**(2), 19 (2018). https://doi.org/10.3390/batteries4020019
3. Farag, M.: Lithium-Ion batteries, modeling and state of charge estimation. Master's Thesis. McMaster University of Hamilton, Hamilton, ON, Canada (2013)
4. Plett, G.L.: Extended Kalman filtering for battery management systems of LiPB-based HEV battery packs: Part 2. Modeling and identification. J. Power Sources **134**, 262–276 (2004)
5. Lakkis, M.E., Sename, O., Corno, M., Bresch, P.D.: Combined battery SOC/SOH estimation using a nonlinear adaptive observer. In: Proceedings of the European Control Conference, Linz, Austria, pp. 1–6 (2015)

Properties of One Method for the Spline Approximation

I. O. Astionenko[1(✉)], P. I. Guchek[1,2], A. N. Khomchenko[3], O. I. Litvinenko[1], and G. Ya. Tuluchenko[1]

[1] Kherson National Technical University, Kherson, Ukraine
{astia, tuluchenko.galina}@ukr.net, phuchek@gmail.com, mmkntu@gmail.com
[2] Nalecz Institute of Biocybernetics and Biomedical Engineering, Warsaw, Poland
[3] Petro Mohyla Black Sea State University, Mykolayiv, Ukraine
khan@kma.mk.ua

Abstract. In the article the influence of the basic functions used to represent a polynomial from the current spline link on the approximating properties of the semi-local smoothing spline, proposed by D.A. Silaev, is studied. When constructing splines of this type, a recurrence formula that binds the group of coefficients of a polynomial from a previous spline link with a similar group of polynomial coefficients from the current spline link, is used. Silayev D.A. were studied the properties of the spline using only a power basis.

It is shown that the study of the magnitude of the eigenvalues of the matrix of stability, which is used in the algorithm for constructing the investigated spline, is not enough to predict the approximation properties of this spline. The accuracy of the approximation is also significantly influenced by the number of conditionality of the matrix, which is a block from the traditional matrix of the least squares method. It is shown that the transition to polynomials, which are presented in the form of certain Hermite polynomials, is expedient. When using semi-local splines, the number of spline units decreases in comparison with the interpolation spline on the same grid. But a significant reduction in the description of the spline does not lead to a marked deterioration in the accuracy of the solutions of boundary value problems solved with the help of splines of the investigated species. The obtained theoretical results are confirmed at solving practical problems.

Keywords: Recurrent spline · Stability matrix · Hermite basic polynomial

1 Introduction

In the works [1–3], the existence and uniqueness of semi-local smoothing splines of arbitrary degrees and the order of smoothness on uniform grids is proved. The peculiarity of constructing such splines is to divide the coefficients into a polynomial that describes the current link spline into two groups. The coefficients of the first group are determined by the conditions of the smoothness of the joints of the spline units. The coefficients of the second group are determined using the least squares method. The

J. Świątek et al. (Eds.): ISAT 2018, AISC 853, pp. 49–60, 2019.
https://doi.org/10.1007/978-3-319-99996-8_5

stability of the algorithm is ensured by finding the optimal ratio of the previous Mh and the final mh length of the current spline link (h is the grid step). The ratio is the optimal one in which the largest unit of the own number of the matrix of stability is the smallest among all possible values for a fixed value M. The stability matrix binds the value of the coefficients to the polynomial from the first group for the current and next spline nodes.

To describe the spline series in the works [1–3], only polynomials are used in the power bases. In our previous work [4], it has been shown that it is expedient to use also Hermite polynomials for cubic splines of zero and first order of smoothness. When solving, by means of splines, of boundary problems with differential equations in second order partial derivatives, it is logical to use splines of second order of smoothness. In this paper, we restrict ourselves to considering half-plane splines of the third and fifth degrees of second order of smoothness.

We shall show that the approximation properties of these splines can be significantly improved by converting to bases based on Hermite polynomials.

2 Main Results of the Study

2.1 Construction of a Semi-local Spline of the Third Degree of Second Order of Smoothness

The Algorithm for Spline Construction

Let a uniform net Δ with a $(N + 1)$ node in step h is be given. Initially, the length of the spline link will be assumed to be Mh. With each link spline we will connect it's local coordinate system, combining its beginning with the first knot of the link spine. Thus, in the local coordinate system, each link spline coincides with the segment $[0; Mh]$. Let us expand in a local coordinate system a polynomial which describing the current link spline, for polynomial basis functions of degree no greater than the third:

$$N = DX, \tag{1}$$

where N – vector of basic functions; D – matrix of coefficients of basic functions; X – vector column of power monomials $X = \begin{pmatrix} 1 & x & x^2 & x^3 \end{pmatrix}^T$.

Then the polynomial from the current link spline can be described as follows:

$$P = CN, \tag{2}$$

where $C = \begin{pmatrix} c_0 & c_1 & c_2 & c_3 \end{pmatrix}$ – vector-row of weight coefficients of the basis functions.

By general formulas from [1–3] it is easy to obtain in an explicit form a system of equations for determining the spline of the third degree of second order of smoothness. In this case, the three coefficients of the polynomial (2) will be determined by the conditions of smooth bonding of the spline links to the second order inclusive. The conditions of smooth bonding of spline links lead to a system of equations:

$$\begin{cases} P^l(mh) = P^{l+1}(0) \\ \left.\frac{dP^l(x)}{dx}\right|_{x=mh} = \left.\frac{dP^{l+1}(x)}{dx}\right|_{x=0} \\ \left.\frac{d^2P^l(x)}{dx^2}\right|_{x=mh} = \left.\frac{d^2P^{l+1}(x)}{dx^2}\right|_{x=0} \end{cases}, \quad (3)$$

where l – number of the current link spline.

The system (3) in the matrix form can be written as follows:

$$\left.\left(D\tilde{X}\right)^T\right|_{x=mh} \cdot C_l^T = \left.\left(D\tilde{X}\right)^T\right|_{x=0} \cdot C_{l+1}^T, \text{ where } \tilde{X} = \begin{pmatrix} 1 & 0 & 0 \\ x & 1 & 0 \\ x^2 & 2x & 2 \\ x^3 & 3x^2 & 6x \end{pmatrix}. \quad (4)$$

Note that the matrix $\left(D\tilde{X}\right)^T$ has three rows and four columns. Regarding the grouping of coefficients from different conditions, the left-hand side of equality (4) will be rewritten in the form of two terms, and the right-hand side of equality is obliquely:

$$\left.\begin{pmatrix} N_1 & N_2 & N_3 \\ \frac{dN_1}{dx} & \frac{dN_2}{dx} & \frac{dN_3}{dx} \\ \frac{d^2N_1}{dx^2} & \frac{d^2N_2}{dx^2} & \frac{d^2N_3}{dx^2} \end{pmatrix}\right|_{x=mh} \cdot \begin{pmatrix} c_0^l \\ c_1^l \\ c_2^l \end{pmatrix} + \left.\begin{pmatrix} N_4 \\ \frac{dN_4}{dx} \\ \frac{d^2N_4}{dx^2} \end{pmatrix}\right|_{x=mh} \cdot c_3^l = \begin{pmatrix} c_0^{l+1} \\ c_1^{l+1} \\ c_2^{l+1} \end{pmatrix}. \quad (5)$$

In the works [1–3] for the matrixes of the system (5) the following notation is introduced:

$$B_0 \cdot \begin{pmatrix} c_0^l \\ c_1^l \\ c_2^l \end{pmatrix} + B_1 \cdot c_3^l = \begin{pmatrix} c_0^{l+1} \\ c_1^{l+1} \\ c_2^{l+1} \end{pmatrix}. \quad (6)$$

The matrix of the least squares method for finding the coefficient c_3^l is formed according to the traditional rules:

$$F = \sum_{i=0}^{M} (P_l(\tilde{x}_i) - \tilde{y}_i)^2 \rightarrow \min\left(c_3^l\right), \quad (7)$$

where $(\tilde{x}_i; \tilde{y}_i)$ – points of the experimental sequence which corresponding to the current spline link.

The solving of the problem for minimization of the functional (7) leads to the equation:

$$\begin{aligned}&\left(\sum_{i=0}^{M} N_1(\tilde{x}_i)N_4(\tilde{x}_i) \quad \sum_{i=0}^{M} N_2(\tilde{x}_i)N_4(\tilde{x}_i) \quad \sum_{i=0}^{M} N_3(\tilde{x}_i)N_4(\tilde{x}_i)\right) \cdot \begin{pmatrix} c_0^l \\ c_1^l \\ c_2^l \end{pmatrix} \\ &+ \left(\sum_{i=0}^{M} N_3(\tilde{x}_i)N_4(\tilde{x}_i)\right) \cdot \left(c_2^l\right) = \left(\sum_{i=0}^{M} \tilde{y}_i \cdot N_4(\tilde{x}_i)\right)\end{aligned} \tag{8}$$

or in accordance with the notation which used in [1–3], the last equation can be written as follows:

$$A_0^l \cdot \begin{pmatrix} c_0^l \\ c_1^l \\ c_2^l \end{pmatrix} + A_1^l \cdot \left(c_3^l\right) = P^l. \tag{9}$$

Note that in the Eq. (8) the matrix consists of one element.

From Eq. (9) we find that

$$\left(c_3^l\right) = \left(A_1^l\right)^{-1} \cdot \left(P^l - A_0^l \cdot \begin{pmatrix} c_0^l \\ c_1^l \\ c_2^l \end{pmatrix}\right). \tag{10}$$

Combining system (6) and Eq. (10), we obtain a recursive formula for calculating coefficients for the link of spline:

$$\begin{aligned}\begin{pmatrix} c_0^{l+1} \\ c_1^{l+1} \\ c_2^{l+1} \end{pmatrix} &= B_1 \cdot \left(A_1^l\right)^{-1} \cdot P^l + \left(B_0 - B_1 \cdot \left(A_1^l\right)^{-1} \cdot A_0^l\right) \cdot \begin{pmatrix} c_0^l \\ c_1^l \\ c_2^l \end{pmatrix} \\ &= B_1 \cdot \left(A_1^l\right)^{-1} \cdot P^l + U \cdot \begin{pmatrix} c_0^l \\ c_1^l \\ c_2^l \end{pmatrix}\end{aligned} \tag{11}$$

where $U = B_0 - B_1 \cdot \left(A_1^l\right)^{-1} \cdot A_0^l$ – matrix of stability.

In works [1–3] there is a necessary condition for the stability of the algorithm for the construction of a semi-local spline, according to which, when the modules of the eigenvalues of the matrix of stability U must be less than one. We investigate how the modulus of the eigenvalues of the matrix of stability U is affected by the change in the form of representation of the polynomial P (2), that is, the transition to another basis.

Forms of Polynomials Which Are Used in the Construction of a Spline of the Third Degree

For a polynomial in a power basis, the matrix D in Eq. (1) is a unit matrix. Consequently, the polynomial in the power base on the current link spline is described by the formula:

$$P_{Power} = C_{Power}X. \tag{12}$$

Also, we use a different form for the representation of a polynomial, namely, a polynomial in the Hermite form with two knots in which the given: for the first knot – the value of the function and its first two derivatives, for the second knot – the value of the function. As you know [5, 6], its basic functions are from the relation:

$$N_{Hermite}^{(3,1)} = D_{Hermite}^{(3,1)} \cdot X = \left(V_{Hermite}^{(3,1)}\right)^{-1} \cdot X, \tag{13}$$

where $V_{Hermite}^{(3,1)} = \begin{pmatrix} 1 & 0 & 0 & 1 \\ x_i & 1 & 0 & x_{i+1} \\ x_i^2 & 2x_i & 2 & x_{i+1}^2 \\ x_i^3 & 3x_i^2 & 6x_i & x_{i+1}^3 \end{pmatrix}$; $x_i = 0$; $x_{i+1} = Mh$; $i = \overline{0; L}$.

Note that x_i are the knots for the conglutination of the spline links and L – quantity of the spline links.

The coefficients of the Hermite polynomial of this type:

$$P_{Hermite}^{(3,1)} = C_{Hermite}^{(3,1)} \cdot N_{Hermite}^{(3,1)}. \tag{14}$$

have the following geometric meaning:

$$C_{Hermite}^{(3,1)} = \left(P_{Hermite}^{(3,1)}\Big|_{x=0}; \quad \frac{dP_{Hermite}^{(3,1)}}{dx}\Big|_{x=0}; \quad \frac{d^2P_{Hermite}^{(3,1)}}{dx^2}\Big|_{x=0}; \quad P_{Hermite}^{(3,1)}\Big|_{x=Mh} \right).$$

Eigenvalues of Stability Matrix for the Spline of the Fifth Degree with the Second Order of Smoothness

We compute for different combinations of values M and m the values of the modulus of the eigenvalues of the stability matrix U from the recurrence formula (11) for polynomials with a power basis (12) and the basis of Hermite (14). To Table 1, we introduce the smallest values of the modules of the eigenvalues of the stability matrix U for a fixed value M, which are selected among the largest values of the modules of eigenvalues, which correspond to the possible value $m = \overline{1; M}$.

Table 1. Parameters of the stability of the algorithm for constructing a cubic spline of second order of smoothness

M	$\min \max_M \lvert\lambda_i\rvert$			
	m	Polynomial with power basis (12)	m	Hermite Polynomial (14)
5	2	0,890	3	0,623
7	4	0,774	4	0,630
9	5	0,778	5	0,634
11	6	0,781	7	0,636
13	7	0,783	8	0,632
15	8	0,784	9	0,630

For lower smoothness orders, the researched characteristic of the stability matrix does not depend on the choice of the basic functions (1) for the polynomial (2) [4].

Testing of a Semi-local Spline of the Third Degree with the Second Order of Smoothness

Test 1. We evaluate the accuracy of the approximation of the experimental dependence which is generated on the basis of the function $f(x) = \sin x$. The choice of the test function is explained by its subsequent application in the article when solving the boundary value problem. The generated sequence contains 50 points which are generated for $x_i = (i-1)h$, where $i = \overline{1;50}$; $h = 0,1$. The estimation will be carried out by the metric C on all points of the spline net $S_3(x)$:

$$\delta = \max_i |\sin x_i - S_3(x_i)|.$$

When constructing a spline, the values of m were chosen to be optimal for each M of Table 1 depending on the basis which is used.

According to Table 2 it is obvious that the application of the basis for the Hermite polynomial (13) has significant advantages. We also note that in works [1–3] the question of the application of other bases, except for power, for the representation of polynomials in the current spline link is not investigated.

Table 2. Estimation for the accuracy of approximation of the experimental dependence by spline $S_3(x)$ in metric C

M	Basis					
	Powerful (12)			Hermite (13)		
	m	L	δ	m	L	δ
5	2	23	0,022	3	15	$2,5 \cdot 10^{-4}$
7	4	11	0,018	4	11	$9,7 \cdot 10^{-4}$
9	5	9	0,028	5	9	$2,7 \cdot 10^{-3}$
11	6	7	0,041	7	6	$3,8 \cdot 10^{-3}$
13	7	6	0,061	8	5	$8,6 \cdot 10^{-3}$
15	8	5	0,082	9	4	$1,7 \cdot 10^{-2}$

2.2 Construction of a Semi-local Spline of the Fifth Degree with Second Order of Smoothness

The Algorithm for Spline Construction

When constructing splines of the fifth degree, the formulas (1–4) remain unchanged. In accordance with the degree spline, the vectors change:

$$X = \begin{pmatrix} 1 & x & x^2 & x^3 & x^4 & x^5 \end{pmatrix}^T \text{and}\, C = \begin{pmatrix} c_0 & c_1 & c_2 & c_3 & c_4 & c_5 \end{pmatrix}.$$

There is also a change in the basis, which now consists of 6 basic functions. We write the system (5) for a new basis:

$$\left.\begin{pmatrix} N_1 & N_2 & N_3 \\ \frac{dN_1}{dx} & \frac{dN_2}{dx} & \frac{dN_3}{dx} \\ \frac{d^2N_1}{dx^2} & \frac{d^2N_2}{dx^2} & \frac{d^2N_3}{dx^2} \end{pmatrix}\right|_{x=mh} \cdot \begin{pmatrix} c_0^l \\ c_1^l \\ c_2^l \end{pmatrix} + \left.\begin{pmatrix} N_4 & N_5 & N_6 \\ \frac{dN_4}{dx} & \frac{dN_5}{dx} & \frac{dN_5}{dx} \\ \frac{d^2N_4}{dx^2} & \frac{d^2N_5}{dx^2} & \frac{d^2N_6}{dx^2} \end{pmatrix}\right|_{x=mh} \cdot \begin{pmatrix} c_3^l \\ c_4^l \\ c_5^l \end{pmatrix} = \begin{pmatrix} c_0^{l+1} \\ c_1^{l+1} \\ c_2^{l+1} \end{pmatrix}. \tag{15}$$

Consequently, the matrix Eq. (15) now takes the form of:

$$B_0 \cdot \begin{pmatrix} c_0^l \\ c_1^l \\ c_2^l \end{pmatrix} + B_1 \cdot \begin{pmatrix} c_3^l \\ c_4^l \\ c_5^l \end{pmatrix} = \begin{pmatrix} c_0^{l+1} \\ c_1^{l+1} \\ c_2^{l+1} \end{pmatrix}. \tag{16}$$

Matrix of the least squares method for finding coefficients c_3^l, c_4^l, c_5^l is formed according to the traditional rules:

$$F = \sum_{i=0}^{M} (P_l(\tilde{x}_i) - \tilde{y}_i)^2 \to \min\left(c_3^l; c_4^l; c_5^l\right), \tag{17}$$

where $(\tilde{x}_i; \tilde{y}_i)$ – points of the experimental sequence which corresponding to the current spline link.

The solving of the problem for minimizing of the functional (17) leads to the equation:

$$\begin{pmatrix} \sum\limits_{i=0}^{M} N_1(\tilde{x}_i)N_4(\tilde{x}_i) & \sum\limits_{i=0}^{M} N_2(\tilde{x}_i)N_4(\tilde{x}_i) & \sum\limits_{i=0}^{M} N_3(\tilde{x}_i)N_4(\tilde{x}_i) \\ \sum\limits_{i=0}^{M} N_1(\tilde{x}_i)N_5(\tilde{x}_i) & \sum\limits_{i=0}^{M} N_2(\tilde{x}_i)N_5(\tilde{x}_i) & \sum\limits_{i=0}^{M} N_3(\tilde{x}_i)N_5(\tilde{x}_i) \\ \sum\limits_{i=0}^{M} N_1(\tilde{x}_i)N_6(\tilde{x}_i) & \sum\limits_{i=0}^{M} N_2(\tilde{x}_i)N_6(\tilde{x}_i) & \sum\limits_{i=0}^{M} N_3(\tilde{x}_i)N_6(\tilde{x}_i) \end{pmatrix} \cdot \begin{pmatrix} c_0^l \\ c_1^l \\ c_2^l \end{pmatrix}$$
$$+ \begin{pmatrix} \sum\limits_{i=0}^{M} N_4(\tilde{x}_i)N_4(\tilde{x}_i) & \sum\limits_{i=0}^{M} N_5(\tilde{x}_i)N_4(\tilde{x}_i) & \sum\limits_{i=0}^{M} N_6(\tilde{x}_i)N_4(\tilde{x}_i) \\ \sum\limits_{i=0}^{M} N_4(\tilde{x}_i)N_5(\tilde{x}_i) & \sum\limits_{i=0}^{M} N_5(\tilde{x}_i)N_5(\tilde{x}_i) & \sum\limits_{i=0}^{M} N_6(\tilde{x}_i)N_5(\tilde{x}_i) \\ \sum\limits_{i=0}^{M} N_4(\tilde{x}_i)N_6(\tilde{x}_i) & \sum\limits_{i=0}^{M} N_5(\tilde{x}_i)N_6(\tilde{x}_i) & \sum\limits_{i=0}^{M} N_6(\tilde{x}_i)N_6(\tilde{x}_i) \end{pmatrix} \cdot \begin{pmatrix} c_3^l \\ c_4^l \\ c_5^l \end{pmatrix} = \begin{pmatrix} \sum\limits_{i=0}^{M} \tilde{y}_i \cdot N_4(\tilde{x}_i) \\ \sum\limits_{i=0}^{M} \tilde{y}_i \cdot N_5(\tilde{x}_i) \\ \sum\limits_{i=0}^{M} \tilde{y}_i \cdot N_6(\tilde{x}_i) \end{pmatrix} \tag{18}$$

The system (18) in the symbols used in the article is briefly written as follows:

$$A_0^l \cdot \begin{pmatrix} c_0^l \\ c_1^l \\ c_2^l \end{pmatrix} + A_1^l \cdot \begin{pmatrix} c_3^l \\ c_4^l \\ c_5^l \end{pmatrix} = P^l, \tag{19}$$

From Eq. (19) we find that

$$\begin{pmatrix} c_3^l \\ c_4^l \\ c_5^l \end{pmatrix} = \left(A_1^l\right)^{-1} \cdot \left(P^l - A_0^l \cdot \begin{pmatrix} c_0^l \\ c_1^l \\ c_2^l \end{pmatrix}\right). \tag{20}$$

By combining system (16) and Eq. (20), we obtain a recurrent formula whose form completely coincides with formula (11), taking into account the new meaning of the notation.

Forms of Polynomials Which Are Used by the Construction of Spline with Fifth Degree

For a polynomial in the power basis, formula (12) holds with the new content of the vectors X and C.

Let's consider all possible cases of Hermite polynomials of the fifth degree with two and three knots [7]. Their construction differs only in the matrix $V_{Hermite}^{(5,j)}$, where j is the number of the type of the Hermite polynomial, which is considered in this article, and the geometric content of the coefficients $C_{Hermite}^{(5,j)}$:

$$P_{Hermite}^{(5,j)} = C_{Hermite}^{(5,j)} \cdot N_{Hermite}^{(5,j)}, \tag{21}$$

where $N_{Hermite}^{(5,j)} = D_{Hermite}^{(5,j)} \cdot X = \left(V_{Hermite}^{(5,j)}\right)^{-1} \cdot X$.

Consider the following types of bases of Hermite:

(1) for Hermite polynomial $P_{Hermite}^{(5,1)}$ with two knots in which the values of the function and its derivatives of the first two orders are given;

(2) for the Hermite polynomial $P_{Hermite}^{(5,2)}$ with two knots for which the values of the function and its derivatives of the first three orders are given in the first knot, and in the second knot –the values of the function and its first derivative;

(3) for Hermite polynomial $P_{Hermite}^{(5,3)}$ with two knots, for which in the first knot a function value and its derivatives of the first four orders are given, and in the second knot – the value of the function is:

(4) for Hermite polynomial $P_{Hermite}^{(5,4)}$ with three knots $\left(x_i < x_{i,0} < x_{i+1}\right)$ for which in the first knot the values of a function and its derivatives of the first two orders are given, in the second knot – the value of the function, in the third node – the value of the function and its first derivative:

(5) for the Hermite polynomial $P^{(5,5)}_{Hermite}$ with three knots $\left(x_i < x_{i,0} < x_{i+1}\right)$ for which in the first knot the values of the function and its derivatives of the first two orders are given; in the second knot – the value of the function and its first derivative, in the third knot – the value of the function:

$$V^{(5,5)}_{Hermite} = \begin{pmatrix} 1 & 0 & 0 & 1 & 0 & 1 \\ x_i & 1 & 0 & x_{i,0} & 1 & x_{i+1} \\ x_i^2 & 2x_i & 2 & x_{i,0}^2 & 2x_{i,0} & x_{i+1}^2 \\ x_i^3 & 3x_i^2 & 6x_i & x_{i,0}^3 & 3x_{i,0}^2 & x_{i+1}^3 \\ x_i^4 & 4x_i^3 & 12x_i^2 & x_{i,0}^4 & 4x_{i,0}^3 & x_{i+1}^4 \\ x_i^5 & 5x_i^4 & 20x_i^3 & x_{i,0}^5 & 5x_{i,0}^4 & x_{i+1}^5 \end{pmatrix}; \quad (22)$$

$$C^{(5,5)}_{Hermite} = \left(P^{(5,5)}_{Hermite}\Big|_{x=0}; \quad \frac{dP^{(5,5)}_{Hermite}}{dx}\Big|_{x=0}; \quad \frac{d^2P^{(5,5)}_{Hermite}}{dx^2}\Big|_{x=0}; \right.$$

$$\left. P^{(5,5)}_{Hermite}\Big|_{x=\left[\frac{M}{2}\right]\cdot h}; \quad \frac{dP^{(5,5)}_{Hermite}}{dx}\Big|_{x=\left[\frac{M}{2}\right]\cdot h}; \quad P^{(5,5)}_{Hermite}\Big|_{x=Mh} \right).$$

Explicit expressions for matrixes are given only for one kind of Hermite polynomial (22), which showed the best approximation properties and is used later.

Characteristics of Matrixes from the Algorithm for Construction of Spline of the Fifth Degree with the Second Order of the Smoothness

When constructing a recurring spline by D.A. Silaev algorithm matrix of the system for finding the part of coefficients for the spline link using the least squares method (LSM) consists of a block of the traditional matrix for this method. In this case, from a block $M_{3,3}$ which is consisting of the last three rows and columns of the Gram matrix G [8]. As you know, the LSM matrix coincides with the Gram matrix. To calculate its numbers of conditionality using different forms of representation of a polynomial, without losing the universality of the results, let's put $[x_i; x_{i+1}] = [0; 1]$. We will also assume that $x_{i,0} = (x_i + x_{+1})/2$ [9].

The calculation of the eigenvalues of the matrix of stability showed that they do not depend on the choice of the basis functions (1) of the polynomial (2) among the five types of Hermite basic functions (21) and the power basis.

Testing of a Semi-local Spline of the Fifth Degree with the Second Order of Smoothness

Test 2. All conditions for the test 1 will be left unchanged for the spline of the fifth degree spline $S_5(x)$. Taking into account the results of Tables 3 and 4, we will carry out testing with the power base and the basis of Hermite $N^{(5,5)}_{Hermite}$, which we obtain according to the formulas (21–22). The results of the computational experiment are classified in the Table 5.

According to Table 5 it is obvious that the use of the basis of the Hermite polynomial $N^{(5,5)}_{Hermite}$ (21–22) has significant advantages.

Table 3. Number of conditionality of the Gram matrix G and of the its block $M_{3,3}$

Polynomial	Number of conditionality	
	of the Gram matrix G	of the matrix block $M_{3,3}$
$P^{(5,1)}_{Hermite}$	$2,056 \cdot 10^6$	$1,977 \cdot 10^5$
$P^{(5,2)}_{Hermite}$	$8,487 \cdot 10^7$	77268,931
$P^{(5,3)}_{Hermite}$	$2,360 \cdot 10^9$	$3,050 \cdot 10^6$
$P^{(5,4)}_{Hermite}$	$1,102 \cdot 10^6$	2884,545
$P^{(5,5)}_{Hermite}$	$1,450 \cdot 10^6$	15,930

Table 4. Examples of estimating the modules of the eigenvalues of the stability matrix for spline of the fifth degree with the second order of smoothness

M	m	$\min_M \max\lvert\lambda_i\rvert$	M	m	$\min_M \max\lvert\lambda_i\rvert$
6	3	0,267	11	7	0,204
7	4	0,226	12	7	0,221
8	5	0,205	13	8	0,208
9	5	0,236	14	9	0,203
10	6	0,214	15	9	0,213

Table 5. Estimation of the accuracy of the approximation of the experimental dependence by the spline $S_5(x)$ in the metric C

M	m	L	$\delta = \max_i \lvert\sin x_i - S_5(x_i)\rvert$	
			Powerful basis of the fifth degree	Basis of Hermite $N^{(5,5)}_{Hermite}$ (21–22)
6	3	15	0,0016	$1,53 \cdot 10^{-7}$
7	4	11	0,0037	$2,55 \cdot 10^{-7}$
9	5	9	0,0051	$1,0 \cdot 10^{-6}$
11	7	6	0,015	$4,6 \cdot 10^{-6}$
13	8	5	0,016	$1,1 \cdot 10^{-5}$
15	8	5	0,014	$2,1 \cdot 10^{-5}$

2.3 Application of Investigated Splines at Numerical Solution of Boundary Value Problem

In [10], a numerical solution of the boundary value problem for a stationary two-dimensional heat equation for a region in the form of an infinite band is obtained, when the power function of thermal sources is interpolated by a cubic spline. Let's solve this problem, applying the studied splines, and compare the accuracy of the solutions of the boundary value problem.

Let the temperature distribution in an infinite band with a rectangular cross-section dimension $a \times h$ is described by the equation:

$$\lambda \frac{\partial^2 T(x;z)}{\partial x^2} + \lambda \frac{\partial^2 T(x;z)}{\partial z^2} + \frac{I^2 \rho_0}{S^2} \cdot \phi(x) \cdot f(z) = 0 \tag{23}$$

with boundary conditions:

$$\left.\frac{\partial T(x;z)}{\partial x}\right|_{x=0} = \left.\frac{\partial T(x;z)}{\partial x}\right|_{x=a} = 0; \tag{24}$$

$$\begin{gathered} \lambda \left.\frac{\partial T(x;z)}{\partial z}\right|_{z=-h/2} = -\alpha(T_{bond} - T_\infty); \lambda \left.\frac{\partial T(x;z)}{\partial z}\right|_{z=h/2} = -\alpha(T_{bond} - T_\infty) \\ 0 \le x \le a; -\tfrac{h}{2} \le z \le \tfrac{h}{2}. \end{gathered} \tag{25}$$

In the Eq. (23) and boundary conditions (24–25) use the notation: $\phi(x) \cdot f(z)$ – power function of heat sources; I – amperage; ρ_0 – resistivity; S – area of the section; λ – coefficient of thermal conductivity; α – coefficient of heat transfer; T_{bond} and T_∞ – temperature at the boundary of the body and ambient temperature.

For solving of the boundary value problem (23–25) in [10] the transition to dimensionless coordinates: $\xi = \frac{x}{a}$; $\zeta = \frac{z}{h}$, – and the dimensionless temperature difference, which is formed by the characteristic temperatures of the process $T1$ and $T2$: $\Delta\overline{T} = \frac{T - T_1}{T_2 - T_1}$, – is carried out. Also introduced in the use of the Biot number $Bi = \frac{\alpha h}{\lambda}$ and Pomeranzev number $Po = \frac{I^2 \rho_0 a h}{\lambda (T_2 - T_1) S^2}$.

After the introduction of all these symbols, the problem (23–25) takes the form:

$$\frac{h}{a} \cdot \frac{\partial^2 T(\xi;\zeta)}{\partial \xi^2} + \frac{a}{h} \cdot \frac{\partial^2 T(\xi;\zeta)}{\partial \zeta^2} + Po \cdot \phi(\xi) \cdot f(\zeta) = 0 \tag{26}$$

with boundary conditions:

$$\left.\frac{\partial \Delta\overline{T}}{\partial \xi}\right|_{\xi=0} = \left.\frac{\partial \Delta\overline{T}}{\partial \xi}\right|_{\xi=1} = 0; \tag{27}$$

$$\begin{gathered} \left.\frac{\partial \Delta\overline{T}}{\partial \zeta}\right|_{\zeta=-1/2} = -Bi\Delta\overline{T}; \left.\frac{\partial \Delta\overline{T}}{\partial \zeta}\right|_{\zeta=1/2} = -Bi\Delta\overline{T}; \\ 0 \le \xi \le 1; -\tfrac{1}{2} \le \zeta \le \tfrac{1}{2}. \end{gathered} \tag{28}$$

To solve the boundary value problem (26–28), the power function of the sources $\phi(\xi)$ is presented in the form of a trigonometric series:

$$\phi(\xi) = \sum_{s=0}^{\infty} a_s \cos(s\pi\xi). \tag{29}$$

The values of the coefficients for the series (29) will be found approximately from the decomposition in the Fourier series by cosine spline $S(\xi)$, which approximates the function $\phi(\xi)$:

$$a_0 = \int_0^1 S(\xi)d\xi; a_s = 2\int_0^1 S(\xi)\cos(s\pi\xi)d\xi. \quad (30)$$

We have repeated the solving algorithm is given in [5] and we obtain a numerical solution of the test boundary value problem with same known functions and quantities. When using the studied splines with Hermite bases: the third degree (13–14) and the fifth degree (21–22) for the values M = 5..7, the absolute error does not exceed 10^{-2} by all the knots of the grids, along with the solution given in work [10].

2.4 Conclusions

In this paper, we propose, for the prediction of the approximation properties of semi-local splines, which are constructed in works [1–3], additionally use the number of conditionality of a matrix, which is a separate block of the traditional Gram matrix for basic functions. It is shown that among polynomials in the Hermite forms there are those that lead to a significant improvement of both stability characteristics of the computational algorithm for constructing a researched spline: the largest module of eigenvalues of the stability matrix and the number of conditionality of the dedicated block of the Gram matrix. Calculated experiments confirmed the obtained theoretical results. Solutions of problems are find with recommended bases have better accuracy.

References

1. Silaev, D.A.: Polulokal'niye sglazhivayushchie splayny. Trudy Semin. I. G. Petrovsk. **29**, 443–454 (2013)
2. Silaev, D.A.: Polulokal'nye sglazhivayushchie S-splayny. Komp'yuternye Issled. Modelirovanie **2**(4), 349–357 (2010)
3. Silaev, D.A., et al.: Polulokal'nye sglazhivayushchie splayny klassa $C1$. Trudy Semin. Imeni I. G. Petrovsk. **26**, 347–367 (2007)
4. Tuluchenko, G., et al.: Generalization of one algorithm for constructing recurrent splines. East. Eur. J. Enterp. Technol. **2–4**(92), 53–62 (2018)
5. Pineshaninov, F., Pineshaninov, P.: Bazisnye funktsii dlya konechnyh elementov [Electron resource]. http://old.exponenta.ru/soft/mathemat/pinega/a1/a1.asp
6. Astionenko, I.O., et al.: Cognitive-graphic method for constructing of hierarchical forms of basic functions of biquadratic finite element. In: Application on Mathematics in Technical and Natural Science, vol. 1773(1), pp. 040002-1–040002-11 (2016). https://doi.org/10.1063/1.4964965
7. Wang, Y.: Smoothing Splines: Methods and Applications. CRC Press, London (2011)
8. Hatmaher, F.R.: Teoria matrits. Phizmatlit, Moscow (2010)
9. Kalitkin, N.N., Shlyahov, N.M.: Simmetrizatsia globalnyh splainov. Mat. Model. **11**(8), 116–126 (1999)
10. Chernenko, V.P., Kobylskaya, E.B.: Primenenie kubicheskogo splina pri chislennom reshenii kraevoi zadachi. Visnyk KDPU Myhaila Ostrogradskogo **6**(53), 38–40 (2008)

An Effective Algorithm for Testing of O–Codes

Ho Ngoc Vinh(✉)

Vinh University of Technology Education, Vinh, Vietnam
hnvinh.skv@moet.edu.vn

Abstract. An extending approach to the concept of the product of context and ambiguous. In this article, we present the concept of overlap product, where contextual words are inserted among code words and strings reduced by the common overlapping context. Thus, the concept of code on the basis of overlap product is created, also called *O*–code. The initial results on the properties and conditions for decoding are the basis to establish an effective algorithm for testing of *O*–codes with the complexity of n^3.

Keywords: Code theory · Code testing · Overlap product · Zigzag code *O*–code

1 Introduction

Recently, the study on code theory tends to apply the concept of context to extend the concept of product and develop new classes of codes. To enrich the properties of code theory, there are a lot of findings of unambiguous product (proposed by Schützenberger [1]) in relation to automate, algebra, code… because they are the parameters to assess the difficulties in decoding and encoding. In [2], Weil applied unambiguous automat and monoid as a tool to establish a product $X = Y \circ Z$, where Y, Z are finite codes in relation to complex theory of language. In [3, 4], Huy and Van established a result to express ω-regular languages of infinite words which is accepted by nonambiguous Büchi *V*-automata as disjoint finite union of a type of unambiguous products of languages and ω–languages whose syntactic monoids are in *V*, where *V* is a variety of finite monoids closed under Schützenberger product. The concept of +–unambiguous product and alternative code, the even alternative code of two languages *X*, *Y* on A^* and some properties of +–unambiguous product, necessary and sufficient properties so that a pair (*X*,*Y*) is an alternative code, even alternative code are considered by Vinh-Huy-Nam [5, 6].

In common codes, a product of two words is to place them next to each other, and the set *X* is a code if any word only has a unique word factorization from the left. There are many extending study applying different techniques such as two way factorization (*Z*–code), controlling code, alternative code,… In this paper, we mention another extending approach, applying contextual product to study code properties on the basis of new product.

First of all, we review some symbols and concepts presented in [7, 8]. Given that *A* is a finite alphabet. A^* is a free monoid given by *A*, with the product and the unit element of ε (empty word) and $A^+ = A^* - \{\varepsilon\}$. A word $u \in A^*$ is *a factor* (*prefix, suffix*) of a word

J. Świątek et al. (Eds.): ISAT 2018, AISC 853, pp. 61–70, 2019.
https://doi.org/10.1007/978-3-319-99996-8_6

$v \in A^*$ if there exist $x, y \in A^*$ such that $v = xuy$ (resp. $v = uy$, $v = xu$). A factor (prefix, suffix) u of v is proper if $xy \neq \varepsilon$ (resp. $y \neq \varepsilon$, $x \neq \varepsilon$).

The number of appearances of characters in the word u is the length of u, denoted as $|u|$, the convention is that $|\varepsilon| = 0$. Given $X \subseteq A^+, w \in A^*$, *w accepts an factorization by the product of words in X* if there exists a sequence $w_1, w_2, \ldots, w_n$, with $n \geq 1, w_i \in X, \forall i \leq n$ so that $w = w_1 w_2 \ldots w_n$. X is considered a *code* if every word $w \in A^+$ has at most one way to factorization into words in X. Supposing that $X, Y \subseteq A^*$, the left quotient (the right quotient) of X and Y is the language $Y^{-1}X$ (resp. XY^{-1}) defined by $Y^{-1}X = \{w \in A^* \mid yw \in X, y \in Y\}$ and $XY^{-1} = \{w \in A^* \mid wy \in X, y \in Y\}$. Shortly, we denote $(A^*)^{-1}X$ as A^-X and $X(A^*)^{-1}$ as XA^-.

2 *O*–Codes

In [9] the approach to the application of the context Y to the beginning of the massage to create the ambiguity. It is possible to create the ambiguity by inserting contextual words between other words. Instead of the concatenation in the common product, the *overlap product* of words is reduced by the common context. Thus, the concept of code on the basis of overlap product, called *O–code*.

In relation to algebra, the idea of overlap product and code on the basis of overlap product is partially derived from findings of unavoidable set.

Unavoidable Set. Given the alphabet A. The set C infix is called an unavoidable set if $\forall w \in A^*$, $|w| > N_C$, $\exists u \in C : w = xuy$, for x, y, $\in A^*$.

In other words, when a string long enough to be able to be factorized into segments separated by words in the unavoidable set C, it is a C contextual factorization (or *O*–factorization). Here, an unavoidable set is a contextual set (Fig. 1).

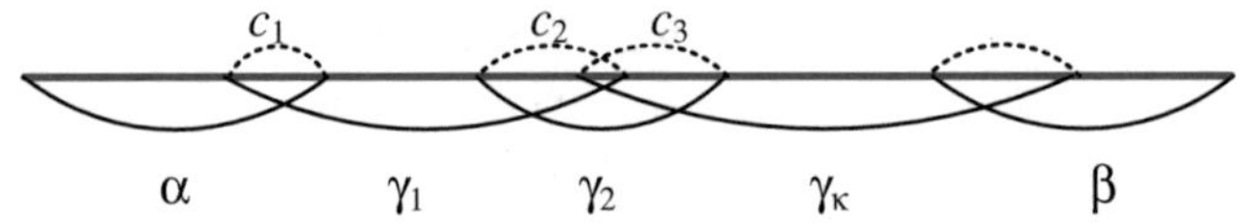

Fig. 1. *O*–factorization.

We classify words with the prefix, suffix or both prefix and suffix belonging to the unavoidable set C into three classes α_i, β_j, γ_k respectively so that no affixes belong to C.

Then, with the basic set $U_C = \{\alpha_i, \beta_j, \gamma_k\}$, every word v: $|v| > N_C$ will be factorized according to the unique *O*–factorization in U_C with $v = \alpha_i \cdot_C \gamma_1 \cdot_C \gamma_2 \cdot_C \ldots \cdot_C \gamma_n \cdot_C \beta_j$. It is shown in the following clause (Fig. 2).

Clause 2.1. *Given the unavoidable set C, the basis U_C. When every word v is long enough, there will be a unique O–factorization in U_C.*

Fig. 2. Classes α_i, β_j, γ_k belonging to C.

Example 2.1. Given the alphabet $A = \{a, b\}$, $C = \{aaa, ab, bb\}$ as an unavoidable set, $N_C = 4$. Actually, if a word begins with a, then the next character is a (if it is b then $ab \in C$), the next character is a or b, both of them belong to C. It is similar with a word beginning with b.

Then, any long enough string can be uniquely factorized with O–factorization, for example, the string $aabaabbaaab$ can be factorized as follows:

$$aabaabbaaab = aab_{\cdot ab}\, abaab_{\cdot ab}\, abb_{\cdot bb}\, bbaaa_{\cdot aaa}\, aaab$$

Definition 2.1. Given $X, C \subseteq A^+$. Then, *O–product* of x, $y \in X$ in the context C, denoted as $x._C y$, is a word with the form $x'cy'$ so that $x = x'c, y = cy' \in X$, where $c \in C$ is the longest word satisfying the above property.

We also denote $x._C y$ with $x._u y$ when u is a specific context.

Example 2.2. Given $X = \{abca, cadb\}$, $C = \{ca, b\}$. When, O–product of $abca$ and $cadb$ in the context C is identified as follows:

$$abca._C\, cadb = abca_{\cdot ca}\, cadb = ab(ca)db.$$

Then, we can give the definition of O–product of two languages. Given $X, Y, C \subseteq A^+$,

$$X._C Y = \{x._C y \,|\, x \in X, y \in Y\}$$

Then $X_1._C X_2._C \ldots ._C X_n = (X_1._C X_2._C \ldots ._C X_{n-1})._C X_n$

We denote: $X^{2C} = X._C X, X^{nC} = X._C \ldots ._C X$ and $X^{+C} = \bigcup_{i \in N} X^{iC}$.

Definition 2.2. Given $X, C \subseteq A^+$. Then, X is considered to have the associative property in the context C if: $\forall x, y, z \in X$, $(x._C y)._C z = x._C(y._C z)$.

In relation to the associative property of *O–product*, we have the following clause:

Clause 2.2. *If C is an infix set then O–product has the associative property.*

Proof. Actually, supposing that $w = x._{C}y = x._{c_1}y$, because C is an infix set, if $w._{C}z = w._{c_1}z$ then the beginning of the word C_2 cannot appear before the beginning of C_1 in w. Therefore, there are two cases (Fig. 3):

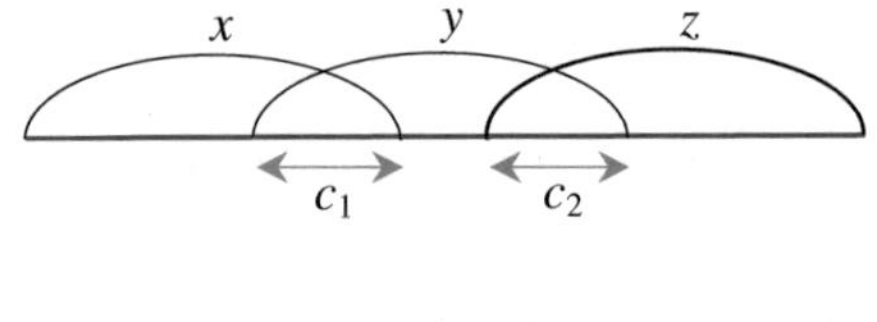

or

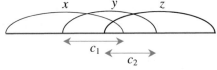

Fig. 3. Two cases of words in C.

Then $(x._{C}y)._{C}z = x._{C}(y._{C}z)$. □

From now on, if there are no special notes, we only consider the contextual infix set.

We can extend the concept of O–product as a common product (in general) as follows:

Definition 2.3. Given $C \subseteq A^+$ and $x, y \in A^+$. Then, the O–product of $x, y \in A^+$ is indentified with:

$$x._{C}y = \begin{cases} x(C \cap Pref(y))^{-1}y & \text{if } x(C \cap Pref(y))^{-1} \neq x \\ xy & \text{otherwise} \end{cases}$$

Then:

$$x_1._{C}x_2._{C}\ldots._{C}x_n = (x_1._{C}x_2._{C}\ldots._{C}x_{n-1})._{C}x_n$$

With two languages X, $Y \subseteq A^+$, we can define

$$X._{C}Y = \{x._{C}y \mid x \in X, y \in Y\}.$$

Similarly, we define $X_1._{C}X_2._{C}\ldots._{C}X_m$.

It can be seen that each factorization according to O–product is an factorization according to common product.

Zigzag code is based on zigzag factorization. The concept of zigzag factorization (two way factorizations) is presented by Anselmo in [10]. Two way factorizations allows extending the concept of common word factorization and developing the class of zigzag code. There are many studies on zigzag code, such as [11, 12].

From the concept *O*–product, if $C = Y$ then the factorization with common product becomes a zigzag factorization.

On the basis of contextual product, we can develop a new concept of code, called *O*–code.

Definition 2.4. Given $X, C \subseteq A^+$, C is an infix set. Then, X is called a *overlap code according to the context* C (shortly, *O–code*) if every word $w \in X^{+C}$ has a unique factorization according to the context C, which means that if $w = x_{0 \cdot C} x_{1 \cdot C} \cdots {}_{\cdot C} x_m = y_{0 \cdot C} y_{1 \cdot C} \cdots {}_{\cdot C} y_n (x_i, y_j \in X)$ then $m = n$ và $x_i = y_i$.

This property is called a unique *O*–factorization of w. Otherwise ($m \neq n$ or $x_i \neq y_i$ with any i), it can be seen that there are two different *O*–factorizations.

According to the above definition, if $C = \{\varepsilon\}$ then *O*–product becomes a common product and *O*–code is a common code. Hence, to some extent, a common code can be seen as a specific case of *O*–code.

The relation between *O*–code and code is shown in the following clause:

Clause 2.3. *There are sets X being O–code but not being code.*

Proof. Actually, with the alphabet $A = \{a,b,c\}$, we consider the set $X = \{cba, abbc, ca, cbaab, bcca\}$ and the context $C = \{a, b, c\}$.

It can be proved that X is a *O*–code. Supposing if there exists a word w with two factorizations then these two factorizations can only begin with *cba* and *cbaab*. We can not further expand because X does not contain words in the form of aaA^*. However, X is not a common code, for example the word $w = cbaabbcca$ has two factorizations: $(cba).(abbc).(ca) = (cbaab).(bcca)$. □

Clause 2.4. *There are sets X being code but not being O–code.*

Proof. Actually, we consider the alphabet $A = \{a, b, c, d\}$, the set $X = \{bac, cd, ba, acd\}$, the context $C = \{a, c\}$. It can be proved that X is a code, but X is not a *O*–code, for example the word $w = bacd$ has two *O*–factorizations: $ac._C\, cd = ba._C\, acd$. □

From Clause 2.3 and Clause 2.4, it is obvious that the class *O*–code and the common code are different.

3 The Algorithm for Testing of *O*–Codes

In this part, we present the procedure for testing of *O*–codes. Because of the specific characteristics of the contextual product, we present the concept of contextual quotient instead of the quotient in conventional code. Then, we develop the word with two factorizations in different contexts.

Definition 3.1. Given $X, Y \subseteq A^+$, $C \subseteq A^+$, $C' = C \cup \{\varepsilon\}$. Then, the quotient in the context C is the set $X^{-C}Y = \{u \mid \exists x \in X, y \in Y, c \in C', x._C u = y, x \neq y\}$.

Similar to common product, we define the exponential m ($m \geq 0$) according to the context C as follows:

$$X^{0C} = X \text{ và } X^{mC} = \{w._C x \mid w \in X^{m-1C}\}.$$

If $s \in X^{mC}$ then $s = x_0._C x_1._C \ldots ._C x_m$, we denote $x_i = s[i]$ (noting that x_i is a specific choice of $s[i]$ in a specific factorization, because s may have many O–factorization).

To test the code property, we will develop two different O–factorizations on the basis of the sets of contextual quotient. The steps of O–code procedure is described in the following recursive formula:

$$\begin{aligned} U_0 &= X^{-C}X \\ U_n &= (U_{n-1}^{-C}X \cup X^{-C}U_{n-1}) \cup U_{n-1}, n \geq 0. \end{aligned} \tag{3.1}$$

Give two strings $s \in X^{mC}$, $t \in X^{nC}$, $s[0] \neq t[0]$, if there are two O–factorizations of s and t then $s = x_0._C x_1._C \cdots ._C x_m, t = y_0._C y_1._C \cdots ._C y_n$, where $x_0 \neq y_0$, $C' = C \cup \{\varepsilon\}$.

Lemma 3.1. *Given $X \subseteq A^+$ and $(U_k)_{k \geq 0}$ is identified according to the formula* (3.1). *If $u \in U_k$ then there exist two numbers m, n such $m + n = k$ and two strings $s \neq t$, $s = x_0._C x_1._C \cdots ._C x_m, t = y_0._C y_1._C \cdots ._C y_n, x_0 \neq y_0$ such $s \in X^{mC}$, $t \in X^{nC}$, $c \in C'$ so that $s._c u = t$.*

Proof. We will inductively prove according to k.

If $k = 0$: from the definition U_0 the result can be inferred.

Supposing that the statement is true with $k \geq 0$, we need to prove that the statement is true with $k + 1$.

Supposing $u \in U_{k+1}$, from the definition $U_{k+1} = U_k^{-C}X \cup X^{-C}U_k$, there are two possibilities: $u \in U_k^{-C}X$ or $u \in X^{-C}U_k$

- In the case $u \in U_k^{-C}X$: then $\exists c \in C'$, $v \in U_k$, $x \in X$ so that $v._c u = x$ where $v \neq x$. According to the inductive hypothesis, because $v \in U_k$ there exist m, n such $m + n = k$, $s \in X^{mC}$, $t \in X^{nC}$, $c' \in C'$ so that $s._{c'} v = t$ and $t \neq s$, $s[0] \neq t[0]$. Then we have $m + 1 + n = k + 1, s' = s._{c'} x \in X^{m+1C}, t \in X^{nC}, t._c u = s'$, because v is a real prefix of x, t is a real prefix of s': $t \neq s'$, $s'[0] = s[0] \neq t[0]$ (see Fig. 4).

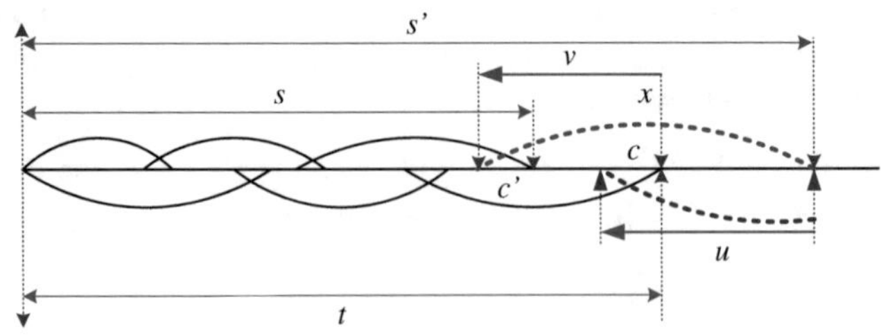

Fig. 4. The case $u \in U_k^{-C}X$.

– In the case $u \in X^{-C}U_k$: there exists $c \in C'$, $v \in U_k$, $x \in X$ so that $x._c u = v$ where $v \neq x$. According to the inductive hypothesis, because $v \in U_k$ there exist m, n such $m + n = k$ and two strings $s \neq t$, $s[0] \neq t[0]$: $s \in X^{mC}, t \in X^{nC}, c' \in C'$ so that $s._{c'}v = t$. Then we have $m + 1 + n = k + 1$, $s' = s._{c'}x \in X^{m+1C}, t \in X^{nC}, s'_{.c}u = t$, because x is a real prefix of u, s' is a real prefix of t : $t \neq s'$, $s'[0] = s[0] \neq t[0]$ (see Fig. 5).

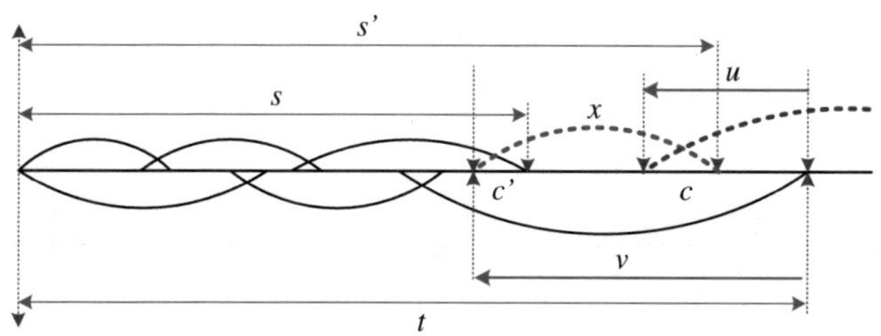

Fig. 5. The case $u \in X^{-C}U_k$

The result can be inferred. □

From Lemma 3.1, we have the following finding.

Clause 3.1. *Given* $X \subseteq A^+$*và* $(U_k)_{k \geq 0}$ *identified with the formula* (3.1). *Then, X is O–code when and only when with every k,* $U_k \cap X = \varnothing$.

Proof. We prove that X is not O–code when and only when there exists U_k such $U_k \cap X = \varnothing$.

(⇐) Supposing that $U_k \cap X = \varnothing$, then we have $u \in U_k$, $u \in X$. From Lemma 3.1 there exist m, n such $m + n = k$ and two strings $s \neq t$, $s = x_0._c x_1._c \cdots ._c x_m, t = y_0._c y_1._c \cdots ._c y_n$, $x_0 \neq y_0$ such $s \in X^{mC}, t \in X^{nC}, c \in C$ so that $s._c u = t$. Then t is a word with two different O–factorizations (because $x_0 \neq y_0$) (Fig. 6).

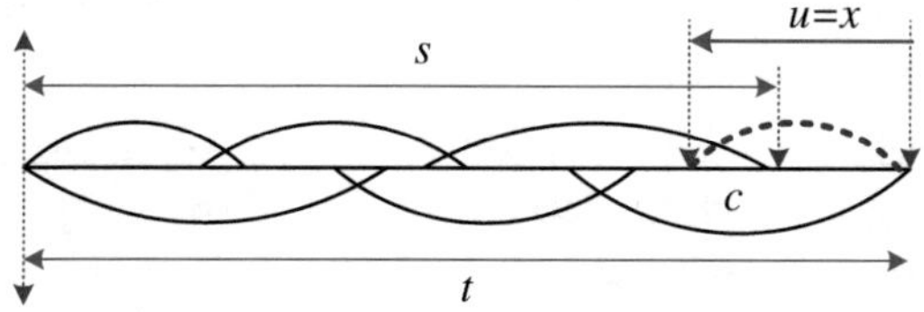

Fig. 6. The case $U_k \cap X = \varnothing$.

(⇒) Otherwise, if X is not O–mã, then there exists a word w with two different O–factorizations. We can show that there exists U_k so that $U_k \cap X \neq \varnothing$. Supposing that

$$w = x_0._c x_1._c \cdots ._c x_m = y_0._c y_1._c \cdots ._c y_n.$$

Without loss of generality, we suppose that $x_0 \neq y_0$, we can show that $U_{m+n-1} \cap X \neq \varnothing$.

Supposing that $|x_0| < |y_0|$. From the definition U_0, we have $u_0 \in U_0$.

Without loss of generality, it can be assumed after some factorizations steps that the upper class x_{k+1} "*overlap*" the right edge y_0 (see Fig. 7). It can be noted that from u_0 to u_k, the right edge matches the right edge of y_0. From u_{k+1} to any u_{k+l} (corresponding to y_l with the right edge crossing x_{k+1}), the right edge matches the right edge x_{k+1}.

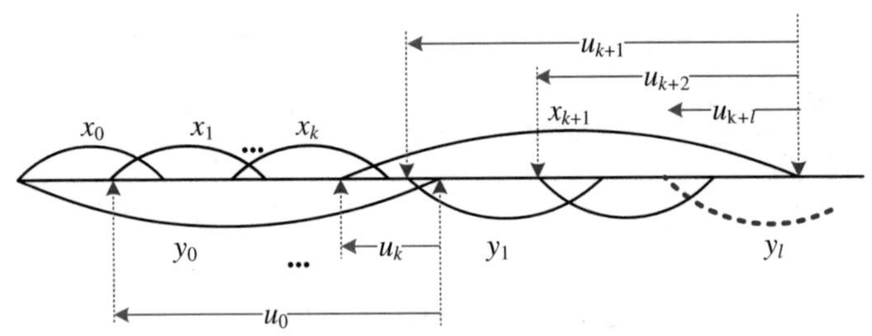

Fig. 7. The case $U_k \cap X \neq \varnothing$.

Because w has two different O–factorizations, at some point, the right edge of x_m will match the right edge of y_n. We consider two cases:

+ Case 1: $y_0 = w$, the right edge of x_{k+1} matches the right edge of y_0 (not crossing). Then $u_k = x_{k+1} \in X$, satisfying the condition $k + 1 + 0 - 1 = k$.

+ Case 2: The right edge of y_l matches the right edge of x_{k+1} and is the right edge of w (not crossing). Then $u_{k+l} = y_l \in X$, satisfying the condition $k + 1 + l - 1 = k + l$.

In general: Such classes as $u_0, \ldots, u_k$ are called the lower classes, $u_{k+1}, \ldots, u_{k+l}$ are called the upper classes. It is easy to see that u_{m+n-1} matches x_m or y_n ($u_{m+n-1} = x_m$ if u_{m+n-1} belongs to the lower class and $u_{m+n-1} = y_n$ if it belongs to the upper class). In all cases, we have $u_{m+n-1} \in X$. The clause is proved. □

Theorem 3.1. *Given* $X \subseteq A^+$ *and* $(U_k)_{k \geq 0}$ *is identified with the formula* (3.1). *If* $U_0 = \varnothing$ *then* X *is a O–code.*

Proof. We prove that, if $U_0 = \varnothing$ then X is a O–code. Indeed, according to the identification of sets $(U_k)_{k \geq 0}$ with the formula (3.1), if there is $U_0 = \varnothing \Rightarrow U_1 = \varnothing \Rightarrow \ldots \Rightarrow U_k = \varnothing$ or $U_k = \varnothing$, with every $k \geq 0$. Therefore, we have $U_k \cap X = \varnothing$, with every $k \geq 0$. According to Clause 3.1, we infer that X is a O–code. □

Theorem 3.2. *Given* $X \subseteq A^+$ *and* $(U_k)_{k \geq 0}$ *is identified with the formula* (3.1). *If* $\exists k \geq 0$ *so that* $U_{k+1} = U_k$ *and* $U_i \cap X = \varnothing$, *with every* $k \geq i \geq 0$ *then* X *is a O–code.*

Proof. Indeed, according to the definition:

$$U_{k+2} = \left(U_{k+1}^{-C} X \cup X^{-C} U_{k+1}\right) \cup U_{k+1}$$

If $U_{k+1} = U_k$ then replacing U_{k+1} with U_k in the above expression:

$$U_{k+2} = \left(U_k^{-C}X \cup X^{-C}U_k\right) \cup U_k = U_{k+1} = U_k$$

Similarly, $U_{k+i} = U_k$, with every $k,i \geq 0$. Because $U_i \cap X = \varnothing$, according to the definition of the U_{k+1} with the formula (3.1), it can be inferred that $U_{k+1} \cap X = \varnothing$, with every $k \geq i \geq 0$.

According to Clause 3.1, it can be inferred that X is a O–code. □

If X is a formal language then the number of sets $(U_k)_{k \geq 0}$ is finite. From Clause 3.1, Theorems 3.1 and 3.2, we have the following algorithm:

* ***The algorithm for testing of O–codes***

Input: Given $X \subseteq A^+$ *as a formal language.*

Output: The conclusion is that X is a O–code or not.

B_1. $U_0 = X^{-C}X, n = 0$,

If $(U_0 = \varnothing)$ *Then goto*B_4,

B_2. *(Loop)*

$U_{n+1} = (U_n^{-C}X \cup X^{-C}U_n) \cup U_n$,

B_3. *If* $(U_{n+1} \cap X \neq \varnothing)$ *Then goto*B_5,

B_4. *If* $(U_{n+1} = U_n)$ *Then goto*B_4,

Else$n = n+1$, *goto*B_2,

B_5. *Write "X is a O–code" and Exit.*

B_6. *Write " X is a O–code" and Exit.*

Remark 3.1. Given X as a formal language, we can develop a morphism monoid φ: $A^* \rightarrow P$, with P being a finite monoid. Let $k = |P|$. Then, the algorithm for testing of O–codes will give the answer with the maximum steps of k.

Indeed, we consider the algorithm for testing of O–codes, calculate the sets $(U_n)_{n \geq 0}$ with the formula (3.1). Because the sets $(U_n)_{n \geq 0}$ satisfied by φ, $U_0 = \varphi^{-1}(K_0), U_1 = \varphi^{-1}(K_1), \ldots, U_n = \varphi^{-1}(K_n)$, with $K_n \subseteq P$.

Because $U_0 \subseteq U_1 \subseteq \ldots \subseteq U_n \subseteq A^*$, we make the following statement: if the sets U_n are separated, with every $n \geq 0$, then: $K_0 \subseteq K_1 \subseteq \ldots \subseteq K_i \subseteq P$. The number of sets K_n is finite and no more than $|P|$ sets (because of inclusion property), it can be inferred that the number of sets U_n is no more than $|P|$ sets. Otherwise, if there is a repetition of the sets U_n then the number of different sets K_n is no more than $|P|$ sets. Therefore, the number of steps of calculate the sets U_n is finite and no more than k steps.

Thus, we can conclude that: if we only consider the number of different sets U_n then the algorithm for testing of O–codes has the complexity of $\mathcal{O}(k)$.

On the other hand, the complexity of each calculating setp U_{n+1} from U_n, regarding each sum on the monoid P, is $(|P|.|P| + |P|.|P|) \approx 2.|P|^2$ and developing the monoid P requires at most $|P|^3$ sums on the monoid P with the flooding algorithm. Therefore, the complexity of the algorithm for testing of O–codes in the worst case is $\mathcal{O}\left(|P|^3\right) \approx \mathcal{O}(k^3)$.

4 Conclusion

Overlap product is an extending study on the approach to applying ambiguous and multiple valued elements on the code concatenation approach that is considered in some studies such as [3, 6, 10, 13],…. The interesting issues that need studying can be: the properties of the contextual set C, subclass O–code, some concepts such as O–formal, O–automat,…

Thanks to the initial findings, we can continue develop such concepts as the decoding lag for O–code, similar to the definition of normal code lag [7] (being a parameter to assess the difficulties of decoding and encoding).

References

1. Schützenberger, M.P.: On a question concerning certain free submonoids. J. Comb. Theory **1** (4), 437–442 (1966)
2. Weil, P.: Groups: codes and unambiguous automata. In: Proceedings of the 2nd Symposium of Theoretical Aspects of Computer Science. Lecture Notes in Computer Science, vol. 182, pp. 351–362. Springer, Heidelberg (1985)
3. Huy, P.T., Van, D.L.: On non-ambiguous Büchi V-automata. In: Proceedings of the Third Asian Mathematical Conference, Philippines. World Scientific, 23–27 October 2000
4. Huy, P.T.: On ambiguities and unambiguities related with ω–Languages. Invited Report in International Conference “Combinatorics and Applications”, Hanoi, 3–5 December 2001
5. Vinh, H.N., Nam, V.T., Huy, P.T.: Codes base on unambiguous products. Lecture Notes in Artificial Intelligence, vol. 6423, pp. 252–262. Springer, Heidelberg (2010)
6. Vinh, H.N., Huy, P.T.: Codes of bounded words. In: Proceedings of the 3rd International Conference on Computer and Electrical Engineering (ICCEE 2010), Chengdu, China, vol. 2, pp. 89–95, 16–18 November 2010
7. Berstel, J., Perrin, D.: Theory of Codes. Academic Press Inc., New York (1985)
8. Gilbert, E.N., Moore, E.F.: Variable length binary encodings. Bell Syst. Tech. J. **38**, 933–967 (1959)
9. Nam, V.T.: Code on the basis of some new product types. Doctoral Dissertation, National Library, Hanoi University of Science and Technology (2007)
10. Anselmo, M.: Sur les codes zig-zag et leur decidabilité. Theory Comput. Sci. **74**, 341–354 (1990)
11. Van, D.L., Saec, B.L., Littovsky, I.: On coding morphism for zigzag code. Theoret. Inform. Appl. **26**(6), 565–580 (1992)
12. Van, D.L., Saec, B.L., Littovsky, I.: Stability for the zigzag submonoids. Theory Comput. Sci. **108**(2), 237–249 (1993)
13. Han, N.D., Vinh, H.N., Thang, D.Q., Huy, P.T.: Quadratic algorithms for testing of codes and ◇-Codes. Fundam. Inform. **130**, 1–15 (2014)

On Transforming Unit Cube into Tree by One-Point Mutation

Zbigniew Pliszka(✉) and Olgierd Unold

Department of Computer Engineering,
Wroclaw University of Science and Technology, Wyb. Wyspianskiego 27,
50-370 Wroclaw, Poland
{zbigniew.pliszka,olgierd.unold}@pwr.edu.pl

Abstract. This work is presenting new properties of vertices of a dimensional unit cube obtained after mutually unambiguous (bijective) transformation of these vertices of a cube into a tree. Some of the presented properties were obtained with the Newton symbol based on an extended definition.

Keywords: Unit cube · Binary tree · Newton symbol
One-point mutation

1 Introduction

In computer science, a unit cube (UC), or more precisely a collection of its vertices (VUC), always plays a basic and primary role. Hence, the tools developed in the earliest years of IT development such as the Hamming [4] measure and the Gray [2] code were used to study the vertices properties of UC. Ultimately, let us list two examples of works that group the topological properties of transformed unit cubes [10,13].

In the topic of transforming UC into a graph, and especially in its unique form of a tree, many works have been created related to optimization problems. In the works concentrating generally on graphs there are presented methods and tools used to solve Hamiltonian paths and cycles, cube edge coloring, polar paths or cycles [2]. With the help of Fibonacci cube there were attempts made not only to create simulators used in architecture, but also to solve problems related to parallel calculations, parallel communication or tolerance of errors [7]. Hence, it would be helpful to mention the special usefulness of graphs when dealing with problems with large data size [14]. Graphs are also a very good object for concurrent or parallel programming [16]. Studies on topographic properties of graphs enabled the development of new computer architectures [6]. Works on trees are devoted to optimization problems such as the minimum spanning tree problem MSTP [5].

A lot of works have also been done on transforming unit cubes into binary trees of different types [1,3,9,11,15]. Also works on parallel programming concern the problem of trees [8].

J. Świątek et al. (Eds.): ISAT 2018, AISC 853, pp. 71–82, 2019.
https://doi.org/10.1007/978-3-319-99996-8_7

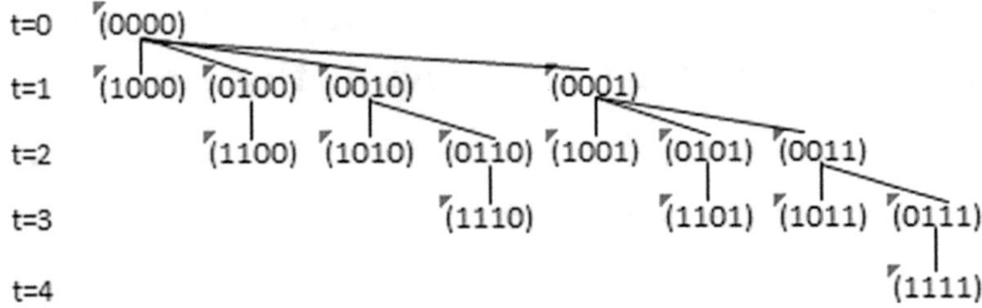

Fig. 1. Space A^n spanned on a tree, binary notation for n = 4.

In earlier work [12], the authors showed the properties of the n-dimensional UCs obtained after using the one-point crossing operator, taken from nature-inspired algorithms. This time they reached the one-point mutation operator. As a result, this work presents new properties of vertices of a UC obtained after mutually unambiguous (bijective) transformation (with the help of the TreeM algorithm) of these vertices of a cube into a (mutation) tree. Some of the presented properties can be introduced thanks to the extended Pascal's triangle.

2 Basic Notions and Definitions

Let us assume, for the uniformity of the formulas in this work, that: the Newton symbol additionally takes $\binom{n}{-1} = 0$ and $\binom{n}{n+1} = 0$ and $\log_2 0 = -1$, whereas at the same time $2^{-1} = 0$. Note that it expands Pascal's triangle on additional values.

Formally, the set of elements of the tree discussed below can be represented as:

$$A^n = \{(a_{n-1}, a_{n-2}, ..., a_0) : \forall i \in \{0, 1, 2, ..., n-1\} \quad a_i \in \{0, 1\}\}$$

This is a set of vertices of a unit cube in n-dimensional space A^n and consists of such elements. In our paper, we will use the terms A^n space and an n-dimensional unit cube interchangeably.

The vertices, in accordance with the TreeM algorithm, can be built in a tree. Throughout the work, let's assume that the root assumes value $0 = (0, \ldots, 0)$ and dimension $n \geq 2$. The example for $n = 4$ is shown in Fig. 1.

The one-point mutation of the vertex is exchange of one vertex value from 0 to 1, and 1 to 0 (in the dual sense, converting the value to the opposite). In the space in question A^n denotes the transition to the neighboring vertex.

Let's assume that a tree obtained with the use TreeM algorithm will be called *a mutation tree*.

3 Algorithm of Transforming Unit Cube into a Mutation Tree

Property 1. Any algorithm, which has the task of reconstructing the entire A^n space from any vertex using only a single point mutation operator, needs to do, at least, $2^n - 1$ mutation.

Proof. Any mutation from any parent vertex gives only one child vertex. Therefore, it remains to perform, at least (assuming optimistic that always after the mutation we will get a new one) as many mutations as the number of remaining elements in A^n after selecting the starting vertex, i.e. $2^n - 1$. ∎

Algorithm, spanning the whole space A^n with the use of single point mutation $2^n - 1$. Figure 1 shows its result for input data $n = 4$ and $a_i = (1, \ldots, 1)$. Number t denotes the level (depth) in the tree (it is equal to the number of nodes, which have to be taken to reach the root). It follows from the previous sentence that the root is at the level $t = 0$. Let's also assume that the phrase "level higher" denotes a decrease in the value of t by 1.

TreeM algorithm

1	Input: n //space dimension
2	Input: a_i //any vertex from A^n (written in binary form)
3	begin
4	T take (absorb) a_i;
5	for $j := n$ downto 1
6	begin
7	Make copies T to $T1$;
8	$T1$ set one level up than T;
9	In $T1$ on all elements a_i, on position j take the opposite value;
10	To root $T1$ hook up root T;
11	New T made of connected $T1$ and T;
12	end;
13	Output: T tree;
14	end.

Before we begin to prove the correctness of the TreeM algorithm, let's note a general property:

Property 2. In the TreeM algorithm, for the vertex at the input $a_i = (a_{i,n}, a_{i,n-1}, ..., a_{i,1}) \in A^n$, and for $j \in \{n, n-1, ..., 1\}$, before the execution of j-th iteration of line 9, all elements of the trees T and $T1$ have the same initial part, namely: $(..., a_{i,j}, a_{i,j-1}, ..., a_{i,1})$.

Proof. The only place in the algorithm where the vertex value and hence, its coordinates, change is exactly line 9 and the change is precisely in the j-th iteration of the value of only j-th coordinates in the tree $T1$. Therefore, since j takes values from a set $\{n, n-1, ..., j+1\}$, then all coordinates values with indices $j, j-1, ..., 1$ have in the already constructed trees T and $T1$ the input value $(..., a_{i,j}, a_{i,j-1}, ..., a_{i,1})$. ∎

Proof. Proof of correctness of the TreeM algorithm.

We will show that TreeM spans the entire space A^n on the tree. In this case, in the obtained tree, each element occurs exactly once. At the input we assume a number n that is the number of coordinates of each vertex and one vertex (assume fixed i) $a_i = (a_{i,n}, a_{i,n-1}, ..., a_{i,1}) \in A^n$. In line 4, the fixed vertex is embedded in the tree T and forms its whole. Lines 7–11 are executed (according to line 5) n times. The rest of the proof will be carried out with the help of mathematical induction. In the first induction step (line 7) for $j = n$, T is composed of one element $a_i = (a_{i,n}, a_{i,n-1}, ..., a_{i,1}) \in A^n$. The same concerns $T1$. In line 8, in terms of content in T and $T1$ nothing changes. In line 9, in the tree $T1$, the only element $a_i = (a_{i,n}, a_{i,n-1}, ..., a_{i,1})$ is replaced with the element $a_s = b_{i,n}, a_{i,n-1}, , ..., a_{i,1})$, where $b_{i,n} = 1$ if $a_{i,n} = 0$, and $b_{i,n} = 0$ if $a_{i,n} = 1$ (opposite in the dual sense). Elements a_i and a_s belong to A^n and are different. Thus, the tree T created by the program executed in lines 10 and 11 consists of two different elements. This ends the first step of induction. In the second induction step, we will show that the next tree T received from the smaller tree of pairs of different elements, as a result of a single execution of line 7–11 of our algorithm, still consists of a tree of pairs of different elements.

For $j \geq 1$ (obviously $j \leq n$) let us assume that at the entrance to the loop (line 7) the tree T consists of pairs of different elements. In line 7 we execute copy of T to $T1$ whereas in line 8 we set the copy $T1$ one level higher. From the induction assumption, all elements of $T1$ (as a copy of T) have different pairs. In line 9, in the tree $T1$, all the elements in the position j will have coordinate $a_{i,j}$ replaced with $b_{i,j}$ opposite in the dual sense to ($b_{i,j} = 1 - a_{i,j}$). This operation preserves the property of the induction assumption for the tree $T1$, otherwise if:

1. there was a pair of equal elements in $T1$ (e.g. $a_s = a_r$), then for this pair, all n equality would have to be satisfied:

$$a_{s,n} = a_{r,n}, a_{s,n-1} = a_{r,n-1}, ..., b_{s,j} = b_{r,j}, ..., a_{s,1} = a_{r,1}$$

   that means that before the operation in line 9, we would have equalities:

$$a_{s,n} = a_{r,n}, a_{s,n-1} = a_{r,n-1}, ..., a_{s,j} = a_{r,j}, ..., a_{s,1} = a_{r,1}$$

   that would also mean that in $T1$ and thus in T there would already be a pair of equal elements what would negate the assumed induction assumption.
2. Further, from the induction assumption all the elements of T would differ in pairs.
3. Also, each element from $T1$ is different from any element of T, otherwise (again indirect proof) if after the operation in line 9 there existed two elements $a_x = (a_{x,n}, a_{x,n-1}, ..., b_{x,j}, ..., a_{x,1}) \in T1$ and $a_y = (a_{y,n}, a_{y,n-1}, ..., b_{y,j}, ..., a_{y,1}) \in T$, and there would be equation $a_x = a_y$, between them, then we would have all the following n equalities:

$$a_{x,n} = a_{y,n}, a_{x,n-1} = a_{y,n-1}, ..., b_{x,j} = a_{y,j}, ..., a_{x,1} = a_{y,1}$$

   which means that before performing the operation in line 9 (performs only operations in T1) in j-th iteration, a_y had the form of $(a_{y,n}, a_{y,n-1}, ..., b_{y,j}, ..., a_{y,1})$.

Thus, on that j-th coordinate it would already have changed value, which means that, contrary to the assumption of a_y ($a_y \in T$), according to remark U1, a_y is not an element of T. Contradiction.

In above points we have exhausted all possible cases. In (1) we have shown that all elements in $T1$ are different in pairs. In (2) the same was stated for T. Finally (in 3) we proved that each element of $T1$ is different from each element of T. We can, undoubtedly, state that all the elements of the set $T \cup T1$ are different when it comes to pairs. And according to the instruction in line 11, $T \cup T1$ is the new T after the completion of j-th iteration. We have thus proved the thesis of the second part of the inductive proof. Using not only the principle of mathematical induction, but also the fact that we did not use any specific property of the number n, we can say that for any n, the treeM algorithm always gives a tree on which there are pairs of different elements of the space A^n are span.

In order to find out if all elements of A^n are placed in the tree, it is enough to calculate: We start with one element (line 4), and according to lines 7 and 10, in each loop recursion we double the number of elements in T. The loop is executed, according to line 5, n times. Hence, after the first loop execution we have 2^1 different elements, after the second 2^2, and so on, until the n-th as the last, we will have 2^n different elements, and this is equal to the power of A^n. Therefore, the whole space as a result of the TreeM algorithm was unrolled in the tree. What proves the correctness of the TreeM algorithm. ■

4 Basic Properties of a Mutation Tree

Property 3. Each mutation tree has $n + 1$ levels.

Proof. Algorithm TreeM in line 4 (which is executed exactly once) creates a tree with one level (the tree contains only one vertex, which obviously is only on one level). Then in loop executed times (loop declaration in line 5), only in line 8 creates one new level (exactly the root of the tree T1 is the vertex passing to the new level). Hence, after the algorithm terminates its operation, the tree will have levels n + 1. What proves the Property 3. ■

Property 4. In the mutation tree, at the input vertex $a_i = (1, \ldots, 1)$ in TreeM algorithm, at each level t, all vertices have the same number of ones (and zeros).

Proof. (reverse induction with respect to n)

1. Assuming in TreeM algorithm at input $a_i = (1, \ldots, 1)$, after the first loop ($j = n$), we get a tree containing one vertex on two levels. Hereby, the new vertex has one number 1 less, which is the result of the execution of line 9 (the only command in the entire TreeM program that causes the change of the coordinates of the vertices). Hence, the property being proven for $j = n$ takes place.

2. Let's assume that the proved property of a tree created with the TreeM algorithm is true from n to $j > 1$. In the next run of the loop after copying the T-tree and moving T1 copy to the higher level (lines 7 and 8), the T1 tree vertices on each level will contain one more 1 than the T-tree. But already in line 9, we remove exactly one number 1 from vertex in T1 tree, which consequently, after executing lines 10 and 11, gives us a tree preserving the proved property. ■

Property 5. In the mutation tree, at the input $a_i = (1, \ldots, 1)$ in the TreeM algorithm, at the level t vertices have t number of ones.

Proof. (a reverse induction with respect to n)

1. The number of ones in vertices at level t, according to the Property 2, can be determined from one vertex at a given level. These elements will be the vertices that are in the process of running of the algorithm with the roots of the T-tree. The algorithm assumes the input vertex $a_i = (1, \ldots, 1)$ and places it in the T-tree (line 4) at the only one, at the moment of operating of algorithm, 0 level.
   The vertex a_i contains n ones, and until the end of the algorithm will be at the lowest level of the tree T. By the end of the program operating, the level indicator in the tree for a_i will be magnified by one n times (line 8 in the loop declared in line 5). That means that once the algorithm running time is completed, a_i will be at a level $0 + n = n$.
2. Let's assume that the proved property is true from n to $t > 1$. In the next loop after copying the T-tree and moving the T1 copy to the higher level (lines 7 and 8), the roots of the trees T1 and T will contain equal number of ones. Next, on line 9, we carefully remove one number 1 from each vertex in the T1 tree, and thus also from the root, which, in the consequence, after execution of lines 10 and 11, gives us again a tree with roots consisting of $t-1$ ones. The loop will be executed $t-1$ times. Hence, this vertex, once the algorithm's operation is completed, will be at the $t-1$ level. What maintains the proved property. ■

Property 6. In the mutation tree obtained after transformation from A^n, according to the TreeM algorithm, the number of vertices at t level is $\binom{n}{t}$.

Proof. (an inductive proof)

1. In the TreeM algorithm, for $n = 1$ this property is obvious. We have only one vertex placed at the only level $t = 0$ and $\binom{1}{0} = 1$.
2. Let us assume that the proved property is true for all pairs of numbers (k, t), where k takes values from 1 to $n-1$, and t takes values from 0 to k. We will show that also for $k = n$ there is Property 6. If in the T-tree for $k = n-1$ for each $t-1$ with the above described constraints, the number of vertices $\binom{k}{t-1} = \binom{n-1}{t-1}$, is equal, then also in T1 tree, being the T copy, the same equality is maintained. But all the vertices in the T1 tree have been shifted by one level up (line 8) and the trees have been merged (lines 10 and 11),

which means that the vertices of the T-tree (prior to merging) occupying the level $t-1$ will be at the same level as the vertices from T1 tree from t level. At the same time in the new tree (after merging) all the vertices of the T-tree from the level $t-1$, will be on the t level. Then it is enough to simply add the number of vertices from $t-1$ level from the T-tree to the number of vertices from the t level from T1 tree to get the number of vertices of the new tree at t level:

$$\binom{n-1}{t-1} + \binom{n-1}{t} = \binom{n}{t}$$

This completes the proof of Property 6. ■

Property 7. In the mutation tree, at the input $a_i = (1, \ldots, 1)$ in the TreeM algorithm, all the leaves form a subset of vertices:

$$L^n = \{(a_{n-1}, a_{n-2}, \ldots, a_0) : a_{n-1} = 1 \wedge \forall i \in 0, 1, 2, \ldots, n-2, a_i \in \{0, 1\}\}$$

And the nodes form a subset:

$$W^n = \{(a_{n-1}, a_{n-2}, \ldots, a_0) : a_{n-1} = 0 \wedge \forall i \in 0, 1, 2, \ldots, n-2, a_i \in \{0, 1\}\}$$

In addition, there are multitude equations:

$$L^n \cap W^n = \emptyset \quad \text{i} \quad L^n \cup W^n = A^n$$

whereas L^n and W^n have the same number of elements, exactly 2^{n-1}.

Proof. After adoption, in the TreeM algorithm, at the input (line 2) vertex $a_i = (1, \ldots, 1)$ and absorbing it into the tree (line 4), in line 5 we start the loop, which will be executed n times (lines from 6 to 12). In the entire algorithm, only on line 9 (contained in the loop), we change the coordinates of the selected vertices. After the first loop execution (for $j = n$) we have a tree composed of two vertices, out of which one is a leaf $(1, 1, \ldots, 1)$, whereas the second a node $(0, 1, \ldots, 1)$ constituting at the same time a temporary root of the tree.

In subsequent iterations of the loop (they will be performed $n-1$ times) the number of vertices in the tree is doubled (line 7), while (we are dealing with a copy) two properties of individual vertices as elements of the tree being a leaf or node are unchanged. On the other hand, in the last $n-1$ iterations of the loop, the vertices coordinates from A^n to $n-1$ of the coordinates are not changed. This means that all copies of leaves are leaves and on n-th coordinate have value 1, and at the same time, all copies of nodes are nodes and on n-th coordinate have a value 0. The disjointness of sets L^n and W^n is the result of their definition (each vertex can have only one of the values 0 or 1 on the $n-1$ coordinate). The conclusion that the sum of sets L^n and W^n constitutes the whole of space A^n follows from the definition of L^n and W^n and from the fact that each vertex belonging to A^n on each of the coordinates must have one of the value, either 0 or 1. The number of elements in sets L^n and W^n will be obtained as the result of TreeM algorithm analysis. After the execution of the first iteration of the loop (declaration on line 5), we have one leaf and one node. The only place in the loop where we the number of leaves and nodes grows is line 7, and there is a copy of an already existing part of the tree (thus doubling the number of leaves and nodes). Such iterations will be executed in our algorithm exactly $n-1$. That means that the quantity, at the same time of the leaves and nodes will be the sum of:

$$1 + \sum_{i=1}^{n-1} 2^{i-1} = 2^{n-1}$$

This completes the proof of Property 7. ■

Let us notice that the Properties 3 and 6 are independent of the vertex given at the input of the TreeM algorithm.

The number of descendents for each node can be obtained from the definition of the Newton symbol. Assuming that n is the dimension of space, the root of the mutation tree will have $\binom{n}{1} = n$ descendants. In subsequent levels (rows), the number of descendents of consecutive nodes will be derived from the distribution of each number of k descendants from the previous level, greater than zero, to consecutive numbers from 0 to $k-1$. The value 0 symbolizes the leaf, thus there are no child elements. Hence, on each t level we will have: $\binom{n}{t}$ elements, $\binom{n-1}{t}$ nodes and $\binom{n-1}{t-1}$ leaves. And from each t level there will be generated $\binom{n}{t+1}$ descendants, where $t \in \{0, 1, \ldots, n\}$. The example is shown in Fig. 2a for $n = 4$.

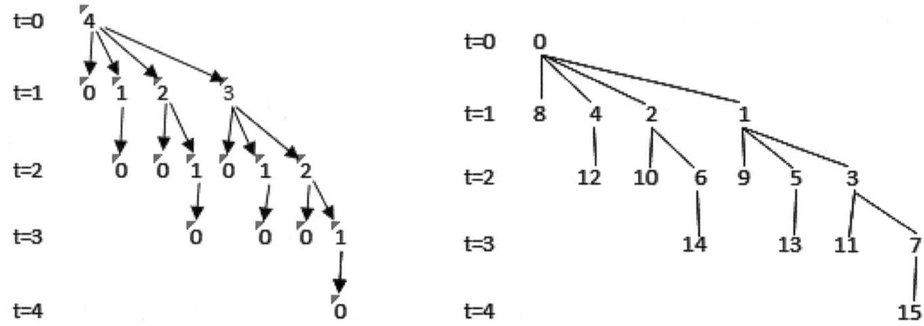

Fig. 2. (a) Number of descendents for nodes in the tree A^n, for $n = 4$. (b) Space A^n built in a tree, a decimal notation for $n = 4$.

Space A^n, can also be formed using the decimal representation (conversion of binary numbers into decimal ones), Fig. 2b.

With the apparent chaos of numbers, we can identify the relations that are characteristic for the mutation tree. Thus, we can show patterns occurring for the value of vertices having the highest values for each non-leaf level.

Excerpt of the table of the highest values (THV) of vertices for the t level, from which the mutation forms the descendent vertices is shown in Table 1.

Let's assume for an element lying at the intersection of a n column and a t line in the THV table designation: T_t^n.

Table 1. Excerpt of the table of the highest values of vertices for the t level, n is a size of space for vertices and t is a depth in a tree $(1 \leq t < n)$.

t	n					
	2	3	4	5	6	7
1	1	2	4	8	16	32
2		3	6	12	24	48
3			7	14	28	56
4				15	30	60
5					31	62
6						63

From the Property 5, we know that at the input $a_i = (1, \ldots, 1)$ in the TreeM algorithm, at the t level vertices have t ones. From the Property 7 we know that nodes have the form defined in the definition of a set W^n. Thus, the largest number in the binary notation for the node at the t level, will be the number:

$$(0, \underbrace{1, ..., 1}_{t} \underbrace{0, ..., 0}_{n-1-t})$$

which in the decimal notation will take the form:

$$T_t^n = 2^{n-1} - 1 - (2^{n-1-t} - 1) = 2^{n-1-t}(2^t - 1)$$

In addition, we have, easy to prove, relationships:

$$T_{n-1}^n = 2^{n-1} - 1$$

$$T_{n-1}^{n+k} = T_{n-1}^n \cdot 2^k = T_{n-1}^n \cdot T_1^{k+2} \qquad for\ 0 \leq k$$

$$T_1^n = 2^{n-2}$$

$$T_t^n = T_{t-1}^{n-1} + 2^{n-2} = T_{t-1}^{n-1} + T_1^n$$

Definition 1. Two vertices $a = (a_{n-1}, \ldots, a_0), b = (b_{n-1}, \ldots, b_0) \in A^n$ are called polar if and only if $\forall i \in \{0, 1, \ldots, n-1\}$ $b_i = 1 - a_i$. These are pairs of maximally distant vertices in the unit cube.

From the above definition and definitions accepted for sets L^n and W^n follows that the polar element for each leaf is a node $1 = l_{n-1} = 1 - 0 = 1 - w_{n-1}$ and vice versa, the polar element for each node is the leaf. Also, if a given vertex is in a tree at the level t, its polar element is located at level $n - t$.

Assuming for the vertex a the number

$$\sum_{i=0}^{n-1} a_i \cdot 2^i.$$

The vertex polar to it b corresponds to the number

$$2^n - 1 - \sum_{i=0}^{n-1} a_i \cdot 2^i = \sum_{i=0}^{n-1} 2^i - \sum_{i=0}^{n-1} a_i \cdot 2^i = \sum_{i=0}^{n-1} (1 - a_i) \cdot 2^i = \sum_{i=0}^{n-1} b_i \cdot 2^i$$

Therefore, for any pair of polar vertices $a = (a_{n-1}, \ldots, a_0), b = (b_{n-1}, \ldots, b_0) \in A^n$ we have the equality:

$$2^n - 1 = \sum_{i=0}^{n-1} a_i \cdot 2^i + \sum_{i=0}^{n-1} b_i \cdot 2^i$$

The position of a given vertex in a tree can be calculated directly from its binary representation. Elements of space a^n always have an even number. If we write successively in the table Tab from the top (depth $t = 0$) and for a set t from the right to the left (for example, for the mutation tree from Fig. 3 the content of Tab is as follows:
$Tab[] = \{\{0\}, \{8, 4, 2, 1\}, \{12, 10, 9, 6, 5, 3\}, \{14, 13, 11, 7\}, \{15\}\})$, then the following relationship would occur between the elements:

$$Tab[2^{n-1} + s] = 2^n - 1 - Tab[2^{n-1} - s + 1]$$

which means that the elements $Tab[2^{n-1} + s]$ and $Tab[2^{n-1} - s + 1]$ are polar elements. This observation allows us to span the tree halfway, the second part is obtained by defining polar elements in reverse order.

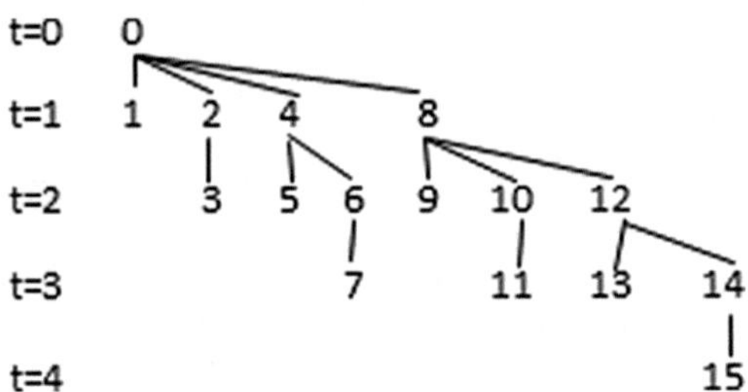

Fig. 3. All possible numbers that are t sums of powers sorted, for $n = 4$.

The number of descendants of the value g is equal to the number of zeros, in the binary representation, at the highest positions standing before one of the highest order. Hence it is: $n - \lfloor log_2 g \rfloor - 1$.

By denoting the j-th descendant of the element with the value g with the symbol $s(g, j)$, for j fulfilling the inequality: $1 \le j \le n - \lfloor log_2 g \rfloor - 1$, the values of the descendants are calculated from the formula:

$$s(g, j) = g + 2^{\lfloor log_2 g \rfloor + j}$$

since the descendant is the parent value plus the value of one, in the binary record, in the next position. Hence, for each element g belonging to a tree (beyond the root), we can also calculate the value of its parent: $g - 2^{\lfloor log_2 g \rfloor}$.

Here, we can also calculate tree leaf values. For any g satisfying the inequality $0 \leq g < 2^{n-1}$ we have:

$$s(g, n - \lfloor log_2 g \rfloor - 1) = g + 2^{n-1}$$

Finally, let us formulate three remarks of the mutation tree:

Remark 1. Each tree element itself is a leaf, or has exactly one leaf. This property could also be seen when the Newton symbol was written down (Fig. 2a).

Remark 2. Nodes have index values from 0 to $2^{n-1} - 1$, whereas leaves from 2^{n-1} to $2^n - 1$.

Remark 3. If we assume that each vertex in our tree is a number that is the sum of the powers of a number 2 (the powers increase from 0 to $n - 1$ from up to down) then each level t, represents all possible numbers that are t sums of powers sorted in ascending order from left to right, Fig. 3.

5 Conclusion

In the paper, we show the properties of a tree obtained by bijective transformation of a unit cube. The transformation was performed using the TreeM algorithm given in Sect. 3 with proof of its correctness. In the tree, called the mutation tree, the descendants of each node are vertices that are adjacent vertices in the cube, and therefore those that differ only in one position in the binary notation. The preliminary investigations showed that the properties of a mutation tree can be useful for some combinatorial optimization problems (like the knapsack problem), which we intend to explore in future work.

Acknowledgements. We would like to thank Prof. Krzysztof Dębicki from University of Wroclaw for giving the Remark 3, which he had noticed during a conversation with ZP. The work was supported by statutory grant of the Wroclaw University of Science and Technology, Poland.

References

1. Abraham, J., Arockiaraj, M.: Wirelength of enhanced hypercubes into r-Rooted complete binary trees. Electron. Notes Discret. Math. **53**, 373–382 (2016)
2. Ammerlaan, J., Vassilev, T.S.: Properties of the binary hypercube and middle level graphs. Appl. Math. **3**(1), 20–26 (2013)
3. Bhatt, S.N., Chung, F.R., Leighton, F.T., Rosenberg, A.L.: Efficient embeddings of trees in hypercubes. SIAM J. Comput. **21**(1), 151–162 (1992)
4. Bookstein, A., Kulyukin, V.A., Raita, T.: Generalized hamming distance. Inf. Retr. **5**(4), 353–375 (2002)

5. Graham, R.L., Hell, P.: On the history of the minimum spanning tree problem. Ann. Hist. Comput. **7**(1), 43–57 (1985)
6. Hwang, K., Jotwani, N.: Advanced Computer Architecture, 3rd edn. McGraw-Hill Education, New York (2011)
7. Klavžar, S.: Structure of Fibonacci cubes: a survey. J. Comb. Optim. **25**(4), 505–522 (2013)
8. Leighton, F.T.: Introduction to Parallel Algorithms and Architectures: Arrays, Trees, Hypercubes. Elsevier, Amsterdam (2014)
9. Liu, Z., Fan, J., Jia, X.: Embedding complete binary trees into parity cubes. J. Supercomput. **71**(1), 1–27 (2015)
10. Nielsen, F.: Topology of interconnection networks. In: Introduction to HPC with MPI for Data Science, pp. 63–97. Springer, Cham (2016)
11. Mulder, H.M.: What do trees and hypercubes have in common? In: Graph Theory, pp. 149–170. Springer, Cham (2016)
12. Pliszka, Z., Unold, O.: On the ability of the one-point crossover operator to search the space in genetic algorithms. In: Rutkowski, L., et al. (eds.) ICAISC 2015, Part I. LNAI, vol. 9119, pp. 361–369. Springer, Cham (2015)
13. Saad, Y., Schultz, M.H.: Topological properties of hypercubes. IEEE Trans. Comput. **37**(7), 867–872 (1988)
14. Sahba, A., Prevost, J.J.: Hypercube based clusters in cloud computing. In: World Automation Congress (WAC), pp. 1–6. IEEE (2016)
15. Wagner, A.S.: Embedding all binary trees in the hypercube. J. Parallel Distrib. Comput. **18**(1), 33–43 (1993)
16. Valiant, L.G.: A scheme for fast parallel communication. SIAM J. Comput. **11**(2), 350–361 (1982)

Pattern Recognition and Image Processing Algorithms

CNN Based Traffic Sign Recognition for Mini Autonomous Vehicles

Yusuf Satılmış, Furkan Tufan, Muhammed Şara, Münir Karslı, Süleyman Eken(✉), and Ahmet Sayar

Computer Engineering Department, Kocaeli University, 41380 Izmit, Turkey
satilmisyusuf58@gmail.com, furkantufan0127@gmail.com, muhammedsara271@gmail.com, munirkarsli@gmail.com, {suleyman.eken,ahmet.sayar}@kocaeli.edu.tr

Abstract. Advanced driving assistance systems (ADAS) could perform basic object detection and classification to alert drivers for road conditions, vehicle speed regulation, and etc. With the advances in the new hardware and software platforms, deep learning has been used in ADAS technologies. Traffic signs are an important part of road infrastructure. So, it is very important task to detect and classify traffic signs for autonomous vehicles. In this paper, we firstly create a traffic sign dataset from ZED stereo camera mounted on the top of Racecar mini autonomous vehicle and we use Tiny-YOLO real-time object detection and classification system to detect and classify traffic signs. Then, we test the model on our dataset in terms of accuracy, loss, precision and intersection over union performance metrics.

Keywords: Autonomous vehicles · Traffic sign recognition
End-to-end deep learning · Intelligent transportation systems
Racecar mini autonomous car

1 Introduction

For the last decades, we have been designing and using microprocessor-based electronic control units in vehicles and with its widespread use came the need for designing and building a safer and more reliable vehicle. Today, software has become the backbone of the automotive industry. The development of autonomous driving technology requires the addition of more sensors or components to the same system, which will do similar tasks, and also controlling/programming them. When a component fails, continuity must be maintained by another component of similar qualities. Lidar [1, 2], radar, ultrasonic, monocular vision, stereo vision, infrared vision are examples of sensors that can support each other. The rapid development of these in-car computing and sensing systems that assist each other, the collection of data, and the rapid developments in the field of computer vision and deep learning [3, 4] will achieve different levels of autonomous driving in the near future.

Compared to other sensors, such as Lidar and ultrasonic sensors, embedded cameras are both cheaper and complementary to the other sensors. Thanks to these mounted cameras, vision-based features can be provided to assist the driver, such as the

J. Świątek et al. (Eds.): ISAT 2018, AISC 853, pp. 85–94, 2019.
https://doi.org/10.1007/978-3-319-99996-8_8

detection and classification of objects on the road, the determination of distances to other vehicles, and mapping the surrounding environment. For safer autonomous vehicles, it is critical to recognize traffic signs. In this paper, we use a Tiny-YOLO [5] real-time object detection and classification system to detect traffic signs. After training the model, we use it in the ROS module. Detecting traffic signs is usually done in two steps in the literature: (i) finding the location of possible signs in a large image obtained from ZED stereo camera mounted on the top of Racecar and (ii) classifying the signs for given a cropped image where only a sign (or maybe nothing relevant) is visible it classifies whether it's a sign or not, and which sign it is.

The remaining of the study is organized as follows: The second section presents the literature on autonomous vehicle steering and traffic sign recognition. The third section presents the used developer Racecar kit. The fourth section presents the proposed methodology for traffic sign recognition. The fifth section first describes how to collect the dataset for training, and second, presents the performance tests. In the last section, we discuss the results.

2 Literature Review

The case for autonomous vehicles navigation in everyday life has gained more importance in recent years with the new developments. Autonomous driving research, such as Eureka Prometheus Project[1] and V-Charge project[2] are big-budget projects supported by the governments. ALVINN (Autonomous Land Vehicle In a Neural Network) was the first project to use neural networks for autonomous vehicles navigation. Compared to network models with hundreds of layers, it is composed of shallow and fully connected layers. It performed well on simple tracks with little obstacles and pioneered the determination of steering angles directly from image pixels. NVIDIA has used modern convolutional networks to extract features from the vehicle cameras frames [6]. Simple real world scenarios like lane keeping, driving on flat unobstructed paths were achieved. Studies on query-efficient learning [7] have been used by Zhang and Cho for autonomous vehicles [8]. Karslı et al. [9] concentrated on training deep network models from front-facing camera data synchronized with the steering angles. They developed three different end-to-end deep learning models and evaluated the success of these models on the racecar mini autonomous vehicle.

In addition to using convolutional neural networks, more advanced approaches involving knowledge of temporal dynamics are also available in the literature [10]. The model consists of fully connected networks (FCN) and long-term memory recursive network (LSTM) architectures. Koutnik et al. [11] trained repetitive neural networks (with over 1 million weights) using reinforced learning method. They aggregated data using TORCS racing car simulator and carried out tests on the same platform. Chi and Mu [12] introduced a deep learning model that effectively combines temporal and spatial information. Alternatively, generative adversarial networks (GANs) have

[1] http://www.eurekanetwork.org/project/id/45.

[2] http://www.v-charge.eu/.

networks that compete against each other to learn representations and subsequently produce accurate instances of learned representations [13]. Kaufler et al. [14] predicted and modeled human driving behavior using GAN. Uçar et al. [15] tested a hybrid model of CNN and SVM for object recognition and pedestrian detection on Caltech-101 and Caltech Pedestrian datasets.

During the last two decades, several methods have been proposed for traffic sign recognition [16]. It incorporates three main steps: (1) Region segmentation to obtain candidate regions with sings in them. We usually take advantage of color features in this step, since signs come in specific colors. (2) Shape analysis to classify signs according to their shapes; circular, triangular or rectangular. And the last, (3) Recognition, in which signs spotted in the previous feature extraction are identified, i.e. its class and meaning is ascertained. We have various classification techniques to pick from for this step specifically; among the most popular ones we can find Artificial Neural Network (ANN) [17], k-Nearest Neighbor (KNN) [18], Support Vector Machine (SVM) [19], and Random Forest (RF) [20]. Jo demonstrated that KNN classifiers are good for this task. However, he come at a very high cost in processing time. On the other hand, linear SVMs have a faster processing time while providing good results. Therefore, a cascade of linear SVM classifiers will be used to implement this system [21]. We use Tiny-YOLO real-time object detection and classification system to detect and classify traffic signs.

3 Features of Used Mini Autonomous Car

In this section, the hardware parts and features of the used car kit will be mentioned. Racecar -the mini autonomous vehicle kit- was originally developed for a competition by MIT in 2015, then updated in 2016 and used for robotics training [22]. The vehicle was developed on a 1/10 scale based on the Traxxas RC Rally Car racing car. The car kit uses an open-source electronic speed controller called VESC [23]. The main processor is the Nvidia Jetson TX2 developer board with 256 CUDA cores.

The hardware platform includes three main sensors: Stereolabs ZED stereo camera, Lidar, and an Inertial measurement unit. ZED stereo camera [24] can extract depth information from two images from two different cameras using standard stereo matching approaches. The Stereolabs SDK implements a semi-global matching algorithm that works on GPU-based computers such as Jetson TX2. This algorithm can also be used for 3D mapping. The second important sensor is the Scanse Sweep Lidar [25]. The Scanse Sweep can collect 1000 samples per second at a distance of 40 m away. The Sweep is a single-plane scanner, i.e. as its head rotates counter clockwise; it records data in a single plane. The beam starts out at approximately 12.7 mm in diameter and expands by approximately 0.5°. The sensor package also includes the Sparkfun Razor 9-DOF IMU [26] as the inertial measurement unit. The 9DoF Razor IMU M0 combines a SAMD21 microprocessor with an MPU-9250 9DoF (9 Degrees of Freedom) sensor to create a tiny, reprogrammable, multipurpose IMU (Inertial Measurement Unit). It can be programmed to monitor and log motion, transmit Euler angles over a serial port or even act as a step-counting pedometer. The 9DoF

Razor's MPU-9250 features three 3-axis sensors—an accelerometer, gyroscope, and magnetometer—that give it the ability to sense linear acceleration, angular rotation velocity and magnetic field vectors.

All these parts are placed on two specially prepared plates cut using a laser cutter. The lower part includes Nvidia Jetson TX2 and IMU. An RGB-D camera is mounted on top. The Lidar and ZED stereo camera are mounted directly on the vehicle chassis. Nvidia Jetson TX2 (our main computer) runs Ubuntu operating system. The main computer also runs the robotic operating system (ROS). ROS environment allows robotic software to be modular. For example, feedback control systems software, motion planning system software, computer vision system software and other detection system software can be divided into their own software modules. Each software module is called a "node". Nodes share information between each other using "messages".

4 Traffic Sign Recognition with CNNs

Traffic sign recognition problem can be treated as a multi-class classification process. The developed system includes two main components, as shown in Fig. 1. The first component is preprocessing step. The second one is Tiny-YOLO model for recognition.

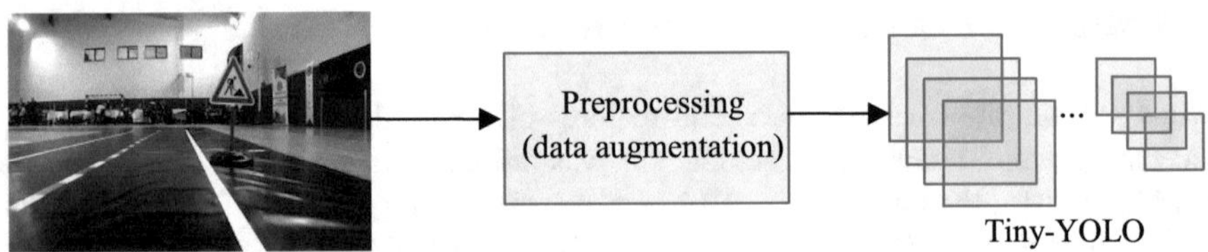

Fig. 1. Traffic sign recognition framework

4.1 Image Augmentation

In order to implement a good image classifier using only a small training dataset, it is necessary to use image augmentation to improve the performance of deep neural networks. Image augmentation is possible by synthesizing training images using at least one of different processing methods, like random rotation, shifts, shear and flips, etc. *imgaug* [27] is a popular library that provides image augmentation for machine learning experiments. Its broad set of features includes the most standard augmentation techniques; these techniques can be applied individually or combined together, to both images and key points/landmarks on images. It has a stochastic interface that is well balanced between simplicity and configurability, and comes with the option to run in background processes to enhance the performance. In this study, we use following augmenters with 50% of all images: GaussianBlur, AverageBlur, MedianBlur, Sharpen, AdditiveGaussianNoise, Dropout, Add, Multiply, ContrastNormalization.

4.2 Convolutional Neural Networks

With recent developments in object classification and detection tasks, we started to rely heavily on Convolutional Neural Networks (CNNs). That was not possible in real world applications in the past because CNNs require tremendous amounts of brute force computing power. A need that had been satisfied only recently with the advancement in GPGPU technologies.

The last few years saw the development of many variations of convolutional neural networks like R-CNN and its modifications Fast R-CNN and Faster R-CNN. Each variation improved on the previous one regarding particularly key criteria like speed and accuracy of the classification. One of the advantages of using convolutional neural networks is that they can perform both object classification and detection simultaneously. All that, without losing on speed, accuracy or the ability to recognize a variety of objects.

Three main types of layers are used to build CNN architectures: Convolutional Layer, Pooling Layer, and Fully-Connected Layer. These layers are stacked together in different ways depending on the architecture of the given CNN. Every layer of a CNN transforms the previous volume of activations to the next using a differentiable function. In our experiments, we settled on using Tiny-YOLO, a variation of the state-of-the-art real-time object classification and detection architecture called YOLO, it features a simpler model architecture and requires a smaller amount of GPU computing resources. We picked Darknet-19 as the pre-trained model to use with it, Darknet-19 consists of 9 convolutional layers, 6 max-pooling layers, 1 average pooling layer and a softmax layer as the last one [28]. Following section first describes how to collect the dataset for training, and second, presents the performance tests.

5 Experimental Analysis

5.1 Experimental Setup

Google Colaboratory, or shortly Colab, [29] is a free cloud service for machine learning education and research, its key advantage is the support for GPU acceleration using an NVIDIA Tesla K80. It provides a Jupyter notebook environment that requires no setup on a local machine to get started. Instead, the code is saved in Google drive and runs on the cloud in a dedicated virtual machine. The VMs are recycled between accounts when idle for a while, and have a maximum lifetime for each user. We use Google Colab for model training and testing in this work.

5.2 Building Dataset

The Logitech F710 joystick controls the speed and the current steering angle of the vehicle through the subscriber from the vehicle's ROS modules. Recorded speed and angle values and images taken from the ZED camera are held in directory.

We collect our own MarcTRdataset by driving the racecar on different tracks. Then we used the "LabelImg" graphical image annotation tool [30] to label the Frames. Using this tool, we encompass the traffic signs in a rectangular bounding box, and then we label the field belonging to the traffic sign class. The Action we make is recorded with each frame while we pass to the next. Annotations are stored as XML files in PASCAL VOC format, the same format used by ImageNet. Each entry holds key information for its respective image: the coordinates, width, height, and class label of the objects in it (see Fig. 2 for an example image and its XML file).

```
▼<annotation>
  <folder>new</folder>
  <filename>park_left_214.jpg</filename>
 ▼<path>
   /park/new/park_left_214.jpg
  </path>
 ▼<source>
   <database>Unknown</database>
  </source>
 ▼<size>
   <width>672</width>
   <height>376</height>
   <depth>3</depth>
  </size>
  <segmented>0</segmented>
 ▼<object>
   <name>park</name>
   <pose>Unspecified</pose>
   <truncated>0</truncated>
   <difficult>0</difficult>
  ▼<bndbox>
     <xmin>252</xmin>
     <ymin>66</ymin>
     <xmax>321</xmax>
     <ymax>141</ymax>
   </bndbox>
  </object>
</annotation>
```


Fig. 2. An example from MarcTRdataset and its XML file

Our dataset contains seven traffic sign classes. Table 1 shows the distribution of types in MarcTRdataset. You can see below a sample of the images in Fig. 3 from the dataset.

Table 1. Distribution of traffic sign types

Type of traffic sign	#
Turn right ahead	423
Turn left ahead	473
No passing	457
End of no passing	451
Road work	447
Pedestrians	454
Parking	859

5.3 Performance Metrics

Dataset is partition into 80% training and 20% test data. The number of test samples is 3566 and model classifies 3564 of those correctly, then the model's accuracy is 99.97%. Loss is calculated on training and validation and its interpretation is how well the model is doing for these two sets. Figure 4 shows the model loss for our dataset. YOLO is used for classification and detection of objects by encompassing them in

Fig. 3. Example images from MarcTRdataset

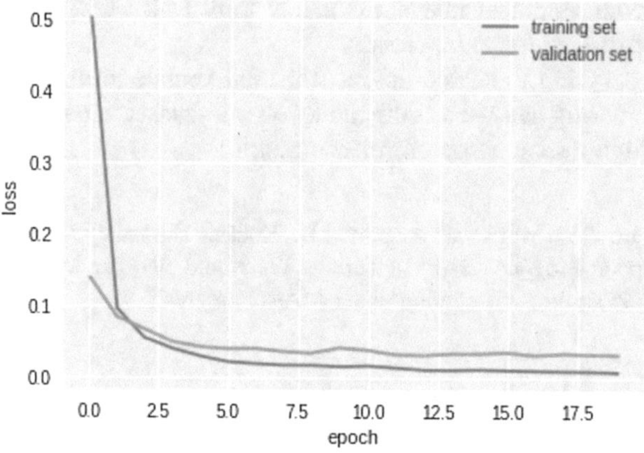

Fig. 4. Model loss

bounding boxes. We use Intersection over Union (IOU) with our model on test dataset, IOU is an evaluation metric used to assess the accuracy of an object detector. Applying IOU for the evaluation of our model requires two sets of bounding boxes: a ground-truth one, that is, hand labeled boxes to specify where in the image our object is, and the predicted one from our model. Table 2 shows performance results of Tiny-YOLO performance on MarcTRdataset.

Table 2. Tiny-YOLO performance on MarcTRdataset

Traffic sign class	Precision	IOU
Turn right ahead	1.00	0.903
Turn left ahead	0.982	0.804
No passing	1.00	0.813
End of no passing	1.00	0.787
Road work	1.00	0.764
Pedestrians	1.00	0.857
Parking	1.00	0.842
Average	*0.999*	*0.824*

6 Conclusion

Traffic sign detection from the raw images taken from a camera mounted on the car, have a very important place in modern autonomous vehicle technology. We have described a traffic sign detection and recognition system, focusing on seven classes of traffic signs and no-passing-signs.

Generally speaking, we provide the following contributions in this paper:

- We collect our own MarcTRdataset by driving the racecar on different tracks and label them with graphical image annotation tool. Dataset will be published after collection of images for more classes.
- We use Tiny-YOLO to enable the racecar mini vehicle to move according to the signs autonomously and we test the model on our dataset in terms of accuracy, loss and intersection over union performance metrics.

Acknowledgement. This work was supported by Kocaeli University Scientific Research and Development Support Program (BAP) in Turkey. We would also like to thank OBSS for their support and OpenZeka for their training under MARC program.

References

1. Yalcin, O., Sayar, A., Arar, O.F., Apinar, A., Kosunalp, S.: Detection of road boundaries and obstacles using LIDAR. In: Proceedings of Computer Science and Electronic Engineering Conference (CEEC), pp. 6–10. IEEE (2014)
2. Yalcin, O., Sayar, A., Arar, O.F., Apinar, A., Kosunalp, S.: Approaches of road boundary and obstacle detection using LIDAR. IFAC Proc. Vol. **46**(25), 211–215 (2013)

3. Goodfellow, I., Bengio, Y., Courville, A.: Deep Learning. MIT Press, Cambridge (2016)
4. Krizhevsky, A., Sutskever, I., Hinton, G.E.: Imagenet classification with deep convolutional neural networks. Commun. ACM **60**(6), 84–90 (2017)
5. Redmon, J., Farhadi, A.: YOLO9000: better, faster, stronger. In: Proceedings of Conference on Computer Vision and Pattern Recognition. IEEE, Honolulu (2017)
6. Bojarski, M., et al.: End to end learning for self-driving cars (2016). arXiv preprint: arXiv: 1604.07316
7. Ross, S., Gordon, G.J., Bagnell, D.: A reduction of imitation learning and structured prediction to no-regret online learning. In: Proceedings of the 14th International Conference on Artificial Intelligence and Statistics (AISTATS), pp. 627–635 (2011)
8. Zhang, J., Cho, K.: Query efficient imitation learning for end-to-end simulated driving. In: Proceedings of the Thirty-First AAAI Conference on Artificial Intelligence, San Francisco, California, USA, pp. 2891–2897 (2017)
9. Karslı, M., et al.: End-to-End Learning Model Design for Steering Autonomous Vehicle, 26. Sinyal İşleme ve İletişim Uygulamaları Kurultayı (2018)
10. Xu, H., Gao, Y., Yu, F., Darrell, T.: End-to-end learning of driving models from large-scale video datasets. arXiv preprint: arXiv:1612.01079
11. Koutník, J., et al.: Evolving large-scale neural networks for vision-based TORCS, pp. 206–212 (2013)
12. Chi, L., Mu, Y.: Deep steering: learning end-to-end driving model from spatial and temporal visual cues. In: Proceedings of the Workshop on Visual Analysis in Smart and Connected Communities, Mountain View, California, USA, pp. 9–16 (2017)
13. Mnih, V.B., et al.: Asynchronous methods for deep reinforcement learning. In: Proceedings of the International Conference on Machine Learning, pp. 1928–1937 (2016)
14. Kuefler, A., Morton, J., Wheeler, T., Kochenderfer, M.: Imitating driver behavior with generative adversarial networks. In: IEEE Intelligent Vehicles Symposium (IV), pp. 204–211 (2017)
15. Uçar, A., Demir, Y., Güzeliş, C.: Object recognition and detection with deep learning for autonomous driving applications. Simulation **93**(9), 759–769 (2017)
16. Vavilin, A., Jo, K.-H.: Graph-based approach for robust road guidance sign recognition from differently exposed images. J. Univ. Comput. Sci. **15**(4), 786–804 (2009)
17. Islam, Kh.T., Raj, R.G.: Real-time (vision-based) road sign recognition using an artificial neural network. Sensors **17**(4), 853 (2017)
18. Han, Y., Virupakshappa K., Oruklu, E.: Robust traffic sign recognition with feature extraction and k-NN classification methods. In: IEEE International Conference on Electro/Information Technology (EIT), pp. 484–848 (2015)
19. Gudigar, A., Jagadale, B.N., Mahesh, P.K., Raghavendra, U.: Kernel based automatic traffic sign detection and recognition using SVM. In: Mathew, J., Patra, P., Pradhan, D.K., Kuttyamma, A.J. (eds.) Eco-Friendly Computing and Communication Systems. Communications in Computer and Information Science, vol. 305. Springer, Heidelberg (2012)
20. Ellahyani, A., ElAnsari, M., ElJaafari, I.: Traffic sign detection and recognition based on random forests. Appl. Soft Comput. **46**, 805–815 (2016)
21. Jo, K.H.: A comparative study of classification methods for traffic signs recognition. In: 2014 IEEE International Conference on Industrial Technology (ICIT), pp. 614–619. IEEE (2014)
22. Karaman, S., et al.: Project-based, collaborative, algorithmic robotics for high school students: programming self-driving race cars at MIT. In: Proceedings of the IEEE Integrated STEM Education Conference, Princeton, NJ, USA, pp. 195–203 (2017)
23. Benjamin's robotics, VESC – Open Source ESC Project. http://vedder.se/2015/01/vesc-open-source-esc/. Accessed 10 May 2018
24. Stereolabs. https://www.stereolabs.com/zed/. Accessed 10 May 2018

25. Scanse Sweep lidar. http://scanse.io/. Accessed 10 May 2018
26. IMU. https://www.sparkfun.com/products/retired/10736. Accessed 10 May 2018
27. imgaug. https://github.com/aleju/imgaug. Accessed 10 May 2018
28. Tashiev, I., et al.: Real-time vehicle type classification using convolutional neural network. In: 1.Ulusal Bulut Bilişim ve Büyük Veri Sempozyumu, pp. 1–5 (2017)
29. Google Colab. https://colab.research.google.com. Accessed 10 May 2018
30. Tzutalin/LabelImg. https://github.com/tzutalin/labelImg. Accessed 10 May 2018

Parallel Processing of Computed Tomography Images

Dawid Połap(✉) and Marcin Woźniak

Institute of Mathematics, Silesian University of Technology,
Kaszubska 23, 44-100 Gliwice, Poland
{dawid.polap,marcin.wozniak}@polsl.pl

Abstract. Medical research is not only expensive but also time-consuming, what can be seen in the queues, and then after the waiting time for the analysis of the effects obtained from tests. In the case of computed tomography examinations, the end result is a series of the described images of the examined object's shape. The description is made on the careful observation of the results.

In this work, we propose a solution that allows to select images that are suspicious. This type of technique reduces the amount of data that needs to be analyzed and thus reduces the waiting time for the patient. The idea is based on a three-stage data processing. In the first one, key-points are located as features of found elements, in the second, images are constructed containing found areas of images, and in the third, the classifier assesses whether the image should be analyzed in terms of diseases. The method has been described and tested on a large CT dataset, and the results are widely discussed.

Keywords: CT images · Image processing · Convolutional neural network

1 Introduction

Computer methods are oriented on improved technologies for more efficient processing of information. Medicine is an example, for which constant development is necessary to keep the highest standards of service. In general along with advances in technology new methods are necessary. This can be achieved by the design of intelligent approaches in which we can use all the science applicable to our task. Computed Tomography (CT) is used to produce an image of the interior organs and tissues of our bodies. CT uses composite projections which depict organs from different directions to create a cross-sectional images of human bodies. In its initial form CT scanner was able to take images of the brain, but with new research and developments next versions were designed to scan other parts of human body. Initially the examination was performed in water and took about 30 min, while after improvements now it much less and the apparatus is fully computerized to make it as easy as possible for the patient.

In recent time we can find many interesting propositions for new applications and methodologies which improve important aspects of CT examinations. In [1] was discussed how the texture analysis methodologies influenced the developments in CT analysis. Authors have presented aspects which still need research and gave examples of these which are in may be hard to improve at the current state of knowledge. Authors

J. Świątek et al. (Eds.): ISAT 2018, AISC 853, pp. 95–104, 2019.
https://doi.org/10.1007/978-3-319-99996-8_9

of [8] presented CT in dental application, while in [9] was presented a discussion on kidney examinations by the use of CT and various segmentation techniques. Authors of [5] proposed a devoted methodology for segmentation of liver tumors. In [2] was presented an algorithm for segmentation of cortex and trabecular bones. In [6] was presented a wide comparison of segmentation techniques, in which deep learning approaches were used for corrections of results from parotid gland segmentation.

Another part of the research is oriented on automatic approaches to classification of diseased tissues or in general for detection of symptoms. In [3] neural networks were presented as classifiers for CT brain images. The authors discussed the complexity of classification and as an efficient technique proposed deep learning. Similarly authors of [4] presented deep learning for neural networks as a solution for breast tissue density classification from non-contrast CT images. This type of screening is also important for classification of lung diseases. In [7] was proposed to use devoted heuristic methodologies as detectors of degenerated tissues from x-ray images, while in [10] was presented a discussion of lung cancer detection by the analysis of various CT screenings. Authors of [12] presented results from the research on CT screenings for the reconstruction of lymph node models, while in [11] was presented how to use posterior screening results for lung nodule detection.

In this article we would like to present results of parallelization on the efficiency of CT screenings examinations. This approach is able to reduce the data processing complexity, since while segmented in parallel several operations can be distributed among devoted processes on each of cores. Therefore we achieve faster comparisons and therefore more efficient processing. The proposed methodology is using the key features localization by the analysis of fundamental elements and after this, results are processed by neural classifier which selects these among initial images which require and additional inspection from doctor. For the proposed methodology we have performed examinations to validate efficiency on open data set.

2 Proposed Technique for CT Images Processing

The patient who undergoes computed tomography awaits for the description of the test results after the examination. It happens that this time is long when only a few doctors deal with these types of things, and the number of made tests is large. Our solution is to simplify this action by reducing the amount of data coming to the technicians describing the images.

2.1 Image Processing

Each images from CT must be processed in such a way that applying a certain feature detector will get the best results. Our proposal is to use a multiple Gaussian blur filter defined by the following matrix (for the size 3×3)

$$\frac{1}{16}\begin{pmatrix} 1 & 2 & 1 \\ 2 & 4 & 2 \\ 1 & 2 & 1 \end{pmatrix}, \tag{1}$$

and two or three times gamma correction which is calculated for each pixel p_{ij} according to

$$p_{ij} = 255 \cdot \left(\frac{p_{ij}}{255}\right)^{2.2}. \tag{2}$$

2.2 Key-Point Search Using SURF Algorithm

One of the most known algorithm for searching key-points on image is Speeded Up Robust Features (SUFR) [13]. Feature detector is based on approximated value of Hessian matrix which define blob detector, and descriptor is defined by Haar's wavelet for a specific pixel. Hessian matrix is defined as

$$H(x,\omega) = \begin{bmatrix} L_{xx}(x,\omega) & L_{xy}(x,\omega) \\ L_{xy}(x,\omega) & L_{yy}(x,\omega) \end{bmatrix}, \tag{3}$$

where the values of the matrix are the convolution of integral image I and derivative using Gaussian kernels $g(\omega)$ and calculated as

$$L_{xx}(x,\omega) = I(x)\frac{\partial^2}{\partial x^2} g(\omega), \tag{4}$$

$$L_{yy}(x,\omega) = I(x)\frac{\partial^2}{\partial y^2} g(\omega), \tag{5}$$

$$L_{xy}(x,\omega) = I(x)\frac{\partial^2}{\partial xy} g(\omega). \tag{6}$$

The determinant of Eq. (3) is represented as

$$det\left(H_{approximate}\right) = D_{xx}D_{xy} - \left(wD_{xy}\right)^2, \tag{7}$$

where w is the weight of the integral image and D_{xx} refers to $L_{xx}(x,\omega)$ (as approximate and discrete kernels). All extremes from determinant are considered as a key-points of the input image. In full version of SURF, there is second stage of performance called key-points description using Haar's wavelet.

2.3 Convolutional Neural Network

Convolutional neural network (CNN) is one of the types of the classifier based on the activity of the human brain, more precisely the primary cortex [14]. The biggest difference between the classic architecture of the neural network is the input and types of layers. The network does not accept data saved in the numerical form and but as a graphic images. In the case of the types, there are three of them – convolutional, pooling and fully connected. The first onc is understood as feature extraction using one, defined filter ω, which has a predetermined matrix of a certain size. Each layer of this type may have a different filter. An example of the filters used is Gaussian blur or sharpening. The filter offset is marked as step *S*. The second type changes the size of the incoming image by applying some minimization based on the choice of one of the neighbors in a given window of the specified size (similar to the previous type). The most frequently used selection is minimization/maximization/average over a certain, specific feature. The last type of the layer named fully connected is similar to classic construction of neural network. There is no input layer, because incoming image from last layer (pooling or convolutional ones) is interpreted as input. Each pixel is understood as one neuron in first, hidden layer. There can be a few layers of these type and in the end, there is output one, which return the results of classification.

As each neural classificatory, these one also needs algorithm to training. The classic one is called backforward propagation described in [15, 16]. The algorithm will minimize the error on the whole network in relation to a certain function *f(.)*. Additionally, output value from neuron *(i,j)* on *l* layer will be marked as a following derivation $\frac{\partial f}{\partial y_{ij}^l}$. The whole learning technique is using chain rule formulated as

$$\frac{\partial f}{\partial \omega_{ab}} = \sum_{i=0}^{N-m} \sum_{j=0}^{N-m} \frac{\partial f}{\partial x_{ij}^l} \frac{\partial x_{ij}^l}{\partial \omega_{ab}} = \sum_{i=0}^{N-m} \sum_{j=0}^{N-m} \frac{\partial f}{\partial x_{ij}^l} y_{(i+1)(j+b)}^{l-1}. \tag{8}$$

On the basis of Eq. (8), the error in current layer *l* (from the end of network) can be defined as

$$\frac{\partial f}{\partial x_{ij}^l} = \frac{\partial f}{\partial y_{ij}^l} \frac{\partial y_{ij}^l}{\partial x_{ij}^l} = \frac{\partial f}{\partial y_{ij}^l} \frac{\partial\left(\sigma\left(x_{ij}^l\right)\right)}{\partial x_{ij}^l} = \frac{\partial f}{\partial y_{ij}^l} \sigma'\left(x_{ij}^l\right), \tag{9}$$

where $\sigma(x)$ is a function that defines the activation of a given neuron. To define a formula for calculating the error in the previous layer, define the gradient for the convolutional layer by the following equation

$$\frac{\partial f}{\partial y_{ij}^{l-1}} = \sum_{a=0}^{m-1} \sum_{b=0}^{m-1} \frac{\partial f}{\partial x_{(i-a)(j-b)}^l} \frac{\partial x_{(i-a)(j-b)}^l}{\partial y_{ij}^{l-1}} = \sum_{a=0}^{m-1} \sum_{b=0}^{m-1} \frac{\partial f}{\partial x_{(i-a)(j-b)}^l} \omega_{ab}. \tag{10}$$

And the above equation is used to define error for convolutional layer (in the case of pooling one, it does not take part in the learning mechanism) as

$$\frac{\partial x_{(i-a)(j-b)}^l}{\partial y_{ij}^{l-1}} = \omega_{ab}. \tag{11}$$

2.4 Proposed Technique

One CT scan results in a series of images. In general this number of items depends on the apparatus which can produce over 100 and more files in high resolution. Graphic processing, for this number of featured images, is time-consuming and heavily computer-loadable both due to resolution of the images and necessary precision of expected results. Together with the development in computer hardware we have new possibility to use more cores of better performance parameters, therefore in this article we suggest using the full capacity of the hardware by performing necessary calculations in parallel on all the cores available in the system. Assume that the computer on which the CT images are stored has *pc* cores. Then each of them can work on one image, by processing the image with various filters, search for possible key-points i.e. by using SURF or other algorithm, and then prepare this image for final classification. This final operation is most crucial for the patient, so in general automated support system shall propose a result to the doctor as a support in consultation if the proposed classification determined correctly the fragment presenting suspect tissues.

In the system which we discuss in this article calculate the positions of key-points for processed CT image. Each of these positions is further analyzed. In the system we use a proposed technique of slice image, where for each of key-points the decision matrix of a size 5x5 pixels is created. The selected key-point is in the middle of this matrix. In case when the point is too close to any of boundaries (defined by the size of input image) and the rest of points are outside these area, the pixels are filled with black color. In this simple way, we prepare fragments of all CT images for evaluation in the final stage. All suspected images are subjected to the classification CNN classifier, which has been previously trained. A model of the proposed operation scheme is presented in Fig. 1.

3 Experiments

Proposed solution was implemented in C# and tested on 6 cores using data provided by Cancer Imaging Archive[1] [17, 18]. The solution was tested under several different parameters – the effectiveness of classification at various network parameters and time of calculations. While in the case of image processing and slice preparing, too many changes were not made, so the classifier was tested in terms of convolutional layers using two different filter configurations. The first one was Gaussian blur and contrast enhancement, and the second one was Gaussian blur and double sharpening. The constructed layers in this way were trained by the backforward propagation algorithm to obtain one of five errors from the set $\{0.01, 0.005, 0.00125, 0.0003, 0.0001\}$. The obtained results for each configuration are presented in Tables 1 and 2. Additionally, confusion matrix for the best accuracy are shown in Figs. 2 and 3.

The network was trained with the *70:30* data proportion, which is training for verification. The results indicate an increase in accuracy for each of the filter sets together with a reduction of the error to which the network was trained. Note that the

[1] http://www.cancerimagingarchive.net/.

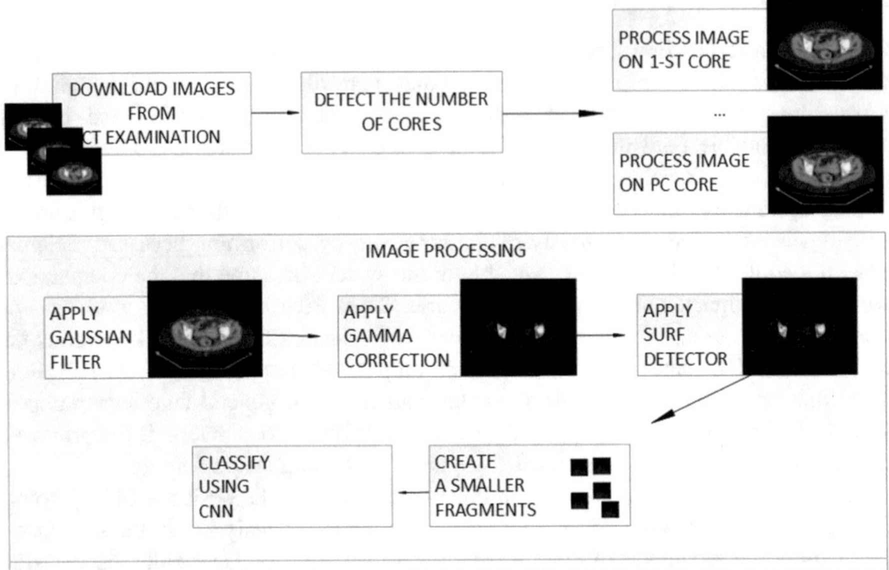

Fig. 1. Proposed decision technique about the need for additional analysis of a given image.

Table 1. The average effectiveness against the obtained learning error for the first set of filters.

Error	0,01	0,005	0,00125	0,0003	0,0001
Classification efficiency	34%	45%	57%	64%	71%

Table 2. The average effectiveness against the obtained learning error for the second set of filters.

Error	0,01	0,005	0,00125	0,0003	0,0001
Classification efficiency	44%	56%	67%	70%	83%

advantage of the second set can be seen, because in the best situation, the efficiency of the network is better by almost *12%* from the other one. When it comes to more accurate measurements, which samples were better classified, it is presented on confusion matrices. In the case of first set (see Fig. 2), the classifier in almost one-third of the correct samples incorrectly attributed them. This is a particularly bad indicator when it comes to the possibility of implementing this technique in industry. Better results were obtained for the second set of filters, where it can be seen that the suspect samples were mostly well recognized. The problem was rather the classification of samples without any signs of disease. If such a trained classifier obtained efficiency over *80%* when it was trained with a set of more than *1600* samples (in the case of the first classifier, there were samples that were not assigned to any of the groups), increasing the accuracy of operation using larger fragments of images, or other set of

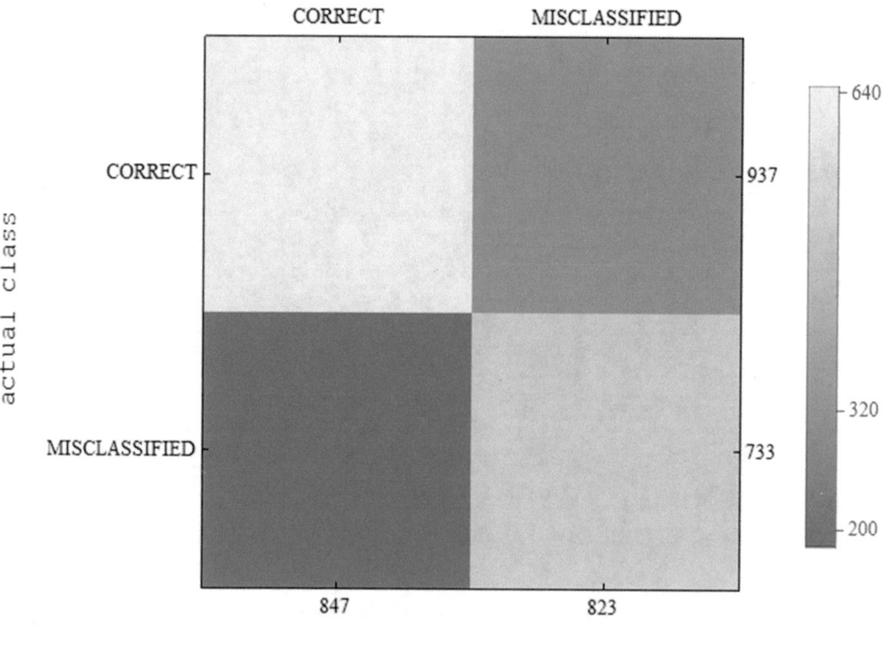

Fig. 2. Confusion matrix for CNN trained for first set of filters to error equal 0.0001.

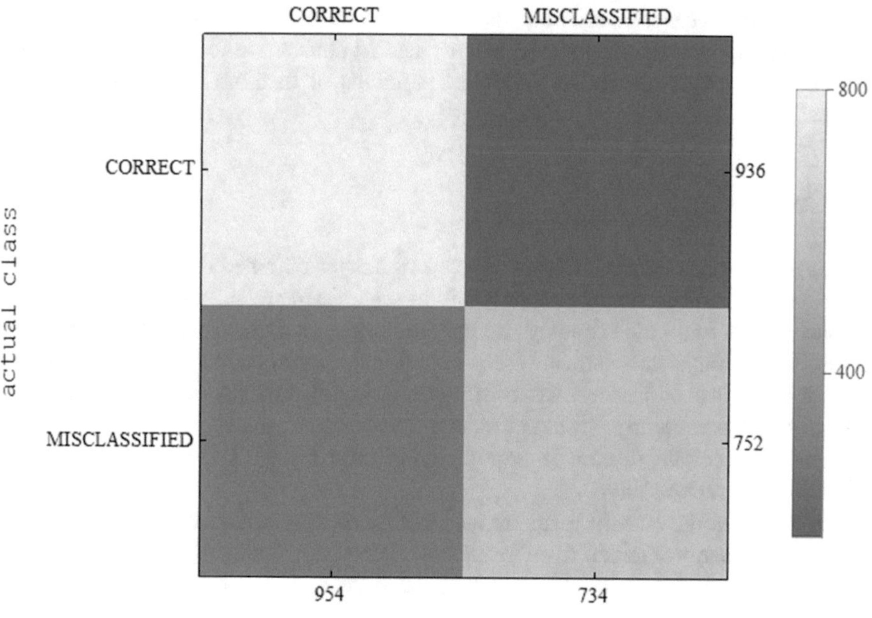

Fig. 3. Confusion matrix for CNN trained for second set of filters to error equal 0.0001.

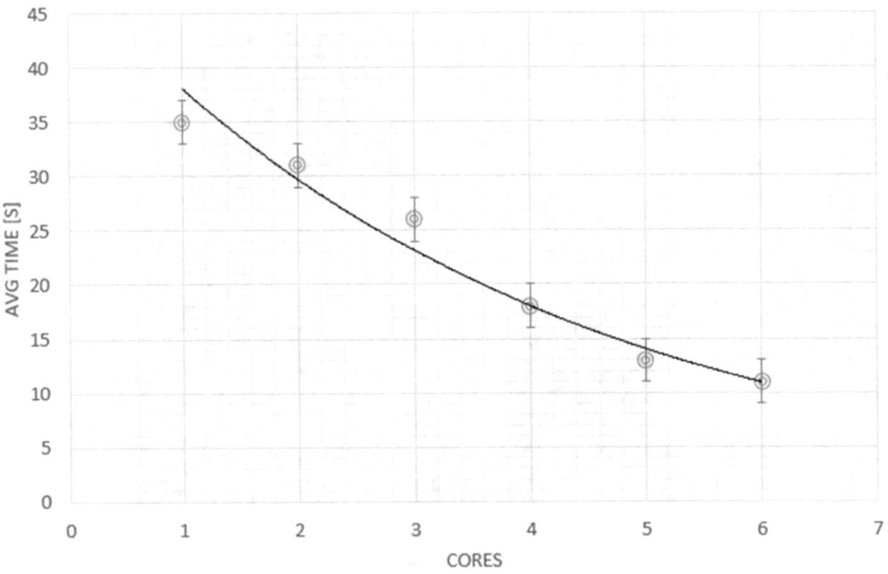

Fig. 4. Graph of time dependence on used cores during processing a set of 10 images.

filters should be subjected to a more detailed analysis. In terms of reducing the amount of time during processing, we illustrated the average measurements in Fig. 4. Operating time decreases almost logarithmically with an increase in the amount of used cores. In two cases, the measurement with the error was not included in the trend line, which means that the method cannot be explicitly named constant in terms of time reduction. On the other hand, the results with such amounts of data are good and indicate its decrease.

4 Conclusions

In this paper, the solution for faster analysis of images obtained during the tests using a CT scanner was described. Our proposition was tested under some parameters. Particularly under two - the accuracy and operation time on a given set with the number of cores. The results clearly indicate the potential in this model of solution, although in the current state, it is still necessary to focus on a few elements so that it can be used in industry. It is particularly worth noting that the automatic analysis of suspected areas of the existence certain disease is important not only for the science, but also for the protection of patients' lives.

In future work, we will focus on increasing accuracy and analyzing the dimension of small fragments obtained from images. An interesting element would be the use of a different algorithm to detect key-points or even the use of hybrid solutions.

Acknowledgments. Authors acknowledge contribution to this project of the "Diamond Grant 2016" No. 0080/DIA/2016/45 from the Polish Ministry of Science and Higher Education.

References

1. Hatt, M., Tixier, F., Pierce, L., Kinahan, P.E., Le Rest, C.C., Visvikis, D.: Characterization of PET/CT images using texture analysis: the past, the present… any future? Eur. J. Nucl. Med. Mol. Imaging **44**(1), 151–165 (2017)
2. He, Y., Shi, C., Liu, J., Shi, D.: A segmentation algorithm of the cortex bone and trabecular bone in Proximal humerus based on CT images. In: 23rd International Conference on Automation and Computing (ICAC), pp. 1–4. IEEE (2017)
3. Gao, X.W., Hui, R., Tian, Z.: Classification of CT brain images based on deep learning networks. Comput. Methods Programs Biomed. **138**, 49–56 (2017)
4. Zhou, X., Kano, T., Koyasu, H., Li, S., Zhou, X., Hara, T., Fujita, H.: Automated assessment of breast tissue density in non-contrast 3D CT images without image segmentation based on a deep CNN. In: Medical Imaging 2017: Computer-Aided Diagnosis, International Society for Optics and Photonics, vol. 10134, p. 101342Q (2017)
5. Sun, C., Guo, S., Zhang, H., Li, J., Chen, M., Ma, S., Qian, X.: Automatic segmentation of liver tumors from multiphase contrast-enhanced CT images based on FCNs. Artif. Intell. Med. **83**, 58–66 (2017)
6. Hänsch, A., Schwier, M., Gass, T., Morgas, T., Haas, B., Klein, J., Hahn, H.K.: Comparison of different deep learning approaches for parotid gland segmentation from CT images. In: Medical Imaging 2018: Computer-Aided Diagnosis, International Society for Optics and Photonics, vol. 10575, p. 1057519 (2018)
7. Woźniak, M., Połap, D.: Bio-inspired methods modeled for respiratory disease detection from medical images. Swarm and Evolutionary Computation *(2018)*
8. Gan, Y., Xia, Z., Xiong, J., Li, G., Zhao, Q.: Tooth and alveolar bone segmentation from dental computed tomography images. IEEE J. Biomed. Health Inform. **22**(1), 196–204 (2018)
9. Torres, H.R., Queirós, S., Morais, P., Oliveira, B., Fonseca, J.C., Vilaça, J.L: Kidney segmentation in ultrasound, magnetic resonance and computed tomography images: a systematic review. Comput. Method Programs Biomed. (2018)
10. Jobst, B.J., Weinheimer, O., Trauth, M., Becker, N., Motsch, E., Groß, M.L., Kauczor, H.U.: Effect of smoking cessation on quantitative computed tomography in smokers at risk in a lung cancer screening population. Eur. Radiol. **28**(2), 807–815 (2018)
11. Teramoto, A., Fujita, H.: Automated lung nodule detection using positron emission tomography/computed tomography. In: Artificial Intelligence in Decision Support Systems for Diagnosis in Medical Imaging, pp. 87–110 (2018)
12. Cooper, L.J., Zeller-Plumhoff, B., Clough, G.F., Ganapathisubramani, B., Roose, T.: Using high resolution x-ray computed tomography to create an image based model of a lymph node. J. Theoret. Biol. (2018)
13. Bay, H., Tuytelaars, T., Van Gool, L.: Surf: Speeded up robust features. In: European conference on computer vision, pp. 404–417 (2006)
14. Krizhevsky, A., Sutskever, I., Hinton, G. E.: Imagenet classification with deep convolutional neural networks. In: Advances in neural information processing systems, pp. 1097–1105 (2012)
15. Oquab, M., Bottou, L., Laptev, I., Sivic, J.: Learning and transferring mid-level image representations using convolutional neural networks. In: IEEE Conference on Computer Vision and Pattern Recognition (CVPR), pp. 1717–1724 (2014)
16. Krizhevsky, A., Sutskever, I., Hinton, G.E.: Imagenet classification with deep convolutional neural networks. In: Advances in neural information processing systems, pp. 1097–1105 (2012)

17. Clark, K., Vendt, B., Smith, K., Freymann, J., Kirby, J., Koppel, P., Tarbox, L.: The Cancer Imaging Archive (TCIA): maintaining and operating a public information repository. J. Digit. Imaging **26**(6), 1045–1057 (2013)
18. Abdoulaye, I.B.C., Demir, Ö.: Mamografi Görüntülerinden Kitle Tespiti Amacıyla Öznitelik Çıkarımı (2017)

Singular Value Decomposition and Principal Component Analysis in Face Images Recognition and FSVDR of Faces

Katerina Fronckova, Pavel Prazak, and Antonin Slaby(✉)

University of Hradec Kralove, Hradec Kralove, Czech Republic
{katerina.fronckova,pavel.prazak,antonin.slaby}@uhk.cz

Abstract. The singular value decomposition (SVD) is an important tool for matrix computations with various uses. It is often combined with other methods or used within specific procedures. The text briefly introduces the SVD and lists its important features and selected elements of the SVD theory. In addition, the text deals with two important issues related to the field of artificial intelligence with extensive practical use. The first is face recognition analysis in relation to face representation using principal component analysis (PCA) and the second is fractional order singular value decomposition representation (FSVDR) of faces. The presented procedures can be used in an efficient real-time face recognition system, which can identify a subject's head and then perform a recognition task by comparing the face to those of known individuals. The essence of the procedures, way of their application, their advantages and shortcomings, and selected results are presented in the text. All procedures are implemented in MATLAB software.

Keywords: Singular value decomposition · Matrix computations
Principal component analysis · Face recognition · Face representation

1 Introduction

The SVD is one of the most important and most versatile matrix computations tools. Its application can be found both in mathematical theory and in many practical areas. The SVD was formulated independently by several authors: Beltrami 1873, Jordan 1874, Sylvester 1889, Autonne 1915, Eckart and Young 1939. Later it was popularized mainly by Golub and reached wider applicability in connection with the development of information technology.

The SVD is related to many other concepts of linear algebra [6,8]. It is possible to use it for example to determine matrix rank, the Frobenius norm or spectral norm of a matrix, the condition number of a matrix, an orthonormal basis for the null space and the column space of a matrix, the approximation of a matrix by a matrix of lower rank. Further large application domain is in statistics in the context of principal component analysis and correspondence analysis.

J. Świątek et al. (Eds.): ISAT 2018, AISC 853, pp. 105–114, 2019.
https://doi.org/10.1007/978-3-319-99996-8_10

Another major application of the SVD is the area of signal processing including compression or data filtering [1,11], also it is used for data registration [2], recognition [14,15], steganography watermarking [10], latent semantic indexing and analysis [4] etc.

The text is organized as follows: Sect. 2 summarizes essential theoretical facts about the SVD. Principal component analysis and its connection to the SVD is introduced in Sect. 3 and then it is used in face recognition. Section 4 presents FSVDR of faces, which can improve recognition accuracy. Finally, Sect. 5 discusses the obtained results.

2 Basic Theoretical Facts About SVD

The following essential theorem states the existence of the SVD for any real matrix [6]. Analogous theorem on the SVD could be formulated for complex matrices as well.

Theorem 1. *Let* $\mathbf{A} \in \mathbb{R}^{m \times n}$ *and* $p = \min\{m, n\}$*. Then there exist orthogonal matrices* $\mathbf{U} \in \mathbb{R}^{m \times m}$*,* $\mathbf{V} \in \mathbb{R}^{n \times n}$ *and a diagonal matrix* $\mathbf{\Sigma} \in \mathbb{R}^{m \times n}$ *with diagonal elements* $\sigma_1 \geq \sigma_2 \geq \cdots \geq \sigma_p \geq 0$ *so that it holds*

$$\mathbf{A} = \mathbf{U}\mathbf{\Sigma}\mathbf{V}^T. \tag{1}$$

Diagonal elements $\sigma_{jj} = \sigma_j, j = 1, \ldots, p$, of the matrix $\mathbf{\Sigma}$ are called the singular values of the matrix $\mathbf{A}$. Let $\mathbf{u}_j$ resp. $\mathbf{v}_j$ denote the jth column of the matrix $\mathbf{U}$ respectively matrix $\mathbf{V}$. Vectors $\mathbf{u}_j, j = 1, \ldots, m$, are called the left singular vectors and vectors $\mathbf{v}_j, j = 1, \ldots, n$, are called the right singular vectors of the matrix $\mathbf{A}$.

The singular values are uniquely determined, and if we in addition to it suppose that they are written in sorted order ($\sigma_1 \geq \sigma_2 \geq \cdots \geq \sigma_p \geq 0$) then the matrix $\mathbf{\Sigma}$ is uniquely determined too. On the other hand, the left singular vectors and the right singular vectors and consequently the matrices $\mathbf{U}$ and $\mathbf{V}$ are not uniquely determined.

The rank r of the matrix $\mathbf{A} \in \mathbb{R}^{m \times n}$ is equal to the number of its non-zero (positive) singular values, i.e. $\sigma_1 \geq \sigma_2 \geq \cdots \geq \sigma_r > 0$, $\sigma_{r+1} = \sigma_{r+2} = \ldots = \sigma_p = 0$.

The non-zero singular values of the matrix $\mathbf{A}$, $\sigma_1, \ldots, \sigma_r$, are square roots of the non-zero (positive) eigenvalues of matrices $\mathbf{A}^T\mathbf{A}$ and $\mathbf{A}\mathbf{A}^T$. The left singular vectors of the matrix $\mathbf{A}$ are the eigenvectors of the matrix $\mathbf{A}\mathbf{A}^T$ and the right singular vectors of the matrix $\mathbf{A}$ are the eigenvectors of the $\mathbf{A}^T\mathbf{A}$. These facts can be used for the calculation of the SVD, see for example [5,6,13] for more details.

3 PCA and SVD in Face Recognition (Eigenfaces)

Principal component analysis is a statistical method, which was developed mainly by Pearson 1901 and Hotelling 1933. It transforms a set of variables to a set of new variables called principal components, which are uncorrelated, and which retain most of the variability present in the original variables. It is commonly used to reduce the dimensionality of multi-dimensional data and serves often as a tool for the initial understanding of data that precedes the actual elaboration of multi-dimensional problems.

The connection of principal component analysis with the SVD matrix decomposition will be demonstrated on the problem of face recognition also known as eigenfaces. We will use the approach introduced and presented by Turk and Pentland in [14,15]. Their work is built on the earlier research of Kirby and Sirovich [7], who examined the use of principal component analysis for representation of face images. The eigenfaces method differs from other methods of computer vision as its essence is not based on the extraction of distinctive characteristics of a face such as, for example, eyes, a nose, etc. but uses for recognizing vectors that are produced via projection of images onto so-called "space of faces". The basis of this space of faces is formed by the eigenvectors of the covariance matrix of the set of faces, i.e. vectors which define the principal components.

Practical demonstration of the described method will use photographs from the ORL (Olivetti Research Laboratory) database [12]. The database contains 10 different face images for each of 40 individuals. Pictures of the same individual differ, for example, by the face expression, lighting conditions, presence of glasses, etc.

We will suppose a training set consisting of n grayscale digital images of faces of different individuals $\mathbf{\Gamma} = \{\mathbf{\Gamma}_1, \mathbf{\Gamma}_2, \ldots, \mathbf{\Gamma}_n\}$ where each image of the dimensions $r \times s = p$ pixels is represented by a vector $\mathbf{\Gamma}_j \in \mathbb{R}^p$ (the elements of the vector represent the intensity values of the individual pixels) or in other words by a point in the p-dimensional Euclidean space. First, the arithmetic mean vector $\mathbf{\Psi}$ of all images is calculated

$$\mathbf{\Psi} = \frac{1}{n}\sum_{j=1}^{n} \mathbf{\Gamma}_j.$$

Subsequently, the differences of each of the images from this mean $\mathbf{\Phi}_j$ are determined

$$\mathbf{\Phi}_j = \mathbf{\Gamma}_j - \mathbf{\Psi}, \quad j = 1, \ldots, n.$$

The following Fig. 1 shows examples of several pictures of the training set and the arithmetic mean of all the training set pictures.

The vectors $\mathbf{\Phi}_j$ are the input information for principal component analysis, which seeks a set of orthonormal vectors $\mathbf{u} \in \mathbb{R}^p$ that best describe the variability distribution of input data. The vector $\mathbf{u}_k$ is chosen, so that the value

$$\lambda_k = \frac{1}{n}\sum_{j=1}^{n} \left(\mathbf{u}_k^T \mathbf{\Phi}_j\right)^2$$

is maximized and at the same time

$$\mathbf{u}_l^T \mathbf{u}_k = \begin{cases} 1 & \text{if } l = k \\ 0 & \text{if } l \neq k. \end{cases}$$

Fig. 1. Several pictures of the training set and the arithmetic mean of all the training set pictures $\boldsymbol{\Psi}$

It is possible to show that $\mathbf{u}_k$ are the eigenvectors and λ_k the eigenvalues of the covariance matrix $\mathbf{C}$, which is defined by the formula

$$\mathbf{C} = \frac{1}{n} \sum_{j=1}^{n} \boldsymbol{\Phi}_j \boldsymbol{\Phi}_j^T = \mathbf{A}\mathbf{A}^T,$$

where $\mathbf{A} = \left(\boldsymbol{\Phi}_1 \; \boldsymbol{\Phi}_2 \cdots \boldsymbol{\Phi}_n \right)$.

The use of the SVD is an appropriate method to determine the eigenvectors and eigenvalues of the covariance matrix. It is used the fact that the eigenvectors of this matrix $\mathbf{C} = \mathbf{A}\mathbf{A}^T$ are at the same time the left singular vectors of the matrix $\mathbf{A}$ and its eigenvalues are the second powers of the singular values of the matrix $\mathbf{A}$. Thus, it is sufficient to calculate the SVD of the matrix $\mathbf{A}$ for finding the vectors $\mathbf{u}_k$

$$\mathbf{A} = \mathbf{U}\boldsymbol{\Sigma}\mathbf{V}^T,$$

and the covariance matrix $\mathbf{C} = \mathbf{A}\mathbf{A}^T$ itself need not be constructed.

The authors assume in the above-mentioned references [14,15] that when solving this problem, it holds $n < p$, and consequently that at most the first $n-1$ eigenvalues are non-zero, and therefore it is possible to omit the other eigenvectors associated with zero eigenvalues. Furthermore, it follows from the nature of principal component analysis that the input set of data could be described using a smaller number of $m < n-1$ eigenvectors (i.e. using m principal components instead of the original p variables). The eigenvalues λ_k express the variability (variance) of the kth principal component defined by the eigenvector $\mathbf{u}_k$. Since the eigenvalues are arranged in descending order, it is clear that the first principal component covers the largest part of the total variability of data, and the influence of the components on the variability decreases with increasing k. One of the possible techniques to determine the appropriate m is to require a certain minimum portion h of the variability of the original data to be preserved, i.e.

$$\frac{\sum_{k=1}^{m} \lambda_k}{\sum_{k=1}^{n} \lambda_k} > h.$$

Subsequently we can just work with the first m eigenvectors $\mathbf{u}_k, k = 1, \ldots, m$. These vectors, also called "eigenfaces" (see Fig. 2), form a basis of the space, also called the space of faces, and the dimension m of this space is considerably lower than the dimension p of the space of the original data.

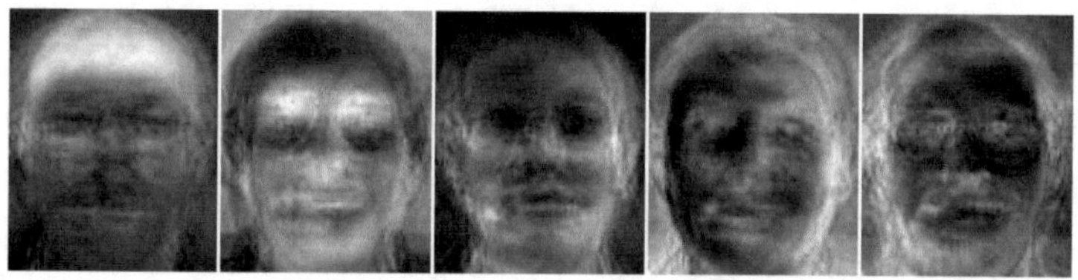

Fig. 2. Visualization of the first five eigenvectors (eigenfaces) $\mathbf{u}_1, \ldots, \mathbf{u}_5$

Now, all input images $\mathbf{\Gamma}_j,\ j = 1, \ldots, n$, are projected onto the space of faces and the vector $\mathbf{\Omega}_j = \left(\omega_{j_1}\ \omega_{j_2}\ \cdots\ \omega_{j_m}\right)^T$ is calculated for each of them. The components of this vector

$$\omega_{j_k} = \mathbf{u}_k^T \left(\mathbf{\Gamma}_j - \mathbf{\Psi}\right) = \mathbf{u}_k^T \mathbf{\Phi}_j \tag{2}$$

represent the individual component scores, which express the contribution of the kth eigenvector to the representation of the jth face image.

Identification of an individual is then converted to the classic recognition problem. Vector $\mathbf{\Omega}_t$ is determined for the new test image $\mathbf{\Gamma}_t$ according to the relation (2), and then the Euclidean distance of this vector from the vectors $\mathbf{\Omega}_j,\ j = 1, \ldots, n$, is calculated for all images of the training set

$$\epsilon_j = \|\mathbf{\Omega}_t - \mathbf{\Omega}_j\|_2.$$

The image is assigned to the individual i if the distance ϵ_i is the minimum of all ϵ_j, and at the same time ϵ_i is smaller than some predetermined recognition threshold θ_ϵ.

Since multiple images that may not even be face images can be projected onto the same vector $\mathbf{\Omega}$, it is appropriate to calculate the distance of the test image from the space of faces

$$\delta = \|\mathbf{\Phi}_t - \mathbf{\Phi}_f\|_2,$$

where $\mathbf{\Phi}_t = \mathbf{\Gamma}_t - \mathbf{\Psi}$ and $\mathbf{\Phi}_f = \sum_{k=1}^{m} \omega_{t_k} \mathbf{u}_k$, and set a certain threshold θ_δ which determines that the image will no longer be considered a face image.

Now one of the following situations may occur:

(i) $\epsilon_i < \theta_\epsilon$ and $\delta < \theta_\delta$: The image is identified as displaying the ith individual.

(ii) $\epsilon_i > \theta_\epsilon$ and $\delta < \theta_\delta$: This case is identified as the face of an unknown individual.
(iii) $\delta > \theta_\delta$: The image is not a face image.

It is obvious that in connection with the first and second option the decreasing value of θ_ϵ will imply the increasing recognition accuracy, but at the same time more faces will remain unrecognized. Calculating the distance from the space of faces and applying the threshold θ_δ allows this method to be used appropriately not only for face recognition but also as a procedure for determining whether a given image is a picture of a face or not. The values of both thresholds θ_ϵ and θ_δ are determined experimentally. Examples of these situations are demonstrated in the following Fig. 3.

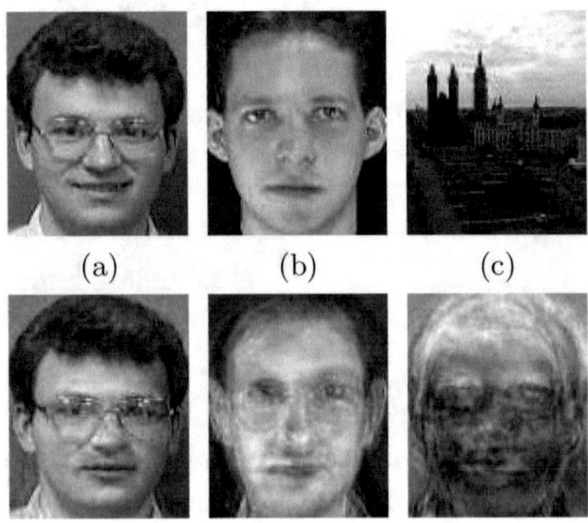

Fig. 3. Three test images and their projections onto the space of faces: (a) an individual included in the training set, (b) an unknown individual who is not in the training set, (c) a non-face image

Table 1 summarizes the values of the distances ϵ and δ for these images. The test image (a) was correctly identified as the individual from the training set having number 5 in the experiment. The minimum distance ϵ of the image (b) is relatively higher compared to the values of the correctly recognized test images of the individuals in the training set, if the threshold θ_ϵ is "suitably" set, this image will be marked as unknown. The high value of the distance δ of the image (c) indicates that this is not a face image.

The presented method for face recognition has been tested by the following simple experiment. The total of 39 individuals from the ORL database were included in the training set. Each of them was represented by the first of ten photos. The test set always contained the second of ten photos of each individual. The threshold values θ_ϵ and θ_δ were not considered, respectively were considered infinite, because all the images in the test set were eligible to be assigned to some individual from the training set. The experiment finished with

Table 1. Values ϵ and δ for the test images of Fig. 3 (rounded to 4 decimal places)

Test image t	$\epsilon = \min_{j=1,\ldots,n} \|\boldsymbol{\Omega}_t - \boldsymbol{\Omega}_j\|_2$	$\delta = \|\boldsymbol{\Phi}_t - \boldsymbol{\Phi}_f\|_2$
(a)	3.6715	8.3403
(b)	11.7247	9.8888
(c)	25.1906	20.3478

the following result: 31 individuals out of the total 39 images in the test set were well-recognized, which indicates about 79.5% success rate.

The face recognition method based on principal component analysis has some shortcomings, which showed up in the experiment. Recognition is sensitive to significant changes in lighting conditions in which the photographs were taken, changes in position and orientation of the face, covering of a part of the face by glasses, etc. The way enabling to eliminate or partially suppress the influence of these effects and thus increase the accuracy of the results obtained is pre-processing (normalization) of all images before performing the method or including multiple different images of each individual in the training set. The eigenface approach is a traditional method for recognizing faces, but since the year 1991, when Turk and Pentland published it, a number of improvements and new methods were proposed.

4 FSVDR of Faces

Liu, Chen and Tan proposed in [9] the way of representing face images called fractional order singular value decomposition representation (FSVDR). Taking advantage of this representation of faces enables to achieve better recognition results, especially in case when the photographs in the training set and the test set differ in lighting conditions or existence of partial coverage of the face. The FSVDR can be used not only with methods based on principal component analysis but also with other face recognition methods.

By default, each grayscale digital image of a face having size $p \times q$ pixels is represented by a matrix containing the intensity values of each pixel. Let this matrix be denoted $\mathbf{G} \in \mathbb{R}^{p \times q}$, and let $\mathbf{G} = \mathbf{U}_G \boldsymbol{\Sigma}_G \mathbf{V}_G^T$ be its SVD. The essence of FSVDR method is to replace the matrix $\mathbf{G}$ by a matrix

$$\mathbf{B} = \mathbf{U}_G \boldsymbol{\Sigma}_G^{\alpha} \mathbf{V}_G^T,$$

$0 \leq \alpha \leq 1$. First, all the images in the training set and the test set are transformed in this way, and then the recognition is performed in a classic way, for example by the eigenfaces method.

Figure 4 illustrates the FSVDR (for different values of parameter α) of two face images that differ in the direction of lighting. It is visually appreciable that the differences between the images are less noticeable with decreasing value of α. The photographs come from the Yale database [3].

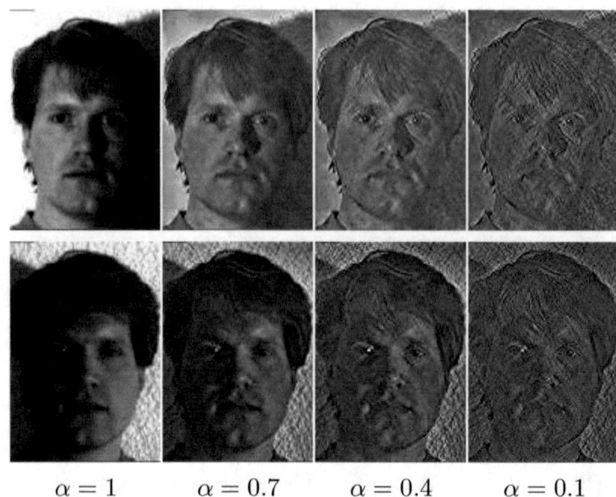

Fig. 4. Two face images with different illumination and their FSVDR ($\alpha = 1$ corresponds to the original image)

The correct choice of the parameter α is the key decision. The value is typically chosen experimentally in accordance with the nature and needs of a specific problem. The authors in [9] state that generally the best results for recognition in connection with methods based on principal component analysis are obtained by the choice $\alpha = 0.1$. We confirmed these findings as described in the following experiment.

The Yale database contains photographs of 15 individuals, each of them is represented by 11 different photographs. Two photographs with different lighting conditions were selected for each individual for the purposes of the experiment performed here. First of them was included into the training set and the second into the test set. All photographs were cropped and their scale was changed. In the first experiment, the standard representation of the photographs was used and recognition was established using the eigenfaces method as was presented in the previous part of this text. Only one person was recognized correctly in this case. In the second experiment, the FSVDR (with parameter $\alpha = 0.1$) was used, following steps of the recognition process remained unchanged. The result of the second experiment was 11 well-identified individuals out of the total number of 15, which indicates 73.3% success rate.

5 Conclusion

The SVD has a variety of uses, some have been known for many years and have been developed by various authors over the course of time, while other applications are related to relatively new domains associated with the development of computing.

The contribution briefly presents the use of the SVD in recognition of faces. We have focused on image recognition and especially face recognition based on the use of synthesis of PCA with the SVD and use of FSVDR. The second approach may also in some sense be seen as an improvement to the first one. The text first summarizes selected essential facts about the SVD, the second part shows the use of the SVD together with PCA for recognition of faces, and the third part is devoted to FSVDR in recognition of faces. The paper presents the essence of both procedures, parameters to be set and their influence on the results, success rates of both approaches, their advantages and disadvantages, and shows results of our experiments. Other outputs could not be included due to the limited scope of the text. All procedures were implemented in MATLAB software.

Acknowledgments. This work and the contribution were supported by a project of Students Grant Agency - Faculty of Informatics and Management, University of Hradec Kralove, Czech Republic. Katerina Fronckova is a student member of the research team.

References

1. Andrews, H.C., Patterson, C.L.: Singular value decompositions and digital image processing. IEEE Trans. Acoust. Speech Signal Process. **24**(1), 26–53 (1976). https://doi.org/10.1109/TASSP.1976.1162766
2. Arun, K.S., Huang, T.S., Blostein, S.D.: Least-squares fitting of two 3-D point sets. IEEE Trans. Pattern Anal. Mach. Intell. **9**(5), 698–700 (1987). https://doi.org/10.1109/TPAMI.1987.4767965
3. Belhumeur, P.N., Hespanha, J.P., Kriegman, D.J.: Eigenfaces vs. Fisherfaces: recognition using class specific linear projection. IEEE Trans. Pattern Anal. Mach. Intell. **19**(7), 711–720 (1997). https://doi.org/10.1109/34.598228
4. Deerwester, S., Dumais, S.T., Furnas, G.W., Landauer, T.K., Harshman, R.: Indexing by latent semantic analysis. J. Am. Soc. Inform. Sci. **41**(6), 391–407 (1990). https://doi.org/10.1002/(SICI)1097-4571(199009)41:6⟨391::AID-ASI1⟩3.0.CO;2-9
5. Golub, G.H., Kahan, W.M.: Calculating the singular values and pseudo-inverse of a matrix. J. Soc. Ind. Appl. Math. Ser. B Numer. Anal. **2**(2), 205–224 (1965). https://doi.org/10.1137/0702016
6. Golub, G.H., Van Loan, C.F.: Matrix Computations, 4th edn. Johns Hopkins University Press, Baltimore (2013)
7. Kirby, M., Sirovich, L.: Application of the Karhunen-Loeve procedure for the characterization of human faces. IEEE Trans. Pattern Anal. Mach. Intell. **12**(1), 103–108 (1990). https://doi.org/10.1109/34.41390
8. Klema, V.C., Laub, A.J.: The singular value decomposition: its computation and some applications. IEEE Trans. Autom. Control **25**(2), 164–176 (1980). https://doi.org/10.1109/TAC.1980.1102314
9. Liu, J., Chen, S., Tan, X.: Fractional order singular value decomposition representation for face recognition. Pattern Recognit. **41**(1), 378–395 (2008). https://doi.org/10.1016/j.patcog.2007.03.027
10. Liu, R., Tan, T.: An SVD-based watermarking scheme for protecting rightful ownership. IEEE Trans. Multimed. **4**(1), 121–128 (2002). https://doi.org/10.1109/6046.985560
11. Sadek, R.A.: SVD based image processing applications: state of the art, contributions and research challenges. Int. J. Adv. Comput. Sci. Appl. **3**(7), 26–34 (2012)

12. Samaria, F.S., Harter, A.C.: Parameterisation of a stochastic model for human face identification. In: Proceedings of 1994 IEEE Workshop on Applications of Computer Vision, pp. 138–142 (1994). https://doi.org/10.1109/ACV.1994.341300
13. Stewart, G.W.: Matrix Algorithms: Volume II: Eigensystems, 1st edn. Society for Industrial and Applied Mathematics, Philadelphia (2001)
14. Turk, M.A., Pentland, A.P.: Eigenfaces for recognition. J. Cognit. Neurosci. **3**(1), 71–86 (1991). https://doi.org/10.1162/jocn.1991.3.1.71
15. Turk, M.A., Pentland, A.P.: Face recognition using eigenfaces. In: Proceedings of 1991 IEEE Computer Society Conference on Computer Vision and Pattern Recognition, pp. 586–591 (1991). https://doi.org/10.1109/CVPR.1991.139758

Model and Software Tool for Estimation of School Children Psychophysical Condition Using Fuzzy Logic Methods

Dmytro Marchuk[1(✉)], Viktoriia Kovalchuk[2], Kateryna Stroj[1], and Inna Sugonyak[1]

[1] Zhytomyr State Technological University, Zhytomyr, Ukraine
[2] Opole University of Technology, Opole, Poland

Abstract. At present, school-age children are regularly exposed to a significant number of negative factors during their school time. The impact of these factors lead to the produce of an organism's response, called stress. Regular stress can in turn cause a deterioration in the health of children.

In this paper we propose a new approach to the use of physical indicators that are tracked using a fitness bracelets. In this work, we obtain and analyze the student's physical activity indicators dynamics. The relationship between the physical condition and the level of psychological loading of schoolchildren during school lessons is analyzed. At the result was developed model for determining stress conditions in school-age children, and a web-service for analyzing and assessing the psychophysiological stress of school-age children is implemented.

Keywords: School-age children · Determine the level of stress · Google fitness API · Decision support systems · Fuzzy knowledge base

1 Introduction

Since ancient times, the issue of our health remains relevant. Improper nutrition, poor sleep, overload of the nervous system, overweight or excessive thinness, unordered schedule of work - all these factors and many others have a great influence on how we feel. This is especially true for children namely pupils, who are in constant motion and change their activity every day. Particularly actual topic for parents is to control the change of physical indicators, namely loss of energy, changes in heart rate, pressure disturbance, and so on. The main objective of the study is to construct a model of control of psychophysical indicators and development of a Web-service for analysis and control of physical activity of pupils.

At the moment, doctors distinguish three main components of health: physical, psychological, social and spiritual. In this paper, the physical component is considered through the collection of physical indicators and psychological components through the measurement of mental activity indicators.

Both boys and girls have a close relationship between psychophysiological indicators and the level of their usual motor activity. For children with a low level of usual

J. Świątek et al. (Eds.): ISAT 2018, AISC 853, pp. 115–124, 2019.
https://doi.org/10.1007/978-3-319-99996-8_11

physical activity are characterized: the average level of situational anxiety, reduced attention, small differences in physical indicators. For children with high levels of usual physical activity are characteristic higher indicators of situational anxiety, increased level of attention, large differences in physical indicators.

Schoolchildren in the process of their life are regularly exposed to a significant number of environmental factors, many of which leads to the development of a corresponding reaction of the body, called stress.

For the first time this concept was introduced by American psychophysiologist Walter Cannon in his work, but the study of stress and related factors is being addressed by another well-known psychophysiologist - Hans Selye [1]. This concept is closely linked with the general adaptive syndrome. Stress can help to make a significant positive impact on the life of the child, being a means of motivation, as well as negatively affect the mood and relationships with others, as well as lead to a number of serious problems of mental and physical health. A large number of children are prone to stressful reactions to external factors due to low living standards, intense school life, poorly planned daily schedule, family problems, and other negative factors.

The problem of parents' monitoring of the health of their children is one of the pressing problems of the present. Nowadays, school-age children spend a lot of time outside the home, which leads to the need to remotely monitor their health. Proceeding from this, the task of developing a decision-support system was developed to control the physical activity of pupils.

2 Main Part

The foundation for this project is a systematic approach: a holistic view of the structure of interoperable elements that are united by the general purpose. The system for analyzing and controlling the physical activity of schoolchildren is a large, complex system for identifying stress periods and their level, which includes a plurality of components whose interaction satisfies the goal.

The composition of the developed decision support system should include the following components:

- subsystem of collection and storage of data on physical indicators;
- subsystem of data processing and formation of conclusions;
- subsystem of tracking over physical indicators in real time;
- subsystem of visualization of the accumulated data.

The DSS should ensure the evaluation of stress periods and their level. At the same time, it could use such output data as blood pressure, pulse, current activity and level of physical and psychological load. Which means the conclusion of the report on stress periods, their limits, types of activities, during which the stress was recorded, the average values of physical indicators.

After analyzing existing methods and approaches for controlling the physical activity of students, the general structure of the system for identifying stressful situations and their periods was developed. In Fig. 1 is a diagram of precedents for better visualization of this functional.

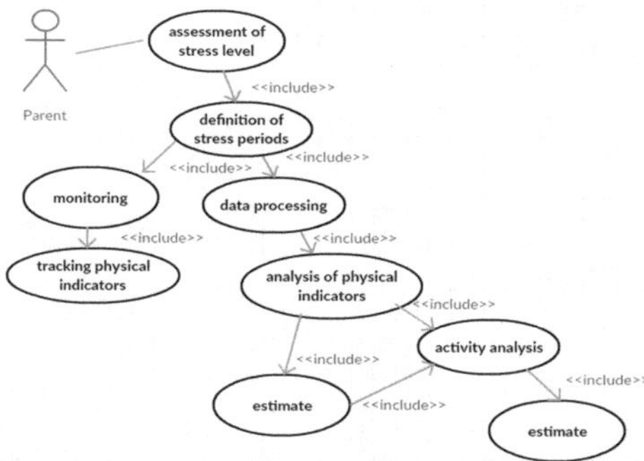

Fig. 1. Diagram of precedents for determining the assessment of the level of stress

Determination of stress levels among schoolchildren. In accordance with the practical studies of data on physical indicators, the degree of stress will be determined at the following levels (from lower to higher): d_1 – light stress level; d_2 – stress below average; d_3 – the average level of stress; d_4 – severe stress; d_5 – the most severe level of stress.

The listed levels $d_1 \div d_5$ are based on the types of diagnoses that need to be recognized. In recognizing the level of stress, we will take into account the following basic parameters available to us, the ranges of which are established by medical studies for the norm. In the round brackets the units of measurement of a parameter are specified: x_1 – level of pulse excess (number of beats per minute); x_2 – duration of excess of pulse (time span); x_3 – level of excess pressure (mmHg); x_4 – duration of excess pressure (time span); x_5 – level of physical activity (score from 1 to 5); x_6 – level of mental load (score from 1 to 5).

These parameters were established by the medical experts [1] during a certain period of studies in children aged 6 to 17 years. The diagnostic task consists in the fact that for each set of values of parameters to match one of the levels: $d_j\left(j = \underline{1,5}\right)$.

Fuzzy Knowledge Base. To implement the algorithm for determining the periods and the level of stress, a mathematical apparatus of fuzzy logic was used, with the help of which a fuzzy knowledge base was created.

Parameters $x_1 \div x_6$, defined above, will be considered as linguistic variables. The child's age will be immediately taken into account when selecting certain limits of the parameters provided. We define the interrelated parameters and introduce the following linguistic variables:

- d – stress assessment, which is measured by one of the levels $d_1 \div d_5$;
- t – estimation of the ratio of the excess of the pulse rate to its duration, which depends on the parameters $\{x_1, x_2\}$;
- y – estimate of the ratio of excess pressure to its duration, which depends on the parameters $\{x_3, x_4\}$;
- z – evaluation of the ratio of mental and physical activity, which depends on the parameters $\{x_5, x_6\}$.

The structure of the model for differential diagnosis of stress levels is similar [2] and is shown in the form of correlations (1)–(4):

$$d = f_d(t, y, z) \tag{1}$$

$$t = f_t(x_1, x_2) \tag{2}$$

$$y = f_y(x_3, x_4) \tag{3}$$

$$z = f_z(x_5, x_6) \tag{4}$$

To evaluate the values of the linguistic variables $x_1 \div x_6$ and also t, y, z, we will use a single scale of qualitative terms: L - low; Ba - below average; A - average; Aa - above average; H - high.

Each of these terms represents a fuzzy set specified by the corresponding membership function.

Fuzzy Logical Equations. Since the formalized knowledge of medical experts under initial conditions may be inadequate, it is foreseen that the knowledge base may be subjected to additional training as experimental data emerge, through the introduction of new rules that will approximate the fuzzy model for identifying stress states, that is, to experimental dependencies. So it is supposed to adapt or customize a fuzzy knowledge base. Taking into account these factors, we obtain the classification of the emergence of different levels of stress in the form of a matrix of knowledge (Table 1), built on such rules using the system of logical expressions 5.

Table 1. Matrix of knowledge

Number the input combination	Input variables			Output variables
	t	y	z	*d*
1.	L	L	L	d_1
2.	Ba	Ba	L	
3.	L	Ba	Ba	
4.	Ba	Ba	Ba	d_2
5.	A	Ba	Ba	
6.	Ba	Ba	A	
7.	A	Ba	A	d_3
8.	Aa	Aa	Ba	
9.	Aa	A	A	
10.	A	H	A	d_4
11.	Aa	Aa	H	
12.	H	Aa	Aa	
13.	H	H	H	d_5
14.	H	Aa	Aa	
15.	H	H	Aa	

1. Dimension of the matrix: (λ +1)N, when (λ +1) – the number of columns that are equal to the number of classification groups of physical indicators;
2. N = $k_1 + k_2 + \cdots + k_m$ – number of rows.
3. First λ matrix columns correspond to the input variables $W_{\lambda i}(t, y, z)$, $i = \underline{1, n}$, and (λ +1) column corresponds to a value d_φ output variable d, $\varphi = \underline{1, m}$.
4. Each row of the matrix represents a certain combination of values of the input variable belonging to one of the possible values of the output variable d. At the same time the first ones k_{φ_1} lines correspond to the value of the output variable d_1, average values k_{φ_2} – values $d_2 \ldots$, the last k_{φ_m} lines – meaning d_m.
5. Variables $W_\lambda(t, y, z)$ are numerical (from 1 to 5). An element of the matrix $\alpha_\lambda^{\varphi\varphi}$ at the intersection of the row with the column corresponds to the linguistic estimation of the parameter W_λ and is involved in determining the possible value of the original variable d, which rank the subjects of the diagnoses according to the principle of changing physical indicators [3].

Categorization of definitions of diagnoses according to the principle of changing physical indicators d = $d = \cup_\varphi d_\varphi$ contains the following classification units:):d_1 – light stress level; d_2 – stress below average; d_3 – the average level of stress; d_4 – severe stress; d_5 – the most severe level of stress.

The introduced matrix of knowledge defines a system of logical expressions such as "IF THAT, ELSE", which binds the values of the input variables $W_1 \div W_n$ to one of the possible solutions, in which case the diagnoses are determined by the principle of changing physical indicators d_φ, $\varphi = \underline{1, m}$.

For the formation of logical conclusions, the corresponding tables (a fragment for the stress level indicator are given in Table 1) are constructed and a formalized system of fuzzy logic equations that will bind the functions of the membership of the diagnoses and the input variables:

The introduction of a matrix of knowledge defines a system of logical statements of the type "IF - THAT, ELSE", which bonds the values of the input variables $x_1 \div x_6$ with one of the possible types of solution $d_j\left(j = \underline{1, 5}\right)$:

$$
\begin{aligned}
&\text{IF } (t = \text{L}) \text{ AND } (y = \text{L}) \text{ AND } (z = \text{L}) \text{ OR } (t = \text{Ba}) \text{ AND } (y = \text{Ba}) \text{ AND } (z = \text{L}) \text{ OR } (t = \text{L}) \\
&\text{AND } (y = \text{Ba}) \text{ AND } (z = \text{Ba}), \text{THAT } d = d_1, \text{ELSE} \\
&\text{IF } (t = \text{Ba}) \text{ AND } (y = \text{Ba}) \text{ AND } (z = \text{Ba}) \text{ OR } (t = \text{A}) \text{ AND } (y = \text{Ba}) \text{ AND } (z = \text{Ba}) \text{ OR } (t = \text{Ba}) \\
&\text{AND } (y = \text{Ba}) \text{ AND } (z = \text{A}), \text{TO } d = d_2, \text{ ELSE} \\
&\text{IF } (t = \text{A}) \text{ AND } (y = \text{Ba}) \text{ AND } (z = \text{A}) \text{ OR } (t = \text{Aa}) \text{ AND } (y = \text{Aa}) \text{ AND } (z = \text{Ba}) \text{ OR } (t = \text{Aa}) \\
&\text{AND } (y = \text{A}) \text{ AND } (z = \text{A}), \text{THAT } d = d_3, \text{ ELSE} \\
&\text{IF } (t = \text{A}) \text{ AND } (y = \text{H}) \text{ AND } (z = \text{A}) \text{ OR } (t = \text{Aa}) \text{ AND } (y = \text{Aa}) \text{ AND } (z = \text{H}) \text{ OR } (t = \text{H}) \\
&\text{AND } (y = \text{Aa}) \text{ AND } (z = \text{Aa}), \text{THAT } d = d_4, \text{ ELSE} \\
&\text{IF } (t = \text{H}) \text{ AND } (y = \text{H}) \text{ AND } (z = \text{H}) \text{ OR } (t = \text{H}) \text{ AND } (y = \text{Aa}) \text{ AND } (z = \text{Aa}) \text{ OR } (t = \text{H}) \\
&\text{AND } (y = \text{H}) \text{ AND } (z = \text{Aa}), \text{TO } d = d_5.
\end{aligned}
\tag{5}
$$

Using the operations ∪ (AND) and ∩ (OR), we write the system of logical expressions for the diagnosis of the parameter du in the following form:

$$\begin{aligned}
\mu^{d_1}(d) &= \left[\mu^{L}(t)\cup\mu^{L}(y)\cup\mu^{L}(z)\right]\cap\left[\mu^{L}(t)\cup\mu^{Ba}(y)\cup\mu^{Ba}(z)\right]\\
&\quad\cap\left[\mu^{Ba}(t)\cup\mu^{Ba}(y)\cup\mu^{L}(z)\right];\\
\mu^{d_2}(d) &= \left[\mu^{Ba}(t)\cup\mu^{Ba}(y)\cup\mu^{Ba}(z)\right]\cap\left[\mu^{A}(t)\cup\mu^{Ba}(y)\cup\mu^{Ba}(z)\right]\\
&\quad\cap\left[\mu^{Ba}(t)\cup\mu^{Ba}(y)\cup\mu^{A}(z)\right];\\
\mu^{d_3}(d) &= \left[\mu^{A}(t)\cup\mu^{Ba}(y)\cup\mu^{A}(z)\right]\cap\left[\mu^{Aa}(t)\cup\mu^{Aa}(y)\cup\mu^{Ba}(z)\right]\\
&\quad\cap\left[\mu^{Aa}(t)\cup\mu^{A}(y)\cup\mu^{A}(z)\right];\\
\mu^{d_4}(d) &= \left[\mu^{Aa}(t)\cup\mu^{A}(y)\cup\mu^{Aa}(z)\right]\cap\left[\mu^{A}(t)\cup\mu^{Aa}(y)\cup\mu^{Aa}(z)\right]\\
&\quad\cap\left[\mu^{Ba}(t)\cup\mu^{Aa}(y)\cup\mu^{Aa}(z)\right];\\
\mu^{d_5}(d) &= \left[\mu^{A}(t)\cup\mu^{H}(y)\cup\mu^{A}(z)\right]\cap\left[\mu^{Aa}(t)\cup\mu^{Aa}(y)\cup\mu^{H}(z)\right]\\
&\quad\cap\left[\mu^{H}(t)\cup\mu^{Aa}(y)\cup\mu^{Aa}(z)\right];
\end{aligned} \tag{6}$$

where $\mu = 1$.

Similar statements are also made for other indicators. The total number of fuzzy logic equations is 15.

The architecture of the development decision support system should be flexible in order to be able to build the functionality quickly and easily, the modules should be as independent as possible, the components of the system should be interchangeable. Access to the DSS must be as mobile as possible.

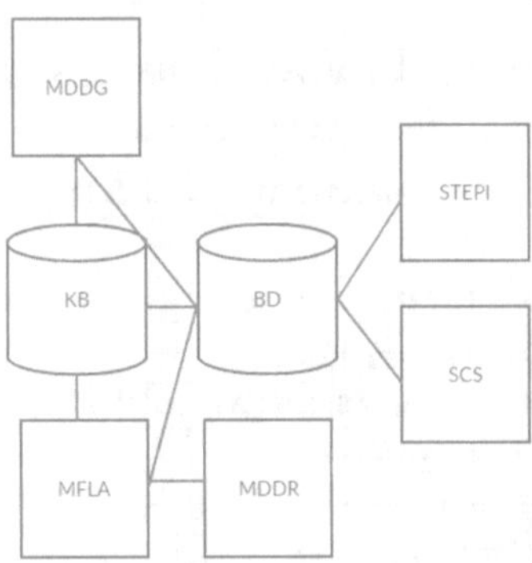

Fig. 2. The general structure of the components of the development DSS

Figure 2 depicts the overall structure of the DSS developed. Within the scope of this master's thesis, the following modules and subsystems were implemented:

- KB is a knowledge base that contains data on the norms of physical parameters for ages 6 to 17 years. It also contains activity data and its characteristics.

- MDDG is a module for displaying dynamic graphs. This module is responsible for displaying graphs that are updated in real time and provide continuous monitoring of the selected physical parameter.
- MDDR is a module for displaying dynamic reports. This module is responsible for the formation and display of reports in stressful periods.
- MFLA is a module for the formation of linguistic assessments. This module is responsible for implementation of algorithm for determination of stress periods and their estimation.
- STEPI is a subsystem of tracking excess physical indicators. This subsystem is responsible for continuous monitoring of physical indicators and informing the user about the excess of these parameters.
- SCS is a subsystem of creating and scheduling, which is responsible for the functioning of creating a schedule, filling its activities.

Based on the goal, the best solution is to use an architecture consisting of a web server, a web service, and an Android application. In this case, the Android application and web service use a shared data store.

To provide an opportunity for data exchange between the Android application and the web server, it was decided to develop a corresponding REST API. In the future, this API can be used to develop applications for other platforms.

Each component has to bear a certain logical load, that is to answer for a certain logical part of the system, to be a completed part of business logic. Such a component must address the list of tasks assigned to it (Fig. 3).

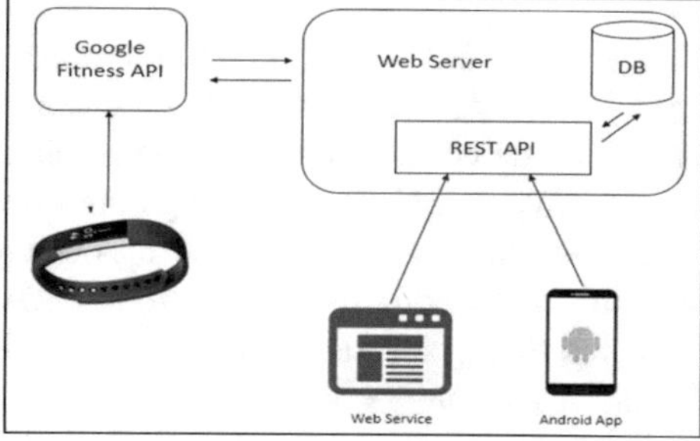

Fig. 3. General architecture of the system under development

In accordance with these requirements, it was decided to create a data warehouse that would be unique and shared. The implemented repository will include user data, schedules, activity types, regulatory physical parameters, custom physical parameters, physical fitness data with the Fitness API.

In addition, it was decided to develop an API for working with a data warehouse, which will allow the user to register and authorize, get their profile, search and add users to the list for tracking, create and fill the activities of the schedule of the day.

On the server, you should develop a synchronization module from the data store with the Google Fitness API.

Based on the requirements for the developed system, a data warehouse was created consisting of 10 relational type tables. The developed structure of the repository provides an opportunity to make a selection of data sets necessary for further analysis and determination of stress periods, as well as the assessment of the level of stress during these periods. Based on the tasks set up a decision support system with the following structure (Fig. 4).

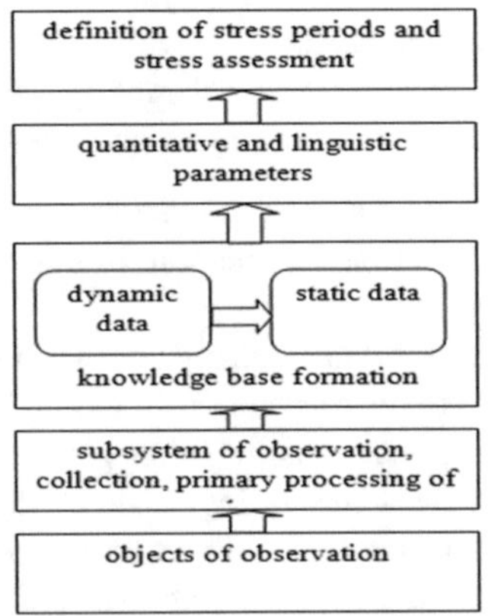

Fig. 4. The structure of the DSS to determine the level of stress

The objects of observation of this system are children aged 6 to 17 years. The subsystem of observation, collection, and primary processing of external data is responsible for obtaining user data when registering, creating and filling out schedules. In addition, this subsystem is responsible for receiving data from the Fitness API.

The formation of the knowledge base takes place by obtaining and transforming data about physical indicators and storing them in the repository. Dynamic data refers to data for each user received through requests to the Fitness API.

After a sample of data for a certain period, the processing of information and the determination of the periods of activity of users, on which the physical indicators exceeded the norm. After assessments of linguistic variables, an assessment of the level of stress is carried out.

The functioning of the developed DSS is not possible without the development of a set of interrelated data processing methods that are necessary for their organization. The developed subsystem should provide high efficiency of work with the accumulated data, if possible eliminate the occurrence of errors. Data processing should take place in a timely manner, to the maximum extent, taking into account all existing factors of influence.

The formation of knowledge base takes place in several stages. Normative data is entered into the database once and are unchanged. In addition, activity data and their characteristics are static. Dynamic data is about users, daily routines, physical indicators at specific moments of time, which are calculated as necessary to display relevant information to the user.

After accumulation of data, the web-service must provide a view of the statistics of physical indicators. These data should be presented in user-friendly form. Therefore, a module was developed to perform a sample of data on physical indicators. Data samples can be represented by parameters such as the type of physical indicators and the period. The resulting data is displayed in graphs.

In order to be able to view data on physical indicators, a user search was developed. The data processing subsystem has the following components: DBMS, DB, Knowledge Base, Query and Reporting Module, Visualization module for data for the selected reporting period.

The Web server interacts with the database through a defined interface to obtain the necessary data. The main interfaces of the service with full information about the physical characteristics of the child are shown in Figs. 5 and 6.

Fig. 5. Profile of the child being tracked

HEALTH + — Home | Watch | Contact | Logout (Parent)

Pulse | Steps | Pressure | Stress

Child stress

Filter by

Start Time	End Time	Stress Level	Avg Pulse	Pulse duration	Activities
Sun, Jan 14, 2018 9:40 AM	Sun, Jan 14, 2018 9:40 AM	1	97	0m	Літературне читання
Sun, Jan 14, 2018 10:10 AM	Sun, Jan 14, 2018 10:50 AM	4	117	40m	Інформатика Перерва між уроками
Sun, Jan 14, 2018 11:40 AM	Sun, Jan 14, 2018 11:50 AM	2	109	10m	Зарубіжна література Перерва між уроками
Sun, Jan 14, 2018 12:30 PM	Sun, Jan 14, 2018 12:30 PM	1	102	0m	Історія України
Sun, Jan 14, 2018 4:50 PM	Sun, Jan 14, 2018 5:40 PM	4	121	50m	Перерва між уроками Геометрія
Sun, Jan 14, 2018 7:20 PM	Sun, Jan 14, 2018 7:30 PM	2	100	10m	Виконання уроків
Sun, Jan 14, 2018 9:10 PM	Sun, Jan 14, 2018 9:30 PM	2	103	20m	Виконання уроків

Fig. 6. Reporting stressful periods

The main functionality is the reporting of the recovered periods of stressful child states in the form of a table. It shows such elements as the actual period itself - the start date and end date, its duration, depending on the analyzed data of physical indicators, the assessment of stress in the child, the average value of the pulse during this period and the activity during which the stressful periods took place.

3 Conclusions

In this work, a thorough analysis of existing methods and solutions for the analysis of psychophysiological indicators and stress situations in children aged 6 to 17 years has been conducted, which has resulted in the design of an algorithm for determining stress periods. Practical result is the development of a system for monitoring the physical activity of students. The developed system consists of two parts - the web-service and the Android-application. The web-service is implemented in the form of a website that provides parents with information on the health of the child.

The basis of constructing an algorithm for determining stress periods was chosen fuzzy logic. The results obtained are useful for preventive medicine and used in schools, lyceums, gymnasia, etc., in which the age group of children reaches the age of 6 to 17 years for the analysis and control of physical activity in order to improve the health status.

References

1. Selye, H.: Stress Without Distress. J. B. Lippincott Co., Philadelphia (1974)
2. Rothstein, O.: Intelligent Identification Technologies: Fuzzy Sets, Genetic Algorithms. Neural Networks. UNIVERSUM-Vinnytsia, Vinnitsa (1999)
3. Barseghyan, A., Barseghyan, A., Kupriyanov, M., Kholod, I.: Analysis of Data and Processes: Studies Manual. BHV-Petersburg, St. Petersburg (2009)

The Artifact Subspace Reconstruction (ASR) for EEG Signal Correction. A Comparative Study

Malgorzata Plechawska-Wojcik[1(✉)], Monika Kaczorowska[1], and Dariusz Zapala[2]

[1] Lublin University of Technology, Lublin, Poland
{m.plechawska,m.kaczorowska}@pollub.pl
[2] Department of Experimental Psychology, The John Paul II Catholic University of Lublin, Lublin, Poland
d.zapala@gmail.com

Abstract. The paper presents the results of a comparative study of the artifact subspace re-construction (ASR) method and two other popular methods dedicated to correct EEG artifacts: independent component analysis (ICA) and principal component analysis (PCA). The comparison is based on automatic rejection of EEG signal epochs performed on a dataset of motor imagery data. ANOVA results show a significantly better level of artifact correction for the ASR method. What is more, the ASR method does not cause serious signal loss compared to other methods.

Keywords: EEG · Artifact correction · ASR · ICA · PCA

1 Introduction

Electroencephalography (EEG) allows for non-invasive neural activity measurement. This technique is widely applied not only in medical application, such as diagnosis and treatment, but also in other areas, such as brain-computer interfaces (BCI), evoked potentials, cognitive load measurement, or other psychological studies.

EEG recording is performed with electrodes placed on the head of an examined person, usually with a dedicated cap. Single recoding might cover from several to even several hundred electrodes.

EEG recordings, however, are often exposed to artifact appearance. Artifacts are unwanted signals of non-cerebral origin registered by EEG electrodes. The presence of artifacts in the signal limits the clinical usefulness of the study and may lead to misdiagnoses, including in particular the detection of non-existent neurological disorders. Artifacts might also impede the signal analysis in both the time and frequency domain, and cause misinterpretations in the phenomenon examined.

Artifacts may have different type, origin and frequency characteristics. Artifacts are usually characterised by the amplitude higher than the rest of the signal. They often take the form of peaks or noised fluctuations. Depending on the artifact type, one or

J. Świątek et al. (Eds.): ISAT 2018, AISC 853, pp. 125–135, 2019.
https://doi.org/10.1007/978-3-319-99996-8_12

many channels might be noised. What is more, if these disturbances are present in most or all of the channels, they often appear in various proportions.

In general, EEG artifact types might be divided into two groups: technical and biological. The first one is related to signal registration, whereas the second one derives from the person examined. Technical artifacts might be caused by inappropriate skin-to-electrode contact caused by, for example, improperly cleansed skin, badly applied gel or a damaged electrode. Other problems might be caused by improper locations of electrodes, not compatible with the accepted standards or inappropriate test conditions, including dissipation of the person examined and inadequately prepared environment. Another important factor is the electric field of external electronic devices, which needs to be filtered.

Biological artifacts are related to physiology, behaviour and motions of the examined person. This group covers participant sweating causing deterioration of the skin-electrode system performance, head and jaw moving, or heart electrical activity. The electrical potential of such muscle tension related artifacts is several times higher than the level of the EEG signal. An important subgroup covers ocular artifacts, related to blinking and eye moving. The movement of the eye during blinking causes a very strong artifact, which can even be seen on all electrodes, including in particular those placed on the prefrontal and frontal leads. The blinking eyes artifacts may cause a potential reaching up to 100 V. For detection of eye artifacts, electrooculography (EOG) recording is often used.

Particular types of artifacts are visible on certain scalp locations. The central scalp sites contain mainly brain activity. Prominent blinks are located in frontal sites, whereas temporo-parietal sites contain temporal muscle artifacts [1]. Detection and correction of such artifacts is an important part of the pre-processing procedure, because high amplitudes of artifactual signal may falsify the results of EEG analysis, in particular event-related potential analysis (consisting in searching for peaks in the averaged signal) or BCI (where the analysis is performed in real-time). Even infrequently occurring artifacts may bias the experiment result as they might have strong influence on the averaged signal.

Although some artifacts might be detected by rejecting single noised electrodes or applying low-order signal statistics such as minimum or maximum, most typical artifacts have irregular character and need more sophisticated detection methods. More adequate are statistical measures of EEG signals, such as linear trend detection, probability of each data epoch or probability distribution of potential values over all epochs, which might help to indicate trials noised with artifacts. What is more, such measures as kurtosis or standard threshold might help in detecting artifacts with specific spectral characteristics.

The aim of this study is to compare the artifact subspace reconstruction (ASR) method with other popular artifact correction methods, such as independent component analysis (ICA) and principal component analysis (PCA). Both mentioned methods, PCA and ICA, are modern artifact correction methods based on signal transformation into new space in order to get independent dimensions. Classical artifact corrections techniques based on blurred data removing or reference adjusting are not considered here.

The comparison criterion is based on the number of automatically rejected epochs of EEG signal. The study analyses motor imagery data taken from 10 subjects.

The originality of the work lies in the comparison of ASR with other artifact correction methods. As the ASR is relatively new technique and it is based on new approach of signal reconstruction with the reference signal fragment, it is worth to check the performance of this solution. The aim of this work is also to check if ASR reconstructs the signal without the significant information loss. Similar analysis for other artifact correction methods were previously performed by other authors [2–4].

This paper is organised as follows. Section 2 provides a review of related research. Section 3 contains a description of the method applied. Section 4 introduces the case study description and the dataset characteristics. Section 5 presents the results. Section 6 is a summary of the paper.

2 Related Work

Detecting and removing EEG artifacts without losing the signal quality is a problematic task. In numerical-based methods dedicated to artifact removing the signal is not removed but corrected. In the case of manual or automatic cleaning, however, contaminated parts of the signal are removed. There is extensive scientific literature on the EEG artifact elimination problem [5–9]. Especially eye movement (including blinks), muscle activity and electrical noise are commonly occurring problems in EEG research.

Among the most popular methods of typical EEG artifact elimination (such as eye blinks or muscle activity) one can find spatial filtering techniques, such as independent component analysis (ICA) [10] or canonical correlation analysis [11]. Among component analysis methods, principal component analysis (PCA) is also widely applied [12]. Among these methods, ICA is considered as the most reliable [13] if there is no prior knowledge about artifacts [14]. These methods, especially ICA, have developed different implementations and several extended versions, such as extended Infomax ICA [15] or Auto-Regressive eXogenous (ARX) [16]. There are also extensions enabling for automation of artifact-related ICA components [17], which originally need to be indicated manually after visual inspection. Other methods, such as linear regression [18] adaptive filtering [19] or Bayes filtering [20], work automatically on a single channel based on estimating the reference channel of artifacts [13].

Modern EEG applications, such as BCIs, need the ability to perform analysis and monitor cortico-cortical interactions in real time [1]. In the case of such real-time analysis ocular artifacts as well as signal drifts are the most problematic signal contamination [21]. Rejection of such artifacts is of great importance; however, it is difficult to implement. What is more, the growing popularity of wearable, mobile EEG poses challenges in reliable real-time modelling of neuronal activity, including fast computation, high modelling performance based on limited amount of data and artifact handling [1].

In [23] movement-related artifacts were applied, prior to Independent Component Analysis. In this study, low walking speed and relatively static nature of movement (treadmill walking) limited artifacts to the level where corrections were possible. For such noised signal, artifact subspace reconstruction (ASR) has been applied [1].

ASR is a recently developed algorithm used by few researchers so far. In [22] the authors apply this method to eliminate high amplitude noise, including movement-related artifacts, before the analysis. In [1] ASR was successfully applied in an ERP study. The authors conducted the analysis with and without ASR and the results confirmed that ASR did not distort the ERP results. In [23] the authors applied ASR to reduce motion artifacts related to treadmill walking. To minimise possible loss of electrocortical signals after ASR, AMICA decomposition was performed and obtained sphering and weighting matrices were applied to the preprocessed EEG dataset. ICA transformations needed in the process of the analysis were applied before ASR cleaning to minimise the possible loss of true cortical activity.

In the literature one can find several papers presenting artifact methods comparison. Among them [2] compares commercially available software applying such criteria as the number of readers able to render assignments, confidence, the intra-class correlation (ICC), in [3] Signal to Interference Ratio (SIR) criterion is used to compare performance of different ICA methods in EOG artifacts removal. In [4] different muscle artifact removal methods based on ICA are compared based on event-related desynchronization (ERD) and dipolarity.

3 Applied Methods

3.1 Artifact Subspace Reconstruction

Artifact subspace reconstruction (ASR) [1] is the EEG artifact correction algorithm available as the EEGLab software plugin. It uses sliding windows of EEG signal, each of which being decomposed with the PCA method. Each obtained EEG fragment is scanned to identify high variance signal exceeding a given threshold. It is done by statistical comparison with data from a derived clean baseline EEG recording containing minimum artifacts. For each sliding window the method searches for principal subspaces significantly deviating from the baseline signal. These fragments are linearly reconstructed by a mixing matrix computed from the calibration data which is a baseline EEG recording [22]. By default, ASR process of artifact removal based on PCA algorithm might also work with ICA component subspace pre-computed on calibration data [24] or estimated with the Online Recursive ICA [25].

3.2 Independent Component Analysis

Independent component analysis (ICA) [10] is related to blind source separation (BSS). This method assumes that the elements of the data are statically independent. The EEG signal gathered from the electrodes placed on the cap can be related to BSS and is also the combination of several independent signals. The components appear after using ICA and the expert should choose components which include artifacts. If artifactual components are removed, an EEG signal is clear.

3.3 Principal Component Analysis

The principal component analysis (PCA) [5] method is based on decomposition of data into components corresponding to various values of variance. The lower the number of components is, the higher is its variance. This algorithm is relayed on the matrix calculus. The calculated components might contain artifacts. The lower the number of a principal component, the more artifacts this component may include. The first and second components are usually removed. As a result, the EEG signal obtained contains a significantly lower number of artifacts.

4 The Case Study

4.1 The Data Set

Ten right-handed subjects (8 females) aged 22–29 (M = 24.60; SD = 2.50) participated in the experiment. All subjects were volunteers and gave a written consent to take part in the study. They also declared that they were neither taking medication nor any psychoactive substances on a permanent basis.

At the end of the whole experimental procedure, participants were paid a remuneration of 20 USD. The study was approved by the Ethics Committee of the Institute of Psychology.

Changes in the brain activity were measured with the GES 300 (Electrical Geodesics, Inc. Eugene, OR, USA) EEG system consisting of a Net Amps 300 amplifier (output resistance 200 MΩ; recording range from 0.01 to 1000 Hz) and a 64-channel cap with active electrodes ActiCAP (Brain Products, Munich, Germany). Electrode impedances were kept below 10 kΩ and the signal was referenced to an FCz channel during registration. Data sampling was defined at 500 Hz and recorded with Net Station 4.4 (EGI, Eugene, OR, USA). The experimental procedure was designed and displayed on a screen with the use of E-Prime, version 2.0 (Psychology Software Tools, Pittsburgh, PA, USA).

During the recording of EEG signals, at the beginning of each trial a grey board divided by vertical line was displayed on the monitor for 3000 ms. Then, on the left or right side, a visual cue appeared (4 s) in the form of a black-and-white checkerboard. Its location indicated which hand movement is to be imagined in a given trial. The subject performed the motor imagery task until the STOP symbol appeared. After the interstimulus interval of random length ranging between 2000–4000 ms, another trial followed. During registration, 180 trials were conducted (90 to imagine the movements of the right and left hands, respectively). For an extensive description of the experimental paradigms see [26] and [27].

4.2 Research Procedure

The data were analysed using EEGLAB, which is a MatLab plugin, and loaded from . set files. The following steps were repeated for every original file:

- The electrodes (up to 2) containing noise or artifacts were removed.
- The signal was filtered using lower and higher edge of the frequency pass band (filter response: 0.5–40 Hz).
- The artifacts were removed from the filtered data using PCA, ICA and ASR algorithms.

The PCA algorithm removed the first and second component. The set of independent components was the result of the ICA algorithm and the expert made a decision which components should be removed. As for ASR, the default parameters were applied. The following steps were carried out for each original file and for created files after using PCA, ICA and ASR:

1. The filtered data was extracted by epochs. EEG data were divided into epochs (segments) matching the period of time from the disappearance of the visual cue to the end of the imagery task (duration 3 s). The following epochs were selected: right, left and relax.
2. The rejecting trials using data statistics were applied to the data from the first step. The abnormal values were used to find and reject epochs with artifacts. Upper and lower limits were defined: ±50 mV.
3. The segments were averaged with reference to individual subjects and conditions (ASR;PCA:ICA:RAW)
4. Segments were subjected to a fast Fourier transform (FFT). The power spectrum (meaning the square root of the sum of the squared real and imaginary parts of the results of the FFT) was calculated and then a decimal logarithm was computed. The procedure was applied to sensorimotor (SMR) rhythms range (8–30 Hz) [28].

An example of applying ASR to the data is presented in Fig. 1.

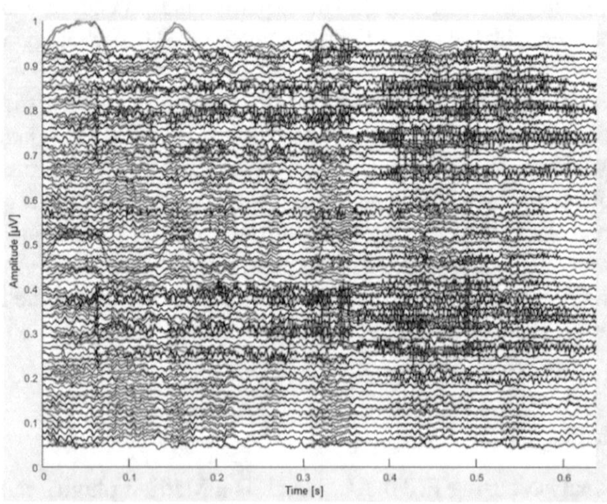

Fig. 1. Example of ASR application. The red signal is the original recording, whereas the blue one is the signal after ASR correction.

Data from a single electrode placed in position C3 was used in the statistical analyses by repeated measure ANOVA. The Dunnett's test [29] with the RAW condition as a control was used for post hoc comparisons. All 64 electrodes were used to visualize the effect of the artifact correction method on the signal distribution on the skull (see Fig. 4A). ANOVA analysis was also performed for number of remaining epochs value (Fig. 3). The procedure of data processing is presented in Fig. 2.

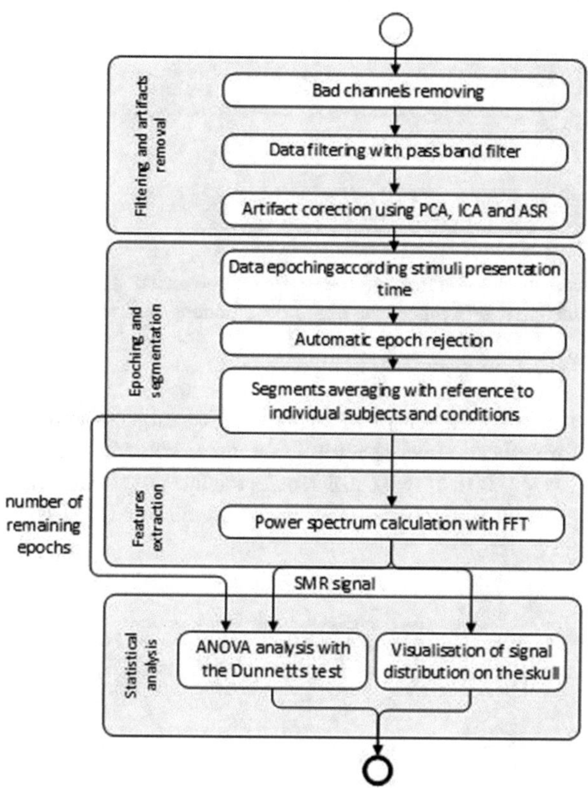

Fig. 2. The procedure of data processing in the present study.

5 Results

The repeated measure analysis of variance with METHOD (RAW; PCA; ICA) and the with-subject factor was performed on a number of removed epochs after automatic rejection. A statistically significant effect was observed ($F(3,27) = 43.305$, $p < 0.001$, $\eta^2 = 0.83$). The Dunnett's post-hoc test with RAW as a control condition showed that the number of the remaining epochs is greatest after ASR procedure ($M = 164.2$, $SE = 6.18$) followed by PCA ($M = 146.2$, $SE = 9.7$) and ICA ($M = 120.2$, $SE = 14.07$) (Fig. 3).

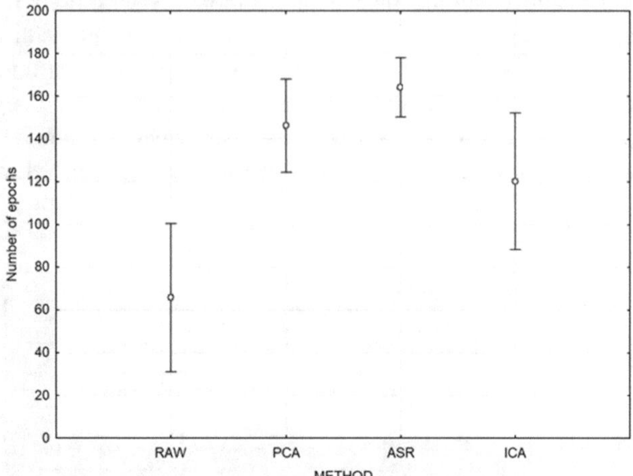

Fig. 3. Differences in the number of the remaining epochs. The vertical bars show 0.95 confidence intervals. All differences in post hoc Dunnetts test are statistically significant at $p < 0.001$.

Also, there was a significant main effect for with-subject factor METHOD (RAW; PCA; ICA) performed on power spectrum of sensorimotor rhythms from C3 electrode ($F(3,27) = 3.51$, $p < 0.03$), $\eta^2 = 0.30$). However, the post hoc Dunnett's test did not confirm any significant differences between the conditions (Fig. 4B).

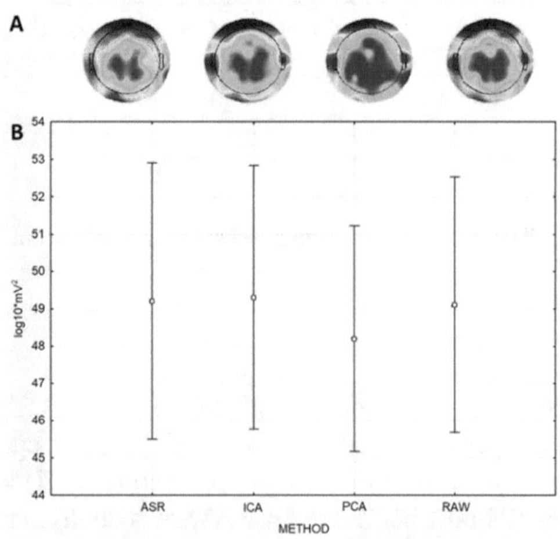

Fig. 4. Power spectrum of sensorimotor rhythms (8–30 Hz): A: distribution on the skull; B: results from C3 electrode. The vertical bars show 0.95 confidence intervals.

6 Summary

The paper presents a case study comparing the recently developed artifact Subspace Reconstruction (ASR) algorithm with other well-known methods dedicated to EEG artifact correction. The ASR method seems to be highly efficient as it can handle heavily noised signal, even artifacts related to movement. The manual analysis of the obtained results performed on several different datasets shows that the ASR method is able to correct artifactual signal in most problematic data sets. The results presented in the paper discuss comparison of ASR to two other methods: independent component analysis (ICA) and principal component analysis (PCA).

The comparison was performed on the dataset of signals taken from 10 right-handed subjects. Each person performed the motor imagery task composed of 180 left and right hand movement imaginary trials. The artifact correction analysing procedure was identical for all three methods. For each method the number of removed epochs after automatic rejection performed in EEGLab software was checked.

Results reveal the ASR method as more effective than the others. The number of remaining epochs after the ASR procedure occur to be significantly greater comparing to other methods and to raw data. This result, obtained in ANOVA analysis, proves that ASR is able to correct a greater number of artifacts than other methods.

Additional calculations were done to test if the artifact correction results affected the EEG data. The power spectrum of sensorimotor rhythms was analysed on the C3 electrode, which should reveal Event-related (De)synchronization (ERD/ERS) associated to imagery of hand movement imagery performed by subjects. Statistical analysis did not confirm any significant differences between results obtained for particular methods. The topplot charts shown illustrate the signal power and its distribution on the skull. The results reveal no impact of the applied artifact correction method on these characteristics.

Acknowledgement. In order to simplify the replication of our results, we have placed a data sets used for analysis in a public repository https://github.com/lareieeg/EEGdata.

References

1. Mullen, T., Kothe, C., Chi, Y.M., Ojeda, A., Kerth, T., Makeig, S., Cauwenberghs, G., Jung, T.P.: Real-time modeling and 3D visualization of source dynamics and connectivity using wearable EEG. In: 35th Annual International Conference on Engineering in Medicine and Biology Society (EMBC). IEEE, pp. 2184–2187 (2013)
2. Weiss, S.A., Asadi-Pooya, A.A., Vangala, S., Moy, S., Wyeth, D.H., Orosz, I., Chang, E.: AR2, a novel automatic muscle artifact reduction software method for ictal EEG interpretation: Validation and comparison of performance with commercially available software. F1000 Research 6 (2017)
3. Kusumandari, D.E., Fakhrurroja, H., Turnip, A., Hutagalung, S.S., Kumbara, B., Simarmata, J.: Removal of EOG artifacts: comparison of ICA algorithm from recording EEG. In: 2nd International Conference on Technology, Informatics, Management, Engineering, and Environment (TIME-E), pp. 335–339 (2014)

4. Frolich, L., Dowding, I.: Removal of muscular artifacts in EEG signals: a comparison of linear decomposition methods. Brain informatics, pp. 1–10 (2018)
5. Berg, P., Scherg, M.: A multiple source approach to the correction of eye artifacts. Electroencephalogr. Clin. Neurophysiol. **90**, 229–241 (1994)
6. Croft, R.J., Barry, R.J.: Removal of ocular artifact from the EEG: a review. Neurophysiol. Clin. **30**, 5–19 (2000)
7. Joyce, C.A., Gorodnitsky, I.F., Kutas, M.: Automatic removal of eye movement and blink artifacts from EEG data using blind component separation. Psychophysiology **41**, 313–325 (2004)
8. Liu, T., Yao, D.: Removal of the ocular artifacts from EEG data using a cascaded spatiotemporal processing. Comput. Methods Progr. Biomed. **83**, 95–103 (2006)
9. Qin, Y., Xu, P., Yao, D.: A comparative study of different references for EEG default mode network: the use of the infinity reference. Clin. Neurophysiol. **121**, 1981–1991 (2010)
10. Delorme, A., Sejnowski, T., Makeig, S.: Enhanced detection of artifacts in EEG data using higher-order statistics and independent component analysis. Neuroimage. **34**, 1443–1449 (2007)
11. DeClercq, W., Vergult, A., Vanrumste, B., VanPaesschen, W., VanHuffel, S.: Canonical correlation analysis applied to remove muscle artifacts from the electroencephalogram. IEEE Trans. Biomed. Eng. **53**, 2583–2587 (2006)
12. Berg, P., Scherg, M.: Dipole modelling of eye activity and its application to the removal of eye artefacts from the EEG and MEG. Clin. Phys. Physiol. Meas. **12**, 49 (1991)
13. Goh, S.K., Abbass, H.A., Tan, K.C., Al-Mamun, A., Wang, C., Guan, C.: Automatic EEG Artifact Removal Techniques by Detecting Influential Independent Components. IEEE Trans. Emerg. Topics Comput. Intell. **1**(4), 270–279 (2017)
14. Uriguen, J.A., Garcia-Zapirain, B.: EEG artifact removal-state of- the-art and guidelines. J. Neural Eng. **12**(3), 031001 (2015)
15. Lee, T.W., Girolami, M., Sejnowski, T.J.: Independent component analysis using an extended infomax algorithm for mixed subgaussian and supergaussian sources. Neural Comput. **11**(2), 417–441 (1999)
16. Wang, Z., Peng, X., TieJun, L., Yin, T., Xu, L., DeZhong, Y.: Robust removal of ocular artifacts by combining Independent Component Analysis and system identification. Biomed. Signal Process. Control **10**, 250–259 (2014)
17. Raduntz, T., Scouten, J., Hochmuth, O., Meffert, B.: EEG artifact elimination by extraction of ICA-component features using image processing algorithms. J. Neurosci. Methods **243**, 84–93 (2015)
18. Wallstrom, G., Kass, R., Miller, A., Cohn, J.F., Fox, N.A.: Automatic correction of ocular artifacts in the EEG: a comparison of regression-based and component-based methods. Int. J. Psychophysiol. **53**(2), 105–119 (2004)
19. Sweeney, K., Ward, T., McLoone, S.: Artifact removal in physiological signals-Practices and possibilities. IEEE Trans. Inf. Tech. Biomed. **16**(3), 488–500 (2012)
20. Gwin, J., Gramann, K., Makeig, S., Ferris, D.: Removal of movement artifact from high-density EEG recorded during walking and running. J. Neurophy. **103**, 3526–3534 (2010)
21. Kilicarslan, A., Grossman, R.G., Contreras-Vidal, J.L.: A robust adaptive denoising framework for real-time artifact removal in scalp EEG measurements. J. Neural Eng. **13**(2), 026013 (2016)
22. Bulea, T.C., Prasad, S., Kilicarslan, A., Contreras-Vidal, J.L.: Sitting and standing intention can be decoded from scalp EEG recorded prior to movement execution. Front. Neurosci. **8**, 376 (2014)

23. Bulea, T.C., Kim, J., Damiano, D.L., Stanley, C.J., Park, H.S.: Prefrontal, posterior parietal and sensorimotor network activity underlying speed control during walking. Front. Human Neurosci. **9**, 247 (2015)
24. Le, Q.V., Karpenko, A., Ngiam, J., Ng, A.Y.: ICA with reconstruction cost for efficient overcomplete feature learning. NIPS, pp. 1017–1025 (2011)
25. Akhtar, M., Jung, T.-P., Makeig, S., Cauwenberghs, G.: Recursive independent component analysis for online blind source separation. IEEE Int. Symp. Circuits Syst. **6**, 2813–2816 (2012)
26. Zapala, D., Francuz, P., Zapala, E., Kopis, N., Wierzgala, P., Augustynowicz, P., Kolodziej, M.: The impact of different visual feedbacks in user training on motor imagery control in BCI. In: Applied Psychophysiology and Biofeedback, pp. 1–13 (2017)
27. Majkowski, A., Kolodziej, M., Zapala, D., Tarnowski, P., Francuz, P., Rak, R.J., Oskwarek, L.: Selection of EEG signal features for ERD/ERS classification using genetic algorithms. In: 18th International Conference on Computational Problems of Electrical Engineering (CPEE), pp. 1–4 (2017)
28. Zapala, D., Zabielska-Mendyk, E., Cudo, A., Krzysztofiak, A., Augustynowicz, P., Francuz, P.: Short-term kinesthetic training for sensorimotor rhythms: Effects in experts and amateurs. J. Mot. Behav. **47**(4), 312–318 (2015)
29. Dunnett, C.W.: A multiple comparison procedure for comparing several treatments with a control. J. Am. Stat. Assoc. **50**(272), 1096–1121 (1955)

The Study of Dynamic Objects Identification Algorithms Based on Anisotropic Properties of Generalized Amplitude-Phase Images

Viktor Vlasenko[1(✉)], Sławomir Stemplewski[2], and Piotr Koczur[1]

[1] Faculty of Nature and Technical Sciences, Opole University, Oleska 48, 45-052 Opole, Poland
vlasenko@uni.opole.pl

[2] Institute of Mathematics and Computer Science, Opole University, Oleska 48, 45-052 Opole, Poland

Abstract. The article presents some results of dynamical objects identification technology based on coincidence matrixes of templates and tested objects' amplitude-phase images (APIm) calculated with discrete Hilbert transforms (DHT). DHT algorithms are modeled on basis of isotropic (HTI), anisotropic (HTA), generalized transforms – AP-analysis (APA) and the difference (residual) relative shifted phase (DRSP-) images to calculate the APIm. The identified objects are recognized as members of classes modeled with 3D templates – images of different types airplanes rotated in space. The dynamic anisotropic properties of APIm causes the increasing of sensitivity to circular angle rotation and make possible effective classification of tested objects at DHT domains. Methods to objects and templates matching accuracy increasing are based on calculations and correlation of intra- and inter-classes coincidence matrixes.

Keywords: Generalized hilbert transforms · Amplitude-phase images
Dynamic object identification

1 Introduction

The complex shape objects (CSO) detection, analysis and recognition procedures are usually the important and resource demanding parts of identification information technologies (IIT) applied to analysis of dynamic scenes including the moving objects images modeling and recognition [1, 2]. Very important factors influencing on effectiveness of recognition are the CSO linear translating and circular rotation moving at field of view causing the distortion of objects' projections and decreasing of effectiveness of recognition based on 2D images as templates. The tasks to these factors elimination in aim to increase of identification effectiveness are still actual and important for digital optic systems at many application areas. As researches show the methodology of digital Hilbert optics (DHO) based on the theory of generalized analytical signals [3] is the prospective area to solving tasks mentioned. This approaches are based on use of discrete Hilbert transforms and hyper-complex signals and generalized Fourier spectra theory [3, 4] for multidimensional images representation [4–6]. The applications of different kind of discrete Hilbert transforms (DHT), generalized

J. Świątek et al. (Eds.): ISAT 2018, AISC 853, pp. 136–144, 2019.
https://doi.org/10.1007/978-3-319-99996-8_13

DHT, Foucault transforms (DH(Fc)T) and others types of hybrid transforms open new directions to increasing the effectiveness of CSO identification. The properties of multidimensional Hilbert-transforms such as circular rotation anisotropy and multi-phase representation [3, 4] based on amplitude-phase fields analysis (APA) need the further investigations in scope of practical usefulness at relevant areas. The article presented below is devoted to studying of possibilities of CSO identification algorithms founded on AP-analysis and evaluations their improving as IIT components based on statistical methods in DHO-domain. The main goal of article is the describing of methods and structures of IIT based on statistical models as matrixes of coincidence (MoC) of multidimensional APA-data images belonged to different classes of dynamical CSO. As dynamic objects (templates and tested referential) for tasks of identification the APA-images are used. As the characteristic features presented as MoC could be used the fields of original images and different Hilbert transforms, fields of generalized amplitudes and phases (AP-images), fields of partial phases, fields of differential residual shifted phase images residual (DRSP-) AP- images. The computing and comparing of MoC facilitates the dynamical objects localization and identification. The article is further developing of previous authors' results in practical IT tasks modeling [6–8]. The conceptual definitions and descriptions such as structural charts of algorithms – block-components of IIT are presented in Sect. 2. Section 3 contains of illustrative pictures showing results of dynamical CSO identification with MoC at spaces of characteristic features such as generalized amplitude and phase fields of 2D hyper-complex images, representing identified objects "in-class" and "inter-class" under conditions of linear translations and circular rotations. Partial phase fields are used as characteristic features of Hilbert-transformed 2D images of CSO for computing of MoC in version of generalized phase splitting (phase_1, phase_2) and in version of DRSP-images (generalized phase domain). Section 4 presents the discussion and conclusions based on analysis of results of computing experiments and modeling.

2 Conceptual Consideration of CSO Identification System Based on DHO-APA Methods: Structures, Functions and Methods

Conceptual chart of APA-methodic based video-information system (VIS) to dynamical CSO identification is presented at Fig. 1. VIS is designed to realize IIT functionality demands [7] with structure consisting of modules (block-components – units). First unit is the library of synthesized 3-D objects – spatial models as the set (Local Data Base LDB) of geometrically transformed (translated, rotated, scaling) 2D templates-Cartesian projections, corresponded to input tested (recognized) objects to be identified. LDB is created as a hierarchic structure of semantics {*Class – Object – Image*}. Next module realizes functions of synthesis of 2-D AP-images – generalized amplitude $A(i,j)$ and phase $\{P(i,j), P1(i,j), P2(i,j)\}$ fields of Hilbert-transformed images of spatial-time objects in process of their moving, and design of LDB based on semantics {*Object – Image – DataArray_AP*}. The third unit realizes procedures of synthesis of statistical models as matrix-histograms (MoC - matrix of coincidence level) of AP-image sets {*hist2*(*A, P1*); *hist2*(*A, P2*); *hist2*(*P1, P2*)} (*P1, P2* – partial phases) to design the template library of models synthesized. Second part of IIT is the module to APA – analysis and identification of tested (recognized - identified objects) images. Analysis is provided on base of AP-image description correlative models (secondary semantics derivative on image, AP-data array and MoC levels).

Procedures of APA are related with such foregoing technological procedures of image acquisition as positioning, space scaling, size and energy of image normalization, targeting, tracking, templates matching, etc. The main method realizes hybrid procedures of templates and tested objects matching and classification decision making is the

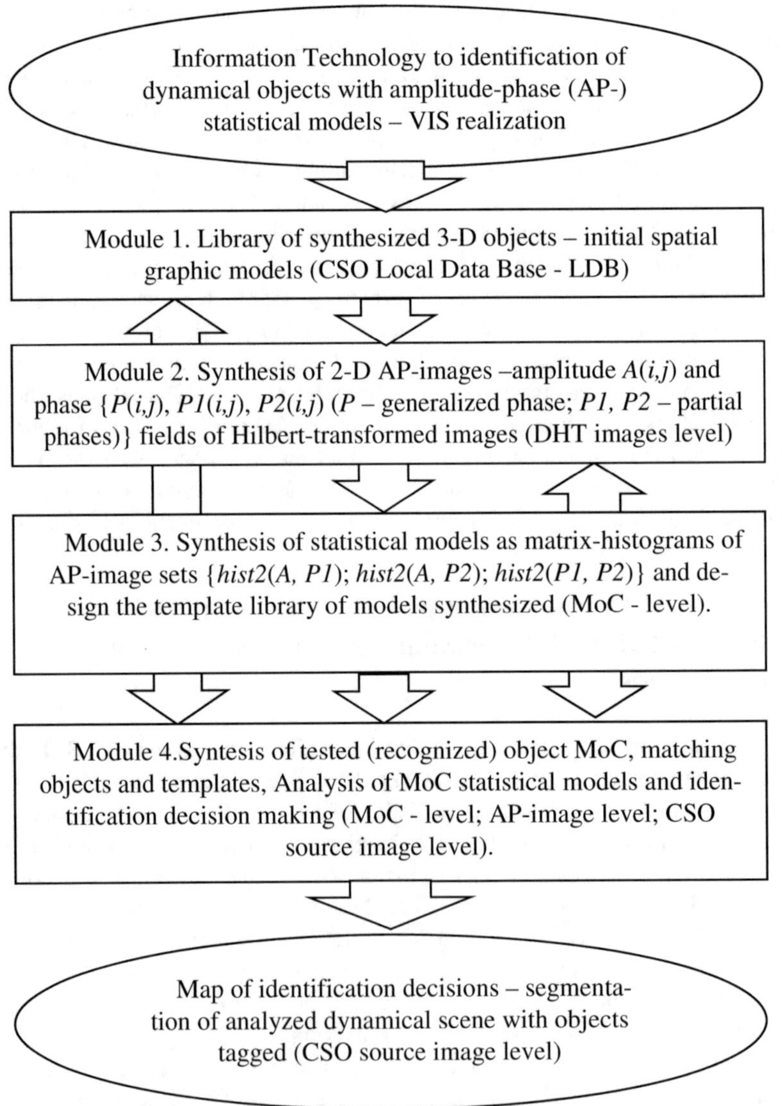

Fig. 1. Conceptual chart of IIT based on dynamical objects AP-images

calculation matrixes of coincidence (MoC) of similar fields amplitude-phase images of objects tested (O1) and templates (O2) ((amplitude_O1 to amplitude_O2, phase_O1 to phase_O2) two objects (O1 and O2). The exact coincidence of fields (this means the full similarity of images morphology, space positioning - localization and scale) is indicated with MoC structure change (from completely filled to quasi-diagonal shape). The block-chart of MoC calculating algorithm is presented on Fig. 2. The functions realized with this algorithm deal with parallel scanning of two AP-images (tested and template), measuring and coding the values of corresponding pixels (with the same indexes (i, j)) of AP-image fields $\{A(i,j), P1(i,j), P2(i,j)\}$ as the indexes of cells $MoC(n, m)$ where the units ("drop-unit") should be added as histogram bins' values (index $n = (0, N_{max})$ is related to the pixel's coded value of the first field (image), index $m = (0, M_{max})$ is related to the pixel's coded value of the second field (image)). After full scanning completing the statistical analysis of MoC content is provided. This content is presented with the statistical moments and coefficient of variance of 2D histograms (in case of different

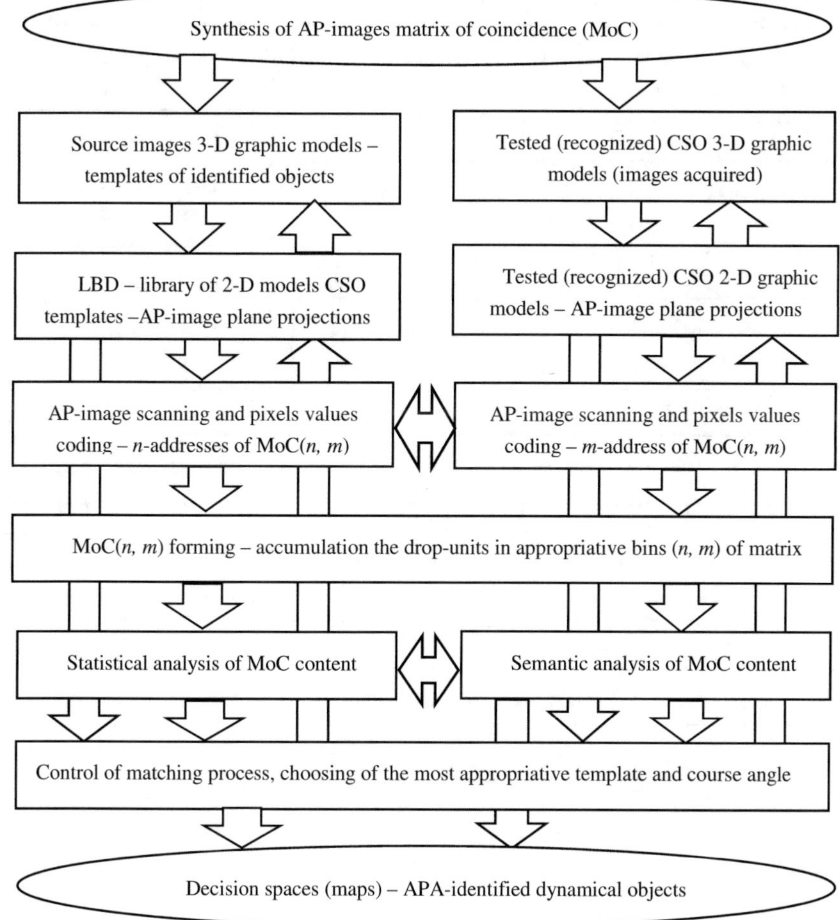

Fig. 2. Conceptual structure of CSO identification algorithm based on MoC synthesis

fields of the AP-image of AP-image analyzed) or matrix of coincidence (in case of the same kind of fields of two compared AP-images analyzed). The full coincidence of images scanning is displayed as "pure" diagonal matrix, whereas the relative changes of objects localization – shifting, translation, rotation, or changes of objects morphology cause the diffusion – spreading of nonzero bins around the matrix main diagonal. As

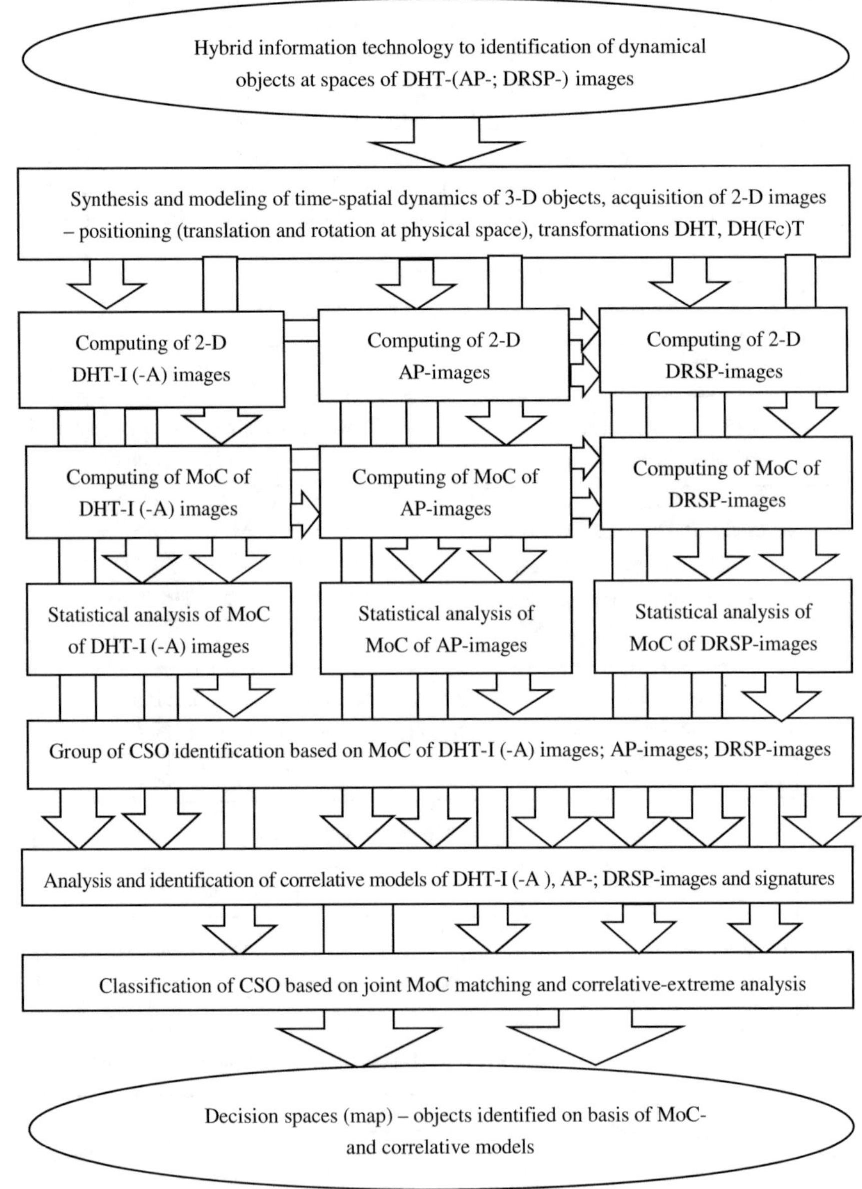

Fig. 3. Conceptual chart of information technology to dynamical objects identification based on methods of AP-analysis, MoC matching and correlative-extreme modeling

investigations show the raising of rotation angle (axis of symmetry) inconsistency or shapes matching decreasing relate to increasing of the norms of difference MoC and decreasing the average of coefficients of variance by rows and columns of MoC. Of Some illustrative examples are presented below (see Sect. 3). This effect is caused by property of DHT anisotropy and shows the high sensitivity of MoC-detectors of OCS localization and shapes changes based on AP-image analysis. The alternative technique could be used for classification decision making is "hybridization" – combining at IIT methods of MoC ("templates – tested objects") synthesis and correlative-extreme analysis in space of AP-images. Structure of algorithm of hybrid technique IIT is presented on Fig. 3. The main functions are calculation of amplitude-phase fields (AP-images) of tested (recognized) CSO, computing the common (in-class and inter-class) 2D histograms, its correlative comparing with semantic models (at level of MoC of templates stored at LDB), detecting the group of most probable ("suspected objects") templates more close fitting to objects tested and on next stages the correlative comparing AP-images with more informative descriptions (on level of AP- and source DHT-images).

3 Examples of Models of Dynamical Objects AP-Images

Modeling of AP-images as illustration of designed IIT functionality has been provided on basis of designed LDB of 3-D airplane models similar to [7] (format "gray" (GR) with specified objects parameters of illumination, ranges of angles rotation φ, ψ, θ and steps of relative angle shift at DRPS-images $\Delta\varphi$, $\Delta\psi$, $\Delta\theta$). Figure 4a and b presents the examples of initial flat (1-D) rotated airplane models with classification tags.

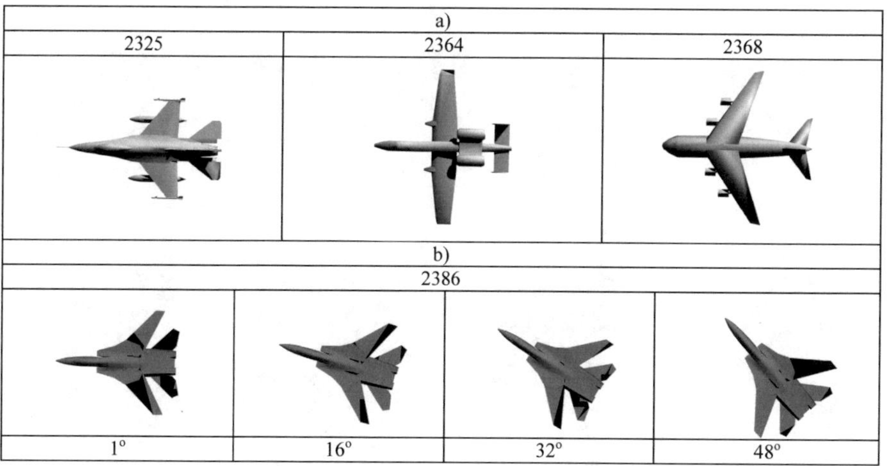

Fig. 4. Examples of tested objects images: initial no rotated (φ, ψ, $\theta = 0°$) – a), 1D rotated – $\varphi = (0\ldots48°)$, ψ, $\theta = 0°$ – b)

Figure 5 presents MoC of AP-images of objects, Fig. 6 – MoC of residual phase DRPS-images corresponding to relative shifts of AP-images $\Delta\varphi = 5^{o}$.

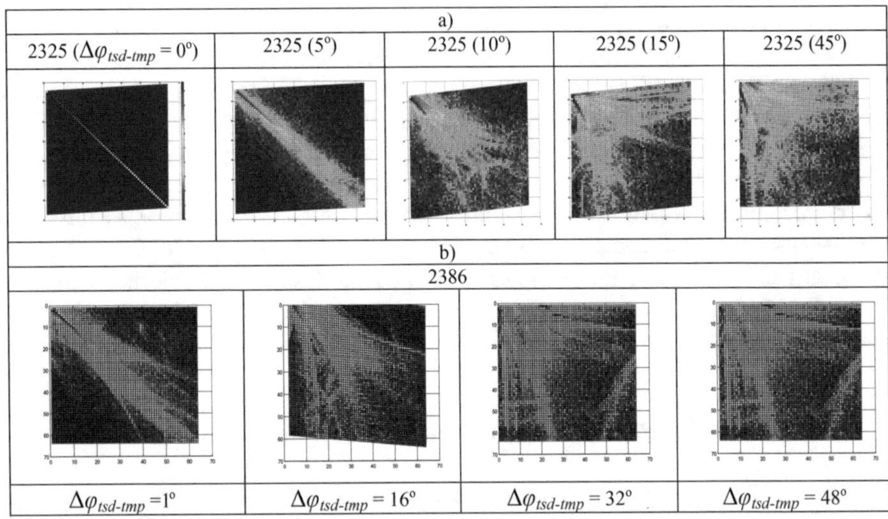

Fig. 5. Examples of MoC corresponding to test (*tsd*) and template (*tmp*) objects images: 2325 - ($\Delta\varphi$, $\Delta\psi$, $\Delta\theta = 0^{o}$; $\Delta\varphi = (5^{o}; \ldots, 45^{o})$ – a), 1D rotated – $\Delta\varphi = (0\ldots48^{o})$, ψ, $\theta = 0^{o}$ – b)

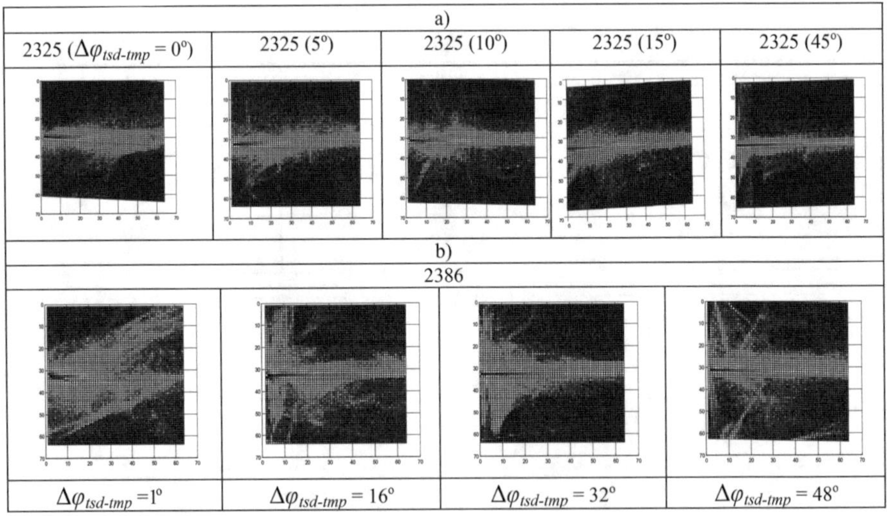

Fig. 6. Examples of MoC of DRPS-images: object 2325 – (a), object 2386 – (b)

Figure 7 presents the coefficients of variance (CoV) of 2D histograms (MoC in-class of phase images (AP-)) vector-columns – a) and the same issue for differential residual phase angle relative shifted images (DRPS-) – b). In case of relative course angle shift (rcas_) is equal 0 the CoV value is equal 8,00, but little changes of rcas_3 (…_16) lead to great CoV decreasing. Figure 7c and d present the coefficients of MoC variance vectors-columns (inter-class 2325-2386) calculated on APIm (c) and DRPS-APIm (d).

Analysis of experimental data shows the advantages of DRPS-APIm MoC-based methods of CSO identification – growth of MoC CoV pick-factor (ratio of CoV full-matched in-class and CoV relatively rotated inter-class images).This methodology could lead to increasing of discrimination abilities and effectiveness of IIT.

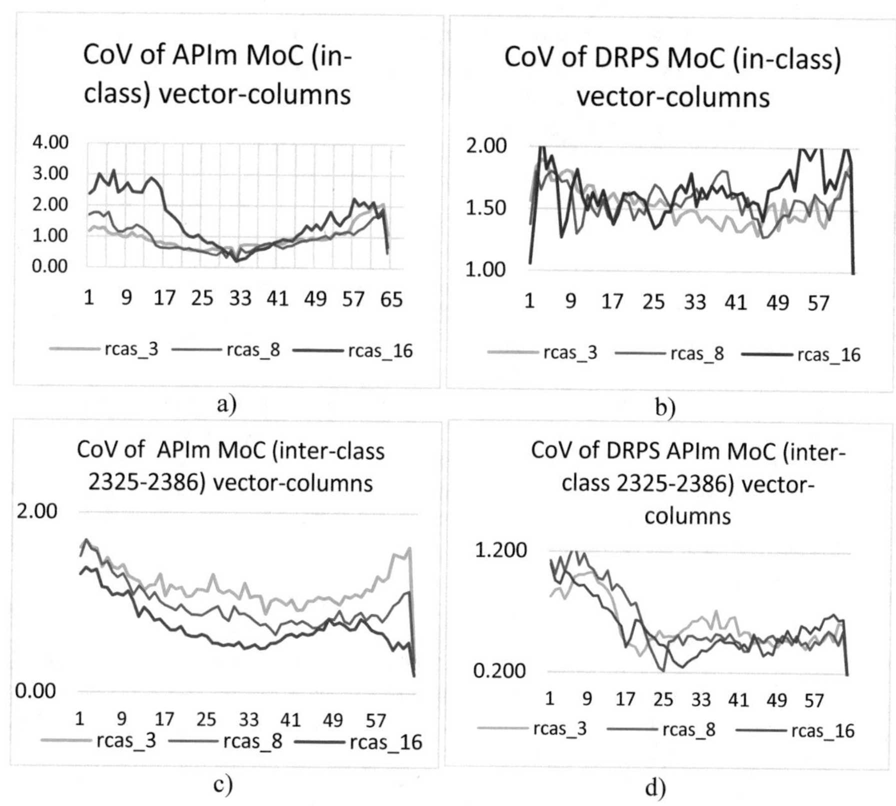

Fig. 7. Analysis of APIm- and DRPS-MoC CoV based methods of CSO matching and identification: APIm of test (identified) object and template images (in class 2325) relatively rotated (rcas = 3; 8; 16°) – a); DRPS APIm MoC (the same conditions) – b); MoC CoV APIm (inter-class 2325-2386) – c): DRPS APIm MoC (inter-class 2325-2386) – d)

4 Summary

Methods of analysis and modeling of IT for complex shape dynamical objects identification based on technics of digital Hilbert optics and used the anisotropic properties of multidimensional generalized Hilbert-Foucault transforms and amplitude-phase images are investigated and verified. Methods are based on calculating of matrixes of coincidence of test (identified) and template amplitude-phase images for matching of images and classification of objects. As measures of convergence and adjacency to improve the discrimination abilities of identification IT its realizing the peak-factors - ratios of coefficients of variance of vector-columns matrix of coincidence of full-matched and arbitrary oriented objects images are proposed. Evaluations of MoC-based models of APIm pointed the significant arising of CoV peak-factor corresponding to class and localization. As alternative the differential residual relative angle shift APIm (DRPS-) method is proposed and elaborated under the same conditions as APIm based method. DRPS-method provides more effective discrimination of localization and shape classification of objects due the measure used is more sensitive to changes of orientation, localization and shape changes.

References

1. Pratt, W.K.: Digital Image Processing: PIKS Inside, 4th edn. Wiley, New York (2010)
2. Duda, R.O., Hart, P.E., Stork, D.G.: Pattern Classification, 2nd edn. Wiley, New York (2001)
3. Hahn, S.L., Snopek, K.M.: Complex and Hypercomplex Analytic Signals Theory and Applications. Artech House, Boston (2017)
4. Hahn, S.L.: Hilbert Transforms in Signal Processing. Artech House, Norwood (1996)
5. Lorenco-Ginori, J.V.: An approach to 2D Hilbert Transform for image processing applications. In: Kamel, M., Campilho, A. (eds.) ICIAR 2007, Montreal, pp. 157–165 (2007)
6. Sudoł, A., Stemplewski, S., Vlasenko, V.: Methods of digital Hilbert optics in modelling of dynamic scene analysis process: amplitude-phase approach to the processing and identification objects' pictures. In: Information Systems Architecture and Technology, pp. 129–138. Politechnika Wrocławska, Wrocław (2014)
7. Vlasenko, V., Stemplewski, S., Koczur, P.: Identification of objects based on generalized amplitude-phase images statistical models. In: Świątek, J., Borzemski, L., Wilimowska, Z. (eds.) Information Systems Architecture and Technology: Proceedings of 38th International Conference on Information Systems Architecture and Technology – ISAT 2017. Advances in Intelligent Systems and Computing, vol. 656, pp. 63–71. Springer, Cham (2018). https://link.springer.com/book/10.1007/978-3-319-67229-8
8. Vlasenko, V., Stemplewski, S., Koczur, P.: Abnormal textures identification based on digital Hilbert optics methods: fundamental transforms and models. In: Świątek, J., Borzemski, L., Wilimowska, Z. (eds.) Information Systems Architecture and Technology: Proceedings of 38th International Conference on Information Systems Architecture and Technology – ISAT 2017. Advances in Intelligent Systems and Computing, vol. 656, pp. 72–79. Springer, Cham (2018). https://link.springer.com/book/10.1007/978-3-319-67229-8

Modeling of Scientific Publications Disciplinary Collocation Based on Optimistic Fuzzy Aggregation Norms

Oleksandr Sokolov[1(✉)], Wiesława Osińska[2], Aleksandra Mreła[3], and Włodzisław Duch[1]

[1] Faculty of Physics, Astronomy and Informatics, Nicolaus Copernicus University in Torun, 5 Grudziadzka, 87-100 Torun, Poland
osokolov@fizyka.umk.pl, wduch@is.umk.pl

[2] Institute of Information Science and Book Studies at Nicolas Copernicus University, Nicolaus Copernicus University in Torun, 1 Wladyslawa Bojarskiego, 87-100 Torun, Poland
wieo@umk.pl

[3] Faculty of Technology, Kujawy and Pomorze Univeristy in Bydgoszcz, 55-57 Torunska, 85-023 Bydgoszcz, Poland
a.mrela@kpsw.edu.pl

https://www.fizyka.umk.pl
https://www.inibi.umk.pl/english
https://www.kpsw.edu.pl

Abstract. Assessment of scientific achievements of scientists is difficult because the science is divided into scientific domains and disciplines. The classification is not a partition, so very often disciplines are related to a few scientific domains. The paper presents the method of calculating scientists' contributions to science, which are based on the number of articles published in journals connected to disciplines which are, in turn, related to scientific domains. The application of fuzzy relations and their composition simplifies the problem of describing these connections. The idea of the scientific contribution unit and the usage of the optimistic fuzzy aggregation norm allows calculating the scientific contribution of each scientist. Since levels of scientific contributions belong to the interval [0,1], there is a possibility to prepare rankings of scientists. The example of the application of this method is supported by the result of the estimation of scientific achievement by the real scientist.

Keywords: Fuzzy logic application · Scientific contribution unit
Parametrization of science · Optimistic fuzzy aggregation norm
Publications collocations

1 Disciplinarity in Science

The problem in science today is too much-fragmented classification system for organizing areas and fields of knowledge, particularly in Poland, where numerous

J. Świątek et al. (Eds.): ISAT 2018, AISC 853, pp. 145–153, 2019.
https://doi.org/10.1007/978-3-319-99996-8_14

disciplines are grouped into scientific fields which form eight academic domains. Such organization obstructs the development of interdisciplinary research as well as the parametrization of scientists and teams specialized in a wide scope of knowledge and elaborating multidisciplinary issues. Disciplinary membership and skills of scientists can be evaluated accordingly to their publishing activity. Selected journals profile match their scientific interests.

Science policy among others consists of journals parametrization, i.e., assigning a score to the defined set of journals and creating the so-called Polish Journal Ranking, which is prepared annually [3]. The ranking is biased in relations to some disciplines [2]. For example, cognitive scientists or psychologists have no many high-scored sources where to publish their works. The inhomogeneous scoring system results from the disproportions of disciplinary measures [5].

2 The Idea of Science Contribution

The idea of science contribution came from the task of classifying researchers and assess their contribution to 8 fields of science according to their articles in magazines.

All journals are related to disciplines, which are assigned in turn to the fields of science. In Poland, there are distinguished 8 scientific domains (Table 1).

Table 1. The scientific domains in Poland [4]

Symbol	Description
SD_1	Social sciences
SD_2	Agricultural, forestry and veterinary sciences
SD_3	Exact sciences
SD_4	Medical, health and sport sciences
SD_5	Humanities
SD_6	Technological sciences
SD_7	Natural sciences
SD_8	The arts

Moreover, 311 disciplines of science are defined. Some disciplines are related to only one scientific domain like, for example, mathematics is related to exact sciences and biocybernetics - to exact sciences and natural sciences. Table 2 presents a part of the relation between scientific domains and disciplines, where 1 indicates that there is a relation and 0 indicates the lack of it.

Nowadays, if the librarians want to estimate the scientific achievement of a scientist in given scientific domain, they sum up the number of all articles, which this scientist has published in journals related to this scientific domain.

Let us consider the exemplary relation between scientists A_1 and A_2 and journals J_1, J_2, J_3 (Table 3). As it can be noticed, it is difficult to estimate these

Table 2. Values of the relation between scientific domains and chosen disciplines

Discipline	SD_1	SD_2	SD_3	SD_4	SD_5	SD_6	SD_7	SD_8
Administration	1	0	0	0	0	0	0	0
Biocybernetics	0	0	1	0	0	0	1	0

Table 3. The exemplary relation between scientists and journals

Scientist	J_1	J_2	J_3
A_1	1	2	1
A_2	2	0	3

scientists' achievements with respect to these domains (Table 1) and prepare the ranking of them (comp. [6]).

The problem of estimating scientists' scientific achievements and possible prepare the ranking of them is more complicated because disciplines are related to one or more scientific domains, and journals are also related to one or more disciplines.

Each published brings new elements to science, and it will be called the scientific contribution. Of course, the more articles in one scientific domain or discipline, the bigger contribution of the scientist in this domain or discipline.

Let A be a unit of a contribution of scientists into scientific domains, which is equal to the value that one scientist adds to science after publishing one paper. Assume that A is a small number, for example, $A = 0.01$. More research is needed to establish value A in such a way to fulfill all bibliometric requirements.

3 Fuzzy Logic

Zadeh [10] proposed the definition of fuzzy sets, namely a fuzzy set $A \subset X$ is a set of pairs $(x, \mu_A(x))$, where $x \in X$ and $\mu_A : X \to [0, 1]$ is a membership function which describes the level of membership of element x to set A.

Let X and Y be two spaces. Then $R \subset X \times Y$ is a fuzzy relation between X and Y if R is a fuzzy set.

Assume that there are two fuzzy relations $R_1 \subset X \times Y$ and $R_2 \subset Y \times Z$. Let T denotes a T-norm and S - S-conorm. Then $R_3 \subset X \times Z$ is a $S-T$ composition (comp. [1]) with the membership function defined in the following way:

$$\mu_3(x, z) = S_{y \in Y}[\mu_1(x, y) T \mu_2(y, z)] \text{ for } x \in X, z \in Z.$$

4 Optimistic Fuzzy Aggregation Norms

The most important feature of aggregation norms is that when the scientist publishes a new article, the contribution of this scientist is always higher.

Definition 1. *Let $I = [0,1]$. Then $S : I \times I \to I$ is called an optimistic fuzzy aggregation norm if it fulfills the following conditions:*

$$S(0,0) = 0 \tag{1}$$

$$S(x,y) = S(y,x) \tag{2}$$

$$S(x,y) \geq \max\{x,y\} \tag{3}$$

Notice that from (3) we can easily deduce

$$S(x,0) \geq x \tag{4}$$

The most important features of optimistic aggregation norm are (1) and (3). Condition (1) shows that if a researcher has not published any paper yet in the given scientific domain, their contribution to this domain is 0. Condition (3) shows that if the level of a contribution to the given domain is positive and if a researcher publishes the new paper related to this domain, the level of contribution will be at least on the same level or higher.

Moreover, condition (4) indicates that if the level of a contribution to this domain is positive and if the researcher has not published any paper, then the level of a contribution is not reduced.

Let S be a well-know S-norm (comp. [9])

$$S(x,y) = x + y - xy \quad \text{for } x, y \in [0,1]. \tag{5}$$

We will show that S is an example of the optimistic fuzzy aggregation norm.

Theorem 1. *Let $S(x,y) = x + y - xy$. Then S is an optimistic fuzzy aggregation norm.*

Proof. Of course, the range of S is $[0,1]$. Now we prove that S fulfills all properties stated in Definition 1.

(1) It is obvious that $S(0,0) = 0$.
(2) This conditions is fulfilled because S is an S-norm.
(3) Let $x, y \in [0,1]$, then $S(x,y) = x + y - xy = y(1-x) + x \geq x$. Similarly, $S(x,y) \geq y$. Hence, $S(x,y) \geq \max\{x,y\}$. □

Figure 1 presents the graph of the optimistic fuzzy aggregation norm.

There are also more examples of optimistic fuzzy aggregation norms:

$$S(x,y) = \ln\left(\frac{(e-1)\sqrt{x^2+y^2}}{\sqrt{1+\left(\frac{\min\{x,y\}}{\max\{x,y\}}\right)^2}} + 1\right) = \ln\left((e-1)\cdot\max\{x,y\} + 1\right)$$

Definition 2. *Let $x_1, x_2, \ldots, x_{N+1} \in [0,1]$ and N is a natural number. Then, the iterations of optimistic fuzzy aggregation norm S are defined as follows:*

$$\begin{aligned} S^1(x_1, x_2) &= S(x_1, x_2), \\ S^2(x_1, x_2, x_3) &= S(S(x_1, x_2), x_3), \\ S^N(x_1, x_2, \ldots x_{N+1}) &= S(S^{N-1}(x_1, x_2, \ldots, x_N), x_{N+1}). \end{aligned}$$

Thus, we can use S to develop the fuzzy model of publications collocations.

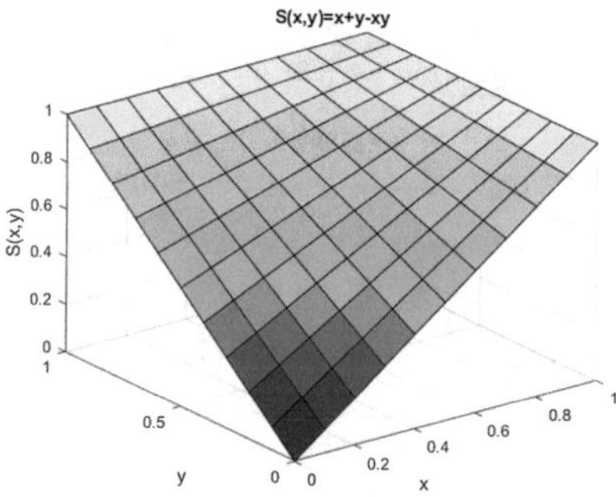

Fig. 1. The graph of optimistic fuzzy relation aggregation norm S

5 Model of Publications Collocations

Let X be a set of all 311 scientific disciplines and Y be a set of all scientific domains. Then, the membership of fuzzy relation R_1 presents the level of relation between disciplines and scientific domains.

Example 1. Let $X = \{Administration, Biocybernetics\}$ and Y is a set of all scientific domains (Table 1), then fuzzy relation $R_1 \subset X \times Y$ can be defined in the following way (Table 4).

Table 4. The exemplary relation between scientists and journals

Discipline	SD_1	SD_2	SD_3	SD_4	SD_5	SD_6	SD_7	SD_8
Administration	1	0	0	0	0	0	0	0
Biocybernetics	0	0	0.8	0	0	0	0.5	0

Indeed, discipline *Administration* is related to only one scientific domain, so

$$R_1(Administration, SD_1) = 1 \;\; R_1(Administration, SD_i) = 0, \;\; i = 2, 3, \ldots, 8.$$

In the case of $Biocybernetics(B)$, it is related to two scientific domains, and exact sciences (SD_3) is first choice and natural sciences (SD_7) is the second one, so $R_1(B, SD_3) = a$, $R_1(B, SD_7) = b$ and $R_1(B, SD_i) = 0$, $i = 1, 2, 4, 5, 6, 8$, where $a, b \in (0, 1)$, for example $a = 0.8$ (discipline of first choice) and $b = 0.5$ (discipline of second choice). □

Let Z be a set of all journals, which number is equal to 5156. The librarians prepared relation R_2 between disciplines and journals in such a way that its membership function shows the level of connection between given scientific journal and discipline. Using S-T composition, fuzzy relation R_3 between scientific domains and journals is built.

Let exemplary fuzzy relation R_1 between journals and disciplines be presented in Table 5 (I), exemplary relation R_2 between disciplines and scientific domains - in Table 5 (II). Moreover, exemplary relation R_3, which values are calculated using the S-T composition, between scientific domains is presented in Table 5 (III).

Table 5. The exemplary relations R_1 (**I**), R_2 - (**II**) and R_3 - (**III**)

I	Discipline			**II**	Scientific domain			**III**	Scientific domain		
Journal	D_1	D_2	D_3	Discipline	SD_1	SD_2	SD_3	Journal	SD_1	SD_2	SD_3
J_1	1	0.8	0	D_1	1	0.4	0	J_1	1	0.8	0
J_2	0	0.5	1	D_2	0	1	1	J_2	0	0.5	1
J_3	0	0	1	D_3	0	0	1	J_3	0	0	1

For each examined scientist, the librarians prepare relation R_4 between scientists and journals, which was presented in Table 3. To proceed, the tables where each line presents the scientific contribution of one journal by each scientist are presented in Table 6 (**I**, **II**). Thus, using the S-T composition, we can get the contribution of each scientist to each scientific domain and present it in Table 6 (**III**).

Now, we are going to calculate the scientific contribution to each scientific domain using the optimistic fuzzy aggregation norm.

Table 6. The exemplary relation between scientists and journals: (**I**) - indicating the article; (**II**) - indicating the scientific contribution and (**III**) - the relation between scientists and scientific domains

I	Journal			**II**	Journal			**III**	Scientific domain		
Scient.	J_1	J_2	J_3	Scient.	J_1	J_2	J_3	Scient.	SD_1	SD_2	SD_3
A_1	1	0	0	A_1	A	0	0	A_1	A	0.8A	0
A_1	0	1	0	A_1	0	A	0	A_1	0	0.5A	0
A_1	0	1	0	A_1	0	A	0	A_1	0	0.5A	0
A_1	0	0	1	A_1	0	0	A	A_1	0	0	A
A_2	1	0	0	A_2	A	0	0	A_2	A	0.8A	0
A_2	1	0	0	A_2	A	0	0	A_2	A	0.8A	0
A_2	0	0	1	A_2	0	0	A	A_2	0	0	A
A_2	0	0	1	A_2	0	0	A	A_2	0	0	A
A_2	0	0	1	A_2	0	0	A	A_2	0	0	A

Indeed, we calculate one of the values of fuzzy relation R_5 between scientists and scientific domains, which are presented in Table 7. We assume that all values of the relation R_5 were zeros before the estimations of scientific contribution.

Table 7. The result relation between scientists and scientific domains, where **I** - formulas with scientific contribution unit, **II** - $A = 0.01$

I	Scientific domain			II	Scientific domain		
Scient.	SD_1	SD_2	SD_3	Scient.	SD_1	SD_2	SD_3
A_1	A	$1.8A - 1.05A^2 + 0.2A^3$	A	A_1	0.01	0.018	0.01
A_2	$2A - A^2$	$1.6A - 0.6A^2$	$3A - 3A^2 + A^3$	A_2	0.02	0.016	0.03

Notice that for this optimistic fuzzy aggregation norm, we have $S(x, 0) = x$ for $x \in [0, 1]$. Hence

$$\begin{aligned} R_5(A_1, SD_2) &= S(S(S(S(0, 0.8A), 0.5A), 0.5A), 0) \\ &= S(S(S(0.8A, 0.5A), 0.5A), 0) \\ &= S(S(0.8A + 0.5A - 0.4A^2), 0.5A), 0) \\ &= S(1.3A - 0.4A^2 + 0.5A - (1.3A - 0.4A^2) * 0.5A), 0) \\ &= 1.8A - 1.05A^2 + 0.2A^3 \end{aligned}$$

6 The Application of the Method

In reality, there are 311 disciplines, 8 scientific domains and 5156 journals.

Figure 2 presents the graph of scientific achievement of a chemist.

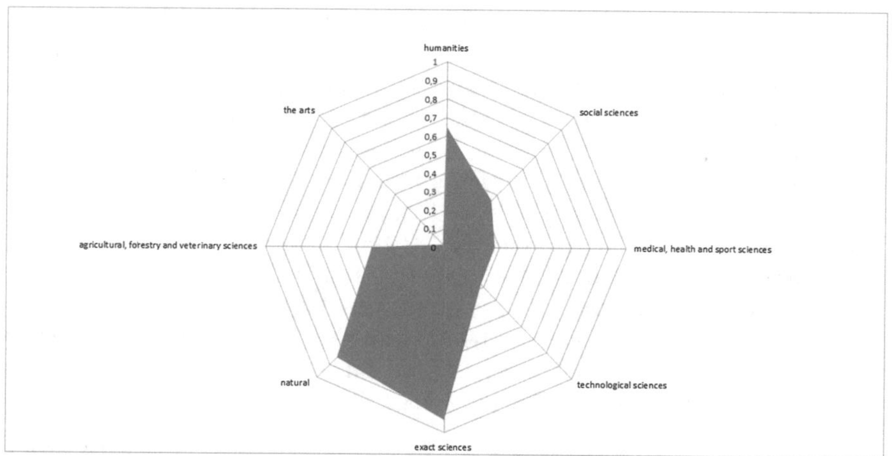

Fig. 2. The diagram of the exemplary scientist's achievements - chemist

Advantages of the application of optimistic fuzzy aggregation norm S and scientific contribution unit A are presented below.

(1) Instead of summing up the quantities of articles to calculate the scientist's scientific contribution, which can cause that the scientific contribution can increase to infinity, we normalize the scientific contribution to the maximum value 1.
(2) The value of scientific contribution unit A might be selected in such a way that the average number of N articles (national, university, units, etc.), would cause that $S^N(A, ...) = 0.5$.
(3) Considering the inclusion of classifications for several different objects (journals, disciplines, and areas), we must create fuzzy relations, so compositions of these relations with the relation between the scientists and articles must also be fuzzy. If, for example, $R_1(journal J_1, discipline D_1) = 0.8$, and then the number of articles of one scientist in this journal is, for example, 23, then using the operation of multiplication, we get 23 articles $\times$ $0.8 = 18.4$. What does the number 18.4 mean? Thus, we cannot use the operation of multiplication in this situation. But using fuzzy relations and optimistic fuzzy aggregation norms let us calculate levels of scientific contributions and compare them.
(4) Using fuzzy relations, we know that scientific contribution always belongs to interval $[0, 1]$. Let us consider the situation when the scientist has achieved the scientific contribution 1, and he publishes one more article? Nowadays, the librarians who estimate the scientific achievements define the threshold - the number of articles above which the scientist becomes recognized in the given discipline, that means they influence in some sense the creation of new paradigms. Above this threshold, it moves to another category - "influential" and counts restart. In our method, these thresholds can also be defined. It can be noticed, that in our method, because all scientific achievements in given scientific domains are numbers belonging to the interval $[0, 1]$, all thresholds may be equal, for example, 0.8, or different for each scientific domain.
(5) For the fuzzy reference value of the author to the scientific domain, it is easy to define the membership function, for example:
   - modest contribution: $\mu(x) = \exp(-((x - 0.2)/\sigma)^2)$,
   - average contribution: $\mu(x) = \exp(-((x - 0.5)/\sigma)^2)$,
   - enormous contribution: $\mu(x) = \exp(-((x - 0.8)/\sigma)^2)$.

   In the same way, the values of relationships between all scientific domains and journals can be evaluated, and we can estimate the contribution of scientists in all these scientific domains.

Notes and Comments. Characterizing scientific journal profile by several disciplines causes the difficulties for automatic classification because there is still no scientometric rules/premises what weights should be given to particular components. Usually, the weights are selected empirically by normalizing the sum to 1 [7]. Fuzzy logic allows calculating disciplinary factors of journals from one side and values of the membership function of researchers to disciplines from the other. Double relations: journals - disciplines, and scientists - disciplines, cause that distribution of disciplines become more continuous. This approach can be useful in describing the ratio of scientists interests, their multidisciplinary input to particular research field and thus the top-down parametrization of individuals or scientific groups.

References

1. Kacprzyk, J.: Wieloetapowe sterowanie rozmyte. Wydawnictwa Naukowo-Techniczne, Warsaw (2001)
2. Kokowski, M.: Jakiej Naukometrii i bibliometrii potrzebujemy w Polsce? Prace Komisji Historii Nauki PAU, no. 14, pp. 135–144 (2015)
3. Kulczycki, E., Rozkosz, A.E.: Does an expert-based evaluation allow us to go beyond the impact factor? experiences from building a ranking of national journals in Poland. Scientometrics **111**(1), 417–442 (2017)
4. List of areas of academic study, academic disciplines and fields of study in the arts and sciences. https://pl.wikipedia.org/wiki/Klasyfikacja_dziedzin_i_dyscyplin_naukowych_w_Polsce
5. Mongeon, P., Paul-Hus, A.: The journal coverage of Web of Science and Scopus: a comparative analysis. Scientometrics **106**(1), 213–228 (2016)
6. Mreła, A., Sokolov, O.: Rankings of students based on experts' assessment and levels of the likelihood of learning outcome acquirement. In: Information and Communication Technologies in Education, Research, and Industrial Applications, pp. 67-88 (2018). https://doi.org/10.1007/978-3-319-76168-8
7. Osińska, V., Bala, P.: New methods for visualization and improvement of classification schemes: the case of computer science. Knowl. Organ. **37**(3), 157–172 (2010)
8. National Center for Biotechnology Information: http://www.ncbi.nlm.nih.gov
9. Wang, X., Ruan, D., Kerre, E.E.: Mathematics of Fuzziness - Basic Issues. Springer, Heidelberg (2009)
10. Zadeh, L.A.: Fuzzy sets. Inf. Control **8**, 338–353 (1965)

Production Planning and Management System

Declarative Modeling of a Milk-Run Vehicle Routing Problem for Split and Merge Supply Streams Scheduling

G. Bocewicz[1(✉)], P. Nielsen[2], and Z. Banaszak[1]

[1] Faculty of Electronics and Computer Science, Koszalin University of Technology, Koszalin, Poland
bocewicz@ie.tu.koszalin.pl

[2] Department of Materials and Production, Aalborg University, Aalborg, Denmark
peter@mp.aau.dk

Abstract. A flow production system with concurrently executed supply chains providing material handling/transportation services to a given set of workstations is analyzed. The considered streams of split and merge supply chains representing all the stages at which value is added to a manufacturing product (including the delivery of raw materials and intermediate components are scheduled under constraints imposed by the solution to an associated milk-run vehicle routing problem. A declarative model of the investigated milk-run delivery principle makes it possible to formulate a vehicle routing and scheduling problem, the solution to which determines the route, the time schedule, and the type and number of parts that different trucks must carry to fulfill orders from various customers/recipients. The goal is to find solutions that minimize both vehicle downtime and the takt time of the production flow. The approach proposed allows to view the above trade-off-like problem as a constraint satisfaction problem and to solve it in the Oz Mozart constraint programming environment.

Keywords: Constraint logic programming · Milk-run · Vehicle routing
Pickup and delivery problem

1 Introduction

Tugger trains [6, 7, 11] have become a popular means of supply in material handling intensive production systems. They can be used in supermarkets to interlink multiple delivery locations along a transport route in a milk run, leading to an efficiency gain (higher transport capacity, reduced labor costs), however, at a cost of more complicated planning and dimensioning, compared to conventionally employed means of transport, such as forklift trucks [1, 9]. Given the advantages following from the milk run schema, we analyzed a flow production system with concurrently executed supply chains [1, 14] providing material handling/transportation services to a given set of workstations.

The streams of split and merge supply chains considered in the study, representing all the stages at which value is added to a manufacturing product, are scheduled under

J. Świątek et al. (Eds.): ISAT 2018, AISC 853, pp. 157–172, 2019.
https://doi.org/10.1007/978-3-319-99996-8_15

constraints imposed by the solution to the associated milk-run vehicle routing problem [2, 10, 13]. In other words, a variety of scheduling, batching, and delivery problems that arise in the assumed set of supply chains, where suppliers make deliveries to several customers, who also make deliveries to succeeding providers/receivers and so on, are being solved with the help of the milk run schema. The goal is to minimize the overall scheduling and delivery cost, which can be achieved by scheduling the jobs and organizing them into batches, each of which is delivered to the next downstream stage as a single shipment. Mathematical models of vehicle routing typically fall into one of the two categories: vehicle flow or set partitioning [2, 12]. Our approach can be described as belonging to the class of set partitioning models, where vehicle routes are defined on the graph of hops, rather than on the graph of customer orders. This postulate is particularly useful when a large number of orders sharing a significantly lower number of pickup and delivery points must be scheduled.

In other words, the model of workstation-to-workstation transport adopted in this study assumes that transport vehicles travel cyclically along routes, servicing workstations; the set of routes, guarantees that all system workstations are serviced, thus ensuring flow of production along established production routes. Understood in this way, the model follows the organization of the milk-run schema, allowing to search for local tugger train routes that minimize the costs of servicing the supply chains. In this context, the present work is a continuation of our previous research related to the design and evaluation of the effectiveness of systems of multimodal processes [3, 4]. By analogy to the milk-run schema, a multimodal process is understood here as a workstation-to-workstation production flow process, whose sections are local, cyclically repeated, milk-run tours. Both problems, i.e. the problem in which cyclic transport routes between workstations are sought for given production routes, as well as the reverse problem in which production routes are sought and transport routes between workstations are given, are combinatorial NP-complete problems [2, 3, 15].

The main achievement of the present study was the formulation of the declarative model of the problem considered which allowed us to view it as a constraint satisfaction problem and to solve it in the Oz Mozart constraint programming environment. The method uses constraint satisfaction to search for feasible solutions, and greedy algorithms to explore it for suboptimal solutions.

The remainder of this paper is organized as follows: Sect. 2 provides a brief overview of related research. Section 3 presents a motivating example introducing the methodology applied and proposes a formulation of a milk-run routing and scheduling problem in the context of constraints imposed by the given supply chain. Section 4 provides results of computational experiments illustrating the proposed approach to milk-run system routing and scheduling. Finally, Sect. 5 offers concluding remarks.

2 Related Work

It can be shown that the total cost spent in a milk-run delivery process is lower than that incurred by applying the direct shipment method [13]. This means that regular shipment/delivery of workpieces by the milk run method is more effective than the use of the direct or the collaborative transportation methods.

Typically, milk-run "trains" consisting of a tugger and three to five trailers use fixed routes. Trains may be shared by multiple suppliers and customers, which means that they collect products at one or more source points and deliver them to the destination points on their way. Of course, the trains need to visit the source points before they visit the destination points [10]. In some cases, they operate on a fixed schedule. The system, therefore, is comparable to a bus system in public transportation [10]. The problems related to the organization and management of milk-run systems derive from the classic vehicle routing, scheduling, and dispatching problems [1, 5, 10, 15]. They are solved, taking into account the specificity of in-plant milk-run solutions, using the methods such as operational research [8], computer simulation [16], and declarative and constraints programming [2, 3, 15]. The most commonly formulated routing problems are those aimed at maximizing the utilization of fleet capacity, finding the best routing and determining the number of parts to be collected from each supplier on each trip. Other frequently encountered routing problems address the questions of "How to assign certain sequences of stops to certain routes?" and "How to configure trains?" [14]. In practice, many restrictions on facility layout, e.g. one-ways or the radius of the curves/turns, as well as different types of trailer configurations have to be taken into account. Apart from choosing the routes which determine the time schedule, one also has to choose the type and number of parts that must be transported by the different trains to fulfill the orders from various customers. In other words, the milk-run scheduling boils down to determining in what time windows parts can be collected from suppliers and delivered to customers along the established routes, so that the cost of transport operations and the size of the inventory in the supply chain are kept at the minimum. In the general case, however, the main point is to simultaneously optimize vehicle routes and dispatch frequency in order to minimize transportation and inventory costs. In that context, the milk-run method seems to be well-suited to solving problems of scheduling and dispatching of inventory in warehouses/supermarkets and production facilities with in-plant transport systems. Another issue that must be considered is that the loading stations can be used by more than one train. This may result in dependencies and blockages between individual trains, e.g. caused by overtaking or stopping. This means that the technical and/or functional constraints on the supply chain distribution systems used in practice require introduction of changes in their production flows. This requires that the conventionally considered problems of finding an optimum supplier schedule and/or an optimal manufacturer schedule be considered together. The objective functions of integrated production and supply flows are the minimization of the total interchange cost and the minimization of the total-interchange-plus-buffer-storage cost [1]. This issue, which takes into account the specific character of milk-run systems, is discussed in the present study as continuation of our previous work on the leveling of multi-product batch production flows [4] and declarative modelling framework for routing and scheduling of Unmanned Aerial Vehicles [3].

3 Modelling

3.1 Motivating Example

Consider the shop floor layout shown in Fig. 1. It consists of a warehouse R_w, a supermarket R_s, and five production cells R_1–R_5. The network of transport connections served by two tugger trains TT_1, TT_2 consists of a set of docking stations $M1$–$M4$ for tugger trains which deliver intermediate components from the warehouse to the production cells and a set of docking stations S_1–S_3 for tugger trains which pick up finished goods to supply them to the supermarket.

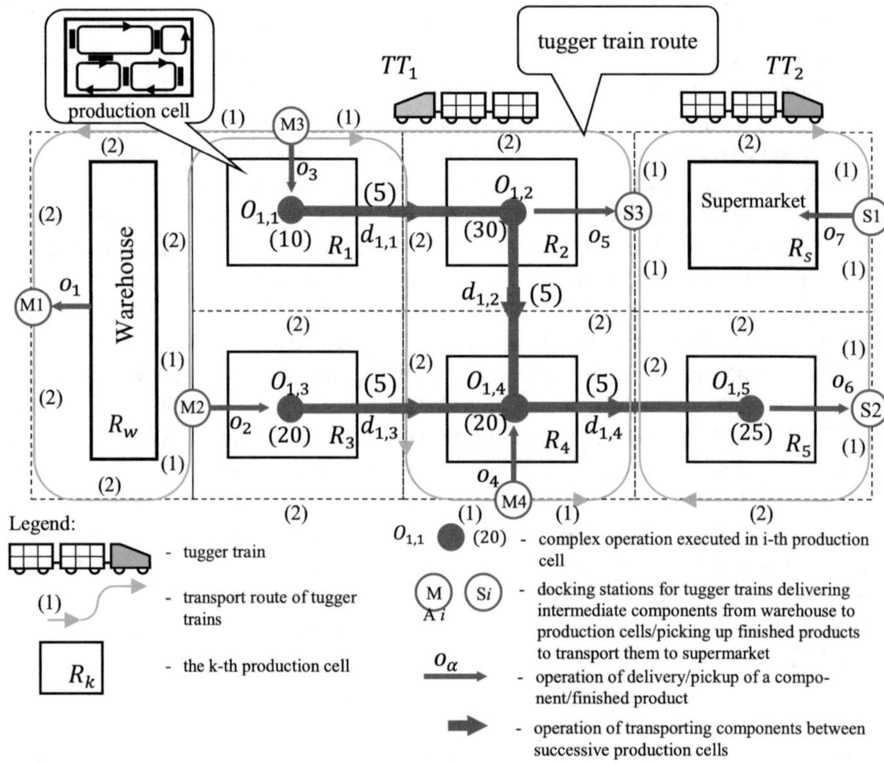

Fig. 1. Layout of the shop floor with marked production flow and milk-run routings.

In the shop floor under consideration, production flow of product J_i (job i) is executed. The technological route for product J_1 is marked in red (see Fig. 1). Assuming that so-called complex operations $O_{i,q}$ (i.e. processes that are made up of elementary operations executed by the individual workstations of a production cell) have the following times: $O_{1,1} = 10$ ut (units of time), $O_{1,2} = 30$ ut, $O_{1,3} = 20$ ut, $O_{1,4} = 20$ ut, and $O_{1,5} = 25$ ut, one can determine the value of production takt time

$TP = 30$ governed by the bottleneck resource R_2. A Gantt diagram illustrating production flow in the investigated system is shown in Fig. 2. As it is easy to notice, whether or not production takt time $TP = 30$ can be achieved is conditioned by timely (just-in-time) delivery/pickup of intermediate components/finished products to/from the given tugger train docking stations. In other words, the production flow schedule shown in Fig. 2 determines the schedule of visits to the individual tugger train docking stations. Assuming that the transport routes established by routing are available and the travel times along their individual sections (as in Fig. 1) are known, the following question can be considered for a fleet of two tugger trains TT_1 and TT_2:

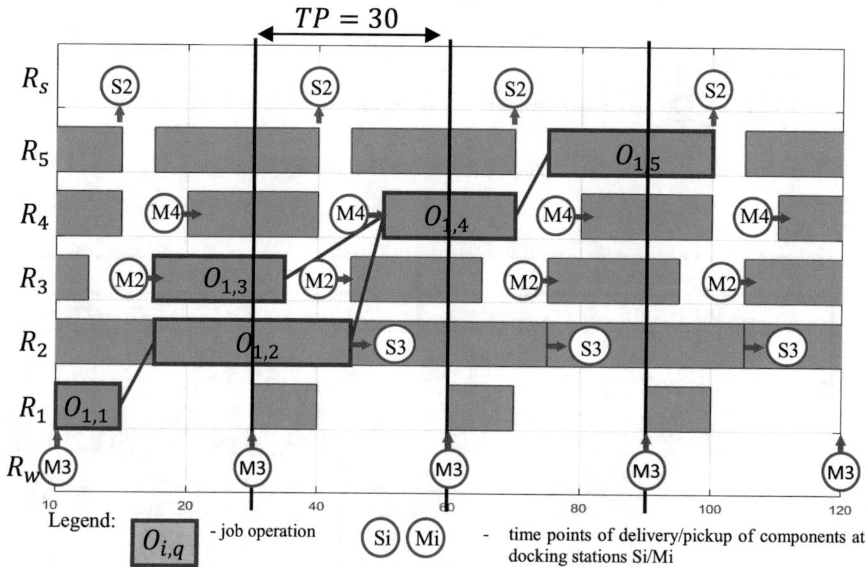

Fig. 2. A Gantt chart of production flow

Do there exist routes for the given tugger train fleet such that items can be moved (delivered/picked up) along them to and from the given docking stations (M)/(S) at time points determined by the production flow schedule from Fig. 2.

Examples of answers to this question are provided by the solutions shown in Figs. 3 and 5. These solutions were obtained assuming that transport between production cells, e.g. R_1 and R_2 or R_2 and R_4 is supported by an overhead transport system. In the first case (Fig. 3), one cyclic transport route to be travelled by two tugger trains was established. A Gantt chart showing how production flows in a system implementing this type of solution is presented in Fig. 4.

In the second case (Fig. 5), two cyclic routes to be travelled by tugger trains were created. A Gantt chart illustrating production flow in a system implementing this type of solution is shown in Fig. 5. In both cases, production takt time increased: In the first case by 6 ut and in the second by 10 ut. It should also be noted that in the first case, the operation cycle of tugger trains spanned two production flow cycles – see Fig. 4.

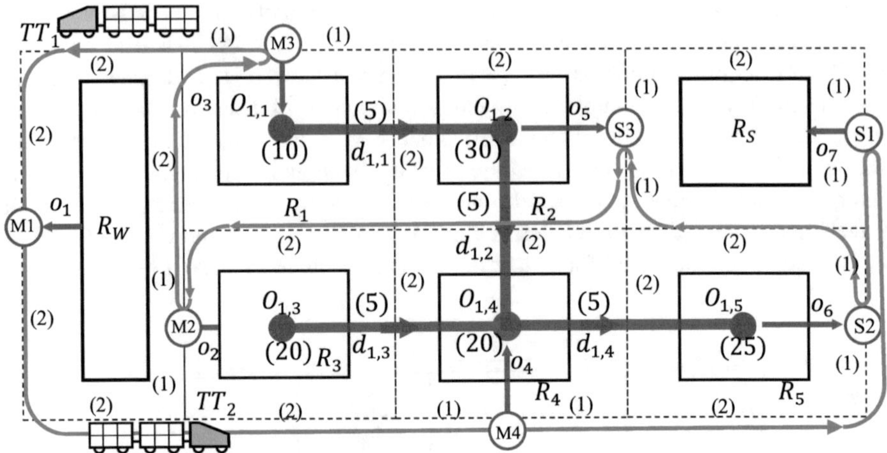

Fig. 3. Transport route for two tugger trains.

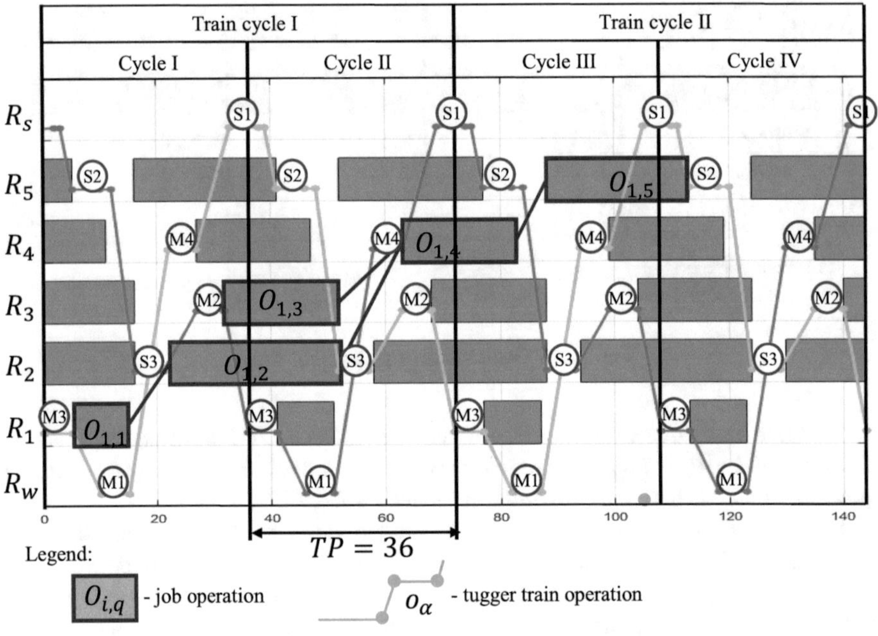

Fig. 4. A Gantt chart of production flow in a system with the milk-run route from Fig. 3

The operation times for tugger trains, which comprised the total times of travel among production cells *TS*, total component delivery/pickup times *TO*, and total train dwelling times *TW*, in one production flow cycle are respectively:

- *TS* = 30 ut; *TO* = 35 ut; *TW* = 7 ut; for the solution from Figs. 3 and 4,
- *TS* = 30 ut; *TO* = 35 ut; *TW* = 15 ut; for the solution from Figs. 5 and 6.

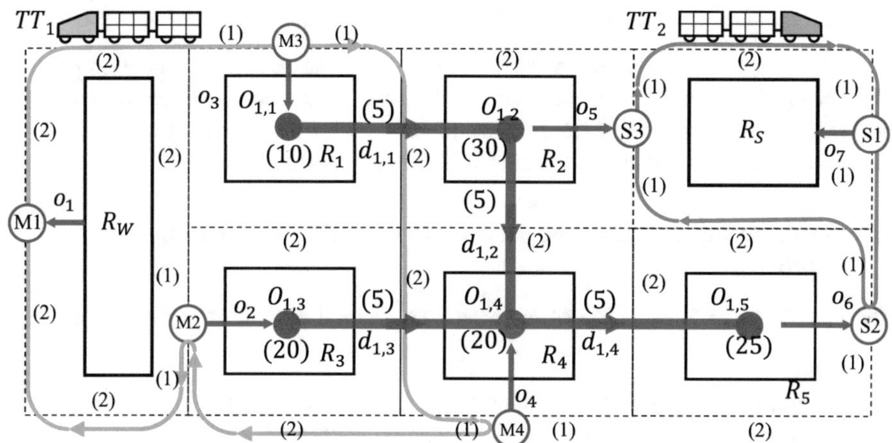

Fig. 5. Transport routes for two tugger trains.

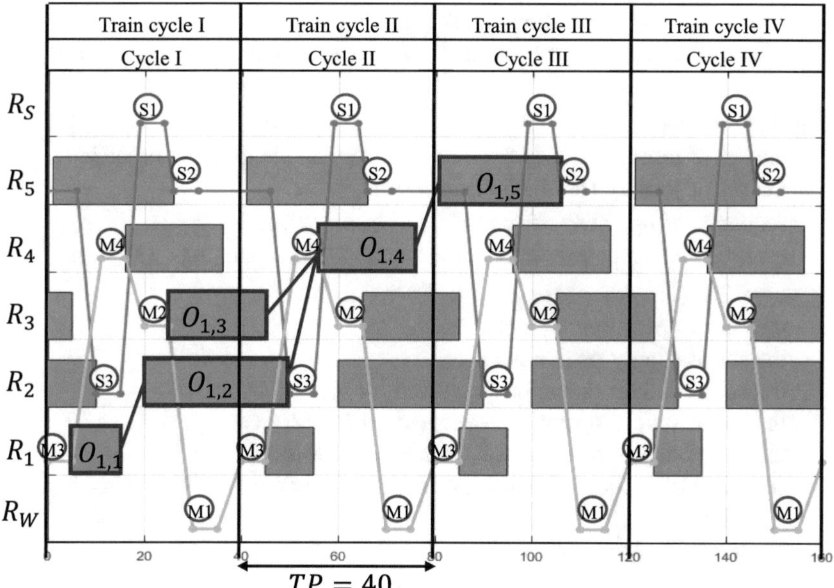

Fig. 6. A Gantt chart of production flow in a system with the milk-run routes from Fig. 5

The solutions obtained result in different sequences of visits to the docking stations. The cyclically repeated sequence has the following form:

- $M1(TT_1) - M2(TT_2) - M3(TT_1) - M4(TT_1) - S1(TT_1) - S2(TT_2) - S3(TT_2) - M1(TT_2) - M2(TT_1) - M3(TT_2) - M4(TT_2) - S1(TT_2) - S2(TT_1) - S3(TT_1) -$ in the first solution – Figs. 3 and 4
- $M1(TT_1) - M2(TT_1) - M3(TT_1) - M4(TT_1) - S1(TT_2) - S2(TT_2) - S3(TT_2) - M1(TT_1) - M2(TT_1) - M3(TT_1) - M4(TT_1) - S1(TT_2) - S2(TT_2) - S3(TT_2) -$ in the second solution – Figs. 5 and 6, $Mj(TT_i)$ – docking station Mj serviced by TT_i.

The tugger train schedules correspond to the steady state of the production process. This means that, generally (in particular in short-run production), the assessment of the degree to which a solution to the organization of a milk-run system allows to utilize the given transport fleet should cover the start-up and shut-down periods.

3.2 Problem Formulation (the Model and the Milk-Run Routing Problem)

The mathematical formulation of the model considered employs the following:

Symbols:

R_k: k-th resource (warehouse, supermarket, production cell);
J_i: job i (production process); $O_{i,q}$: operation q of J_i;
TT_v: transport process v (v-th tuger train);
o_α: α-th supply operation (operation of delivery/pickup of an intermediate component/finished product to/from a production cell);
b_α: index of supply operation which precedes o_α;
f_α: index of supply operation which follows o_α;
ρ: size of production batch (number of jobs executed during one cycle).

Sets and sequences:

R: the set of resources R_k (warehouses, supermarkets, production cells);
J: the set of jobs J_i, (production processes);
O_i: sequence of operations for J_i: $O_i = \left(O_{i,1}, \ldots, O_{i,q}, \ldots, O_{i,lm_i}\right)$;
p_i: route of J_i, sequence of resources on which operations $O_{i,q}$ are executed: $p_i = \left(p_{i,1}, \ldots, p_{i,q}, \ldots, p_{i,lm_i}\right)$, $p_{i,q} \in R$;
Q_k: the set of operations executed on R_k;
$\mathcal{O}$: the set of supply operations o_α;
S_k: the set of pickup operations executed on R_k, $S_k \subseteq \mathcal{O}$;
M_k: the set of supply operations executed on R_k, $M_k \subseteq \mathcal{O}$;
TT: the set of transport means TT_v (transport processes);
B: sequence of predecessor indices of supply operations, $B = (b_1, \ldots, b_\alpha, \ldots, b_\omega)$, $b_\alpha \in \{0, \ldots \omega\}$;
F: sequence of successor indices of supply operations, $F = (f_1, \ldots, f_\alpha, \ldots, f_\omega)$, $f_\alpha \in \{1, \ldots \omega\}$.

Parameters:

m: number of resources;

n: number of jobs, l: number of transport means;

lm_i: number of operations of J_i; ω: number of supply operations,

$t_{i,q}$: operation time of $O_{i,q}$; tr_α: operation time of o_α,

$td_{a,b}$: travel time between the resource of operation $O_{i,a}$ and the resource of operation $O_{i,b}$;

$d_{\alpha,\beta}$: travel time between the resource of operation o_α and the resource of operation o_β;

TP^*: maximum value of production takt time TP.

Variables:

TP: production takt time; $x_{i,q}$: start time of operation $O_{i,q}$,

$y_{i,q}$: end time of operation $O_{i,q}$; xt_α: start time of operation o_α,

yt_α: end time of operation o_α; CU: production utilization rate;

xs_α : the moment the resource occupied by tugger train is released after completion of operation o_α;

b_α: index of the supply operation preceding operation o_α (operations o_{b_α} and o_α are executed by the same tugger train); $b_\alpha = 0$ means that o_α is the first operation of the system cycle;

f_α: index of the supply operation following o_α, (operations o_α and o_{f_α} are executed by the same tugger train).

Constraints:

I. For job operations (production processes):

$$y_{i,q} = x_{i,q} + \rho \times t_{i,q}, q = 1 \ldots lm_i, \forall J_i \in J \quad (1)$$

$$y_{i,a} + td_{a,b} \leq x_{i,b}, \text{when } O_{i,a} \prec O_{i,b}, \forall J_i \in J, \quad (2)$$

$$y_{i,q} \leq x_{i,q} + TP, q = 1 \ldots lm_i, \forall J_i \in J, \quad (3)$$

$$(y_{i,a} \leq x_{j,b}) \vee (y_{j,b} \leq x_{i,a}), \text{when } O_{i,a}, O_{j,b} \in Q_k, i \neq j, \forall R_k \in R, \quad (4)$$

$$CU = \rho / TP. \quad (5)$$

II. For tugger trains (transport process operations):

$$yt_\alpha = xt_\alpha + tr_\alpha, \alpha = 1, 2, \ldots, \omega, \quad (6)$$

$$b_\alpha = 0, \forall \alpha \in BS, BS \subseteq BI = \{1, 2, \ldots, \omega\}, |BS| = l, \quad (7)$$

$$b_\alpha \neq b_\beta \forall \alpha, \beta \in BI \backslash BS, \alpha \neq \beta, \tag{8}$$

$$f_\alpha \neq f_\beta \forall \alpha, \beta \in BI, \alpha \neq \beta, \tag{9}$$

$$(b_\alpha = \beta) \Rightarrow (f_\beta = \alpha), \forall b_\alpha \neq 0, \tag{10}$$

$$[(b_\alpha = \beta) \wedge (b_\beta \neq 0)] \Rightarrow (yt_\beta + d_{\beta,\alpha} \leq xt_\alpha), \alpha, \beta = 1, 2, \ldots, \omega, \tag{11}$$

$$[(f_\alpha = \beta) \wedge (b_\beta = 0)] \Rightarrow (yt_\alpha + d_{\alpha,\beta} \leq xt_\beta + TP), \alpha, \beta = 1, 2, \ldots, \omega, \tag{12}$$

$$xs_\alpha \geq yt_\alpha, \alpha = 1, 2, \ldots, \omega, \tag{13}$$

$$[(f_\alpha = \beta) \wedge (b_\beta \neq 0)] \Rightarrow (xs_\alpha = xt_\beta - d_{\alpha,\beta}), \alpha, \beta = 1, 2, \ldots, \omega, \tag{14}$$

$$[(f_\alpha = \beta) \wedge (b_\beta = 0)] \Rightarrow (xs_\alpha = xt_\beta - d_{\alpha,\beta} + TP), \alpha, \beta = 1, 2, \ldots, \omega, \tag{15}$$

$$\begin{aligned} &[(xs_\alpha < yt_\beta) \wedge (xs_\beta - TP < yt_\alpha)] \vee [(xs_\beta < yt_\alpha) \wedge (xs_\alpha - TP < yt_\beta)], \\ &\forall o_\alpha, o_\beta \in S_k \cup M_k, k = 1, \ldots, m. \end{aligned} \tag{16}$$

III. For transport and production processes (linking tugger trains with jobs)

$$y_{i,q} = xt_\alpha + c \times TP, c \in \mathbb{N}, \forall o_\alpha \in S_k, \forall O_{i,q} \in Q_k, k = 1, \ldots, m. \tag{17}$$

$$x_{i,q} = yt_\alpha + c \times TP, c \in \mathbb{N}, \forall o_\alpha \in M_k, \forall O_{i,q} \in Q_k, k = 1, \ldots, m. \tag{18}$$

Question: Do there exist routes (represented by sequences *B*, *F*) for the given tugger train fleet (set *TT*) and batch size ρ, which ensure the existence of a production schedule $(x_{i,q}, xt_\alpha)$ that allows the achievement of the given capacity utilization rate $CU \geq CU'$?

3.3 Method

Looking for the answer to the question formulated above, we assumed that key importance should be attributed to production efficiency understood as the production capacity utilization rate *CU*. Usually, the production capacity is defined as the number of product items that can be manufactured within a given time. For the purposes of this work, production capacity utilization rate has been defined with reference to production takt time *TP*, as the ratio of batch size ρ to production takt time *TP* in accordance with (5). In practice, technical or technological constraints do not allow bottleneck capacity to be fully exploited. The assumption adopted here entails the use of a milk-run routing and scheduling procedure that comprises the following steps:

1. determine the bottleneck in the production flow (the size of production batch ρ is known) and the associated production takt time *TP*, than calculate the capacity utilization rate CU'
2. determine the time points at which tugger trains will be docked at the given storage/collection stations, such that the established value of *CU* is maintained;
3. determine milk-run routes for the given fleet of tugger trains *TT* – if this set is empty, go to step 4, if not, go to step 5,
4. if this step is repeated for the ϕ-th time, go to step 7, if not, go to step 6,
5. check whether the obtained milk-run schedule for the system $(x_{i,q}, xt_\alpha)$ allows you to maintain the established capacity utilization rate CU' – if so, go to step 7, if not, then go to step 6.
6. increase the size of the production batch ρ and, accordingly, increase the production takt time, and go to step 2,
7. stop – if the stop condition of step 5 has been met, the solution obtained corresponds to the maximum capacity utilization rate CU' – if not, the solution obtained does not guarantee admissible capacity utilization rate $CU > CU'$. If the stop condition of step 4 is met, then there is no admissible solution.

It is easy to see that the iterative procedure described above uses the following sequential scheme: the determination of the bottleneck makes it possible to determine production takt time *TP* (and CU') – designation of *TP* makes it possible to determine the time points of docking tugger trains in the given storage/collection stations – the established docking time points are a condition for finding routes for the tugger train fleet *TT*, ensuring full utilization of its capacity CU'. To describe this methodology in detail, one obviously must take into account the adopted assumptions regarding possible train blockage, permissible buffer capacities which impose limits on production batches, the deployment of storage/collection points, etc. Due to the limited scope of this study, these issues have been omitted.

3.4 Milk-Run Routing and Scheduling Subject to Supply Chain Constraints

Step 3 of the procedure presented in Sect. 3.3. involves the establishment of routes for the given fleet of tugger trains *TT* and a known size of the production batch ρ. To perform step 3, then, one has to answer the following question: *Does there exist a set of routes (represented by sequences B, F) for a given tugger train fleet (set TT) and a given batch size* (ρ)*, which guarantee the existence of a production schedule* ($x_{i,q}, xt_\alpha$) *that allows the achievement of the assumed capacity utilization rate* $CU \geq CU'$*?*

This kind of decidability problem can be viewed as a Constraint Satisfaction Problem:

$$CS = (\mathcal{V}, \mathcal{D}, \mathcal{C}) \tag{19}$$

where: $\mathcal{V} = \{B, F, X, XT\}$ is a set of decision variables, where $X = \{x_{i,q} | i = 1 \ldots n, q = 1 \ldots lm_i\}$, $XT = \{xt_\alpha | \alpha = 1, 2, \ldots, \omega\}$; $\mathcal{D}$ is a discrete finite set of domains of variables $\mathcal{V}$; $\mathcal{C}$ is a set of constraints describing the following relations: the execution order of job operations (1)–(3) and tugger train operations (17); exclusion of job operations (4) and tugger train operations performed on shared resources (16). These constraints ensure cyclic routes (7)–(10), and determine the order of execution of transport operations (6), (11)–(15) and capacity utilization requests $CU \geq CU'$.

To solve the *CS* problem formulated in this way (19), one must establish such values (determined by $\mathcal{D}$) of decision variables *B*, *F* (tugger train routes) and *X*, *XT* (production schedules and supply operation schedules), for which all the constraints $\mathcal{C}$ (including the mutual exclusion constraint, etc.) will be satisfied. In that context the *CS* problem integrates the issues of tugger trains routing (B, F) and scheduling of transport/production operations (X, XT). These problems are typically solved using constraint programming environments, such as Oz Mozart, ILOG, ECL^{i}PS^{E} [3, 4, 15].

4 Computational Experiments

For the system from Fig. 1, in which the production flow (Fig. 2) determines production takt time $TP = 30$ ut and capacity utilization rate $CU = 1/30$, the goal is to find the number of tugger trains (set *TT*) and batch size ρ which make it possible to service delivery/pickup stations so that the given level of production capacity utilization rate $(CU = 1/30)$ is achieved? The answer to this question is sought assuming that:

- times of complex operations $t_{i,q}$ and travel times between workstations $td_{a,b}$, $d_{a,\beta}$ are as given in Fig. 1,
- delivery/pickup times for all tugger trains are the same at $tr_{\alpha} = 5$ ut.

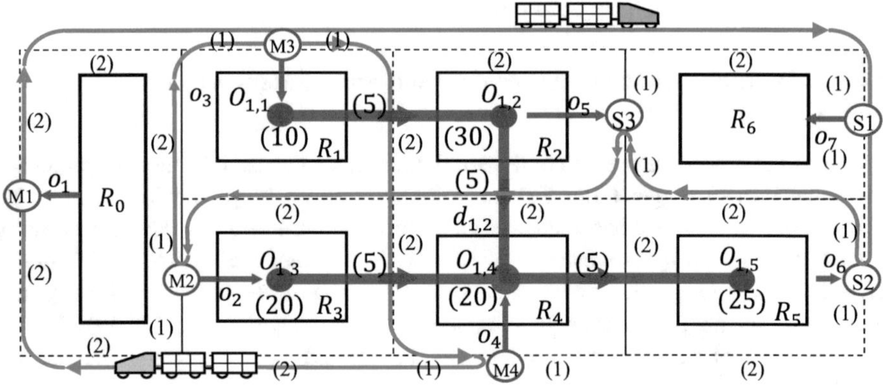

Fig. 7. Cyclic transport routes for two tugger trains.

Two alternative solutions are shown in Figs. 7 and 9. These figures show tugger train routes obtained using the proposed method. Problem *CS* (19) was solved twice (OzMozat system, Intel Core i5-3470 3.2 GHz, 8 GB RAM, calculation time, 2 s).

The first solution, corresponding to a situation in which the available fleet consists of two tugger trains, $TT = (TT_1, TT_2)$ includes one route, shown in Fig. 7. This route results in the schedule illustrated in Fig. 8. It is easy to note that this solution ensures production takt time $TP = 60\text{ut}$ when the size of the production batch is $\rho = 2$. This means that a batch of two product items J_1 is manufactured within one cycle. The capacity utilization rate is the same as in the solution shown in Fig. 2: $CU = 1/30$.

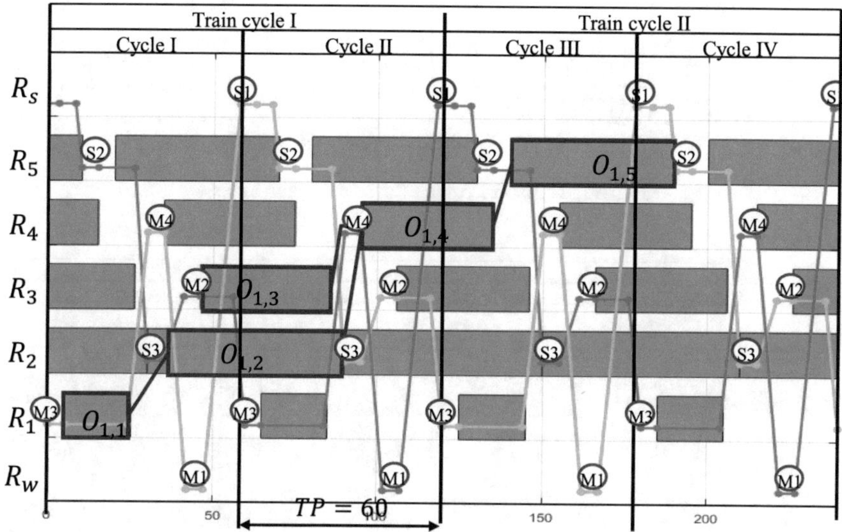

Fig. 8. A Gantt chart of production flow in a system with the milk-run routes from Fig. 7: $\rho = 2$; $TP = 60$; $CU = 1/30$

Other situations can also occur in the system under consideration. For example, the second solution involves the use of a fleet of four tugger trains: $TT = (TT_1, TT_2, TT_3, TT_4)$ when the size of the production batch is $\rho = 1$ – Figs. 8 and 9. As you can see, both solutions yield the same production capacity utilization rate $CU = 1/30$, but they differ in production batch size ρ and the number of trains TT used to ensure uninterrupted production flow.

The different sizes of production batches result in different tugger train utilization rates. Total travel time between production cells TS, total component delivery/pickup times TO, and total train dwelling times TW in one production flow cycle are:

- $TS = 40$ ut; $TO = 35$ ut; and $TW = 45$ ut; for the solution of Figs. 7 and 8,
- $TS = 64$ ut; $TO = 35$ ut; and $TW = 21$ ut; for the solution of Figs. 9 and 10.

By increasing the size of the production batch, one can reduce the number of tugger trains used by a fleet, however, at the cost of reducing their utilization rate: vehicle waiting time TW in the first solution is 24 ut longer than in the second solution.

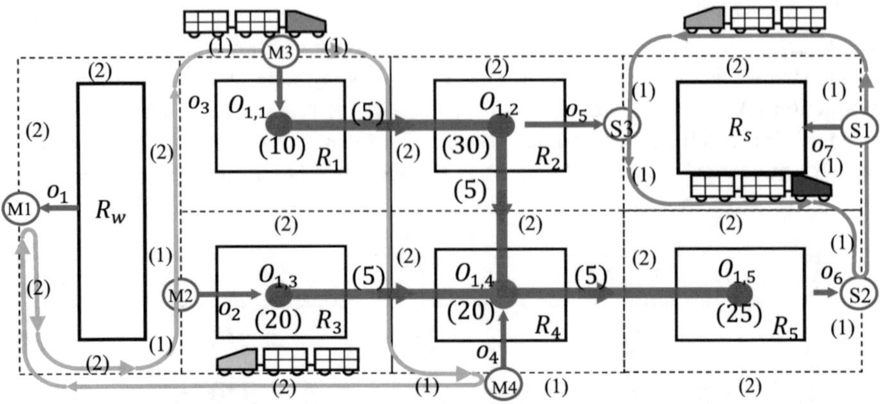

Fig. 9. Transport routes for four tugger trains.

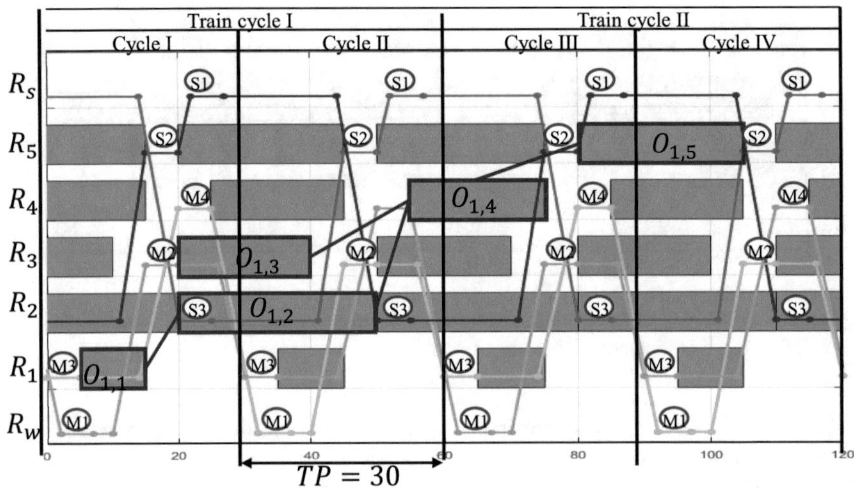

Fig. 10. A Gantt chart of production flow in a system from Fig. 9

Table 1 compares the results of the experiments carried out for $\rho = 1, 2$ and $|TT| = 1$–4, in the system shown in Fig. 1. Production capacity utilization rate $CU = 1/30$ is calculated for the following parameters: ($\rho = 1; |TT| = 4$); ($\rho = 2; |TT| = 2$); ($\rho = 2; |TT| = 3$); ($\rho = 3; |TT| = 4$). It should be noted that the manufacture of larger production batches ρ, on the one hand, means that the preset CU can be maintained with the use of a smaller fleet, but, on the other hand, it leads to reduced utilization of the trains making up the fleet – tugger train waiting times for solutions with $\rho = 2$ (38%–48%) are longer than for solutions with $\rho = 1$ (18%).

Table 1. Experimental results for the system in Fig. 1, where $\rho = 1 - 2$ and $|TT| = 1 - 4$

Production batch size ρ	Number of tugger trains $\vert TT \vert$	Production takt time TP_{min}	Production capacity utilization rate: $CU = \frac{\rho}{TP_{min}}$	% of tugger train utilization $\frac{TS+TO}{\vert TT\vert \times TP_{min}}$	% of tugger train downtime $\frac{TW}{\vert TT\vert \times TP_{min}}$
1	1	64	1/64	98%	2%
	2	36	1/36	90%	10%
	3	31	1/31	87%	13%
	4	30	1/30	82%	18%
2	1	69	1/34,5	94%	6%
	2	60	1/30	62%	38%
	3	60	1/30	54%	46%
	4	60	1/30	52%	48%

5 Conclusions

This paper shows how the milk-run schema can be applied in routing, scheduling and batching, to deal with problems that arise during delivery of products to several production cells, which also make deliveries to other customers involved in an arborescent production flow. The proposed declarative model makes it possible to view the problem under consideration as a constraint satisfaction problem, and solve it with the use of constraint programming platforms. The problem solving procedure implemented in this environment uses the greedy strategy scheme, which makes it possible to analyze practical-scale problems. In our future research, we plan to investigate the conditions imposed on the cyclicity of local processes, the fulfillment of which guarantees the achievement of bottleneck production capacity. Another direction of research we wish to explore is related to the concept of a multimodal production floor model and, in particular, the need to determine the conditions, the fulfillment of which guarantees minimum production start-up and shutdown periods, as well as transition between different steady states.

Acknowledgements. The work was carried out as part of the POIR.01.01.01-00-0485/17 project, “Development of a new type of logistic trolley and methods of collision-free and deadlock-free implementation of intralogistics processes”, financed by NCBiR.

References

1. Agnetisa, A., Hallb, N.G., Pacciarellic, D.: Supply chain scheduling: sequence coordination. Discret. Appl. Math. **154**, 2044–2063 (2006)
2. Badica, A., Badica, C., Leon, F., Luncean, L.: Declarative representation and solution of vehicle routing with pickup and delivery problem. Procedia Comput. Sci. **108C**, 958–967 (2017). International Conference on Computational Science, ICCS 2017

3. Bocewicz, G., Nielsen, P., Banaszak, Z., Thibbotuwawa, A.: Routing and scheduling of unmanned aerial vehicles subject to cyclic production flow constraints. In: Proceedings of 15th International Conference on Distributed Computing and Artificial Intelligence (2018, in print)
4. Bocewicz, G., Nielsen, P., Banaszak, Z., Wojcik, R.: An analytical modeling approach to cyclic scheduling of multiproduct batch production flows subject to demand and capacity constraints. In: Advances in Intelligent Systems and Computing, vol. 656, pp. 277–289 (2017)
5. Bocewicz, G., Muszyński, W., Banaszak, Z.: Models of multimodal networks and transport processes. Bull. Pol. Acad. Sci. Tech. Sci. **63**(3), 635–650 (2015)
6. Droste, M., Deuse, J.: A planning approach for in-plant milk run processes to optimize material provision in assembly systems. In: Proceedings of 4th CIRP CARV 2011, pp. 605–610 (2012)
7. Gyulai, D., Pfeiffer, A., Sobottka, T., Váncza, J.: Milkrun vehicle routing approach for shop-floor logistics. Procedia CIRP **7**, 127–132 (2013)
8. Hall, N.G., Potts, C.N.: Supply chain scheduling: batching and delivery. Oper. Res. **51**(4), 566–584 (2003)
9. Hentschel, M., Lecking, D., Wagner, B.: Deterministic path planning and navigation for an autonomous fork lift truck. IFAC Proc. Vol. **40**(15), 102–107 (2007)
10. Kitamura, T., Okamoto, K.: Automated route planning for milk-run transport logistics with NuSMV model checker. IEICE Trans. Inf. Syst. **E96-D**(12), 2555–2564 (2013)
11. Meyer, A.: Milk Run Design (Definitions, Concepts and Solution Approaches). Dissertation, Karlsruher Institut für Technologie (KIT) Fakultät für Maschinenbau, KIT Scientific Publishing, Karlsruher (2015). https://doi.org/10.5445/ksp/1000057833
12. Parragh, S.N., Doerner, K.F., Hartl, R.F.: A survey on pickup and delivery problems: transportation between pickup and delivery locations. J. Betriebswirtsch. **58**(2), 81–117 (2008)
13. Setiani, P., Fiddieny, H., Setiawan, E.B., Cahyanti, D.E.: Optimizing delivery route by applying milkrun method. In: Conference on Global Research on Sustainable Transport (GROST 2017). Advances in Engineering Research (AER), vol. 147, pp. 748–757 (2017)
14. Schmidt, T., Meinhardt, I., Schulze, F.: New design guidelines for in-plant milk-run systems (2016). http://www.mhi.org
15. Sitek, P., Wikarek, J., A hybrid approach to the optimization of multiechelon systems. Math. Prob. Eng. **2015** (2015). https://doi.org/10.1155/2015/925675
16. Staab, T., Klenk, E., Günthner, W.A.: Simulating dynamic dependencies and blockages in inplant milk-run traffic systems. In: Bye, R.T., Zhang, H. (eds.) Proceedings of the 27th European Conference on Modelling and Simulation (2013)

Energy Consumption in Unmanned Aerial Vehicles: A Review of Energy Consumption Models and Their Relation to the UAV Routing

Amila Thibbotuwawa[1(✉)], Peter Nielsen[1], Banaszak Zbigniew[1], and Grzegorz Bocewicz[2]

[1] Department of Materials and Production, Aalborg University, Aalborg, Denmark
{amila,peter}@mp.aau.dk, Z.Banaszak@wz.pw.edu.pl

[2] Faculty of Electronics and Computer Science, Koszalin University of Technology, Koszalin, Poland
bocewicz@ie.tu.koszalin.pl

Abstract. The topic of unmanned aerial vehicle (UAV) routing is transitioning from an emerging topic to a growing research area with UAVs being used for inspection or even material transport as part of multi-modal networks. The nature of the problem has revealed a need to identify the factors affecting the energy consumption of UAVs during execution of missions and examine the general characteristics of the consumption, as these are critical constraining factors in UAV routing. This paper presents the unique characteristics that influence the energy consumption of UAV routing and the current state of research on the topic. This paper provides the first overview of the current state of and contributions to the area of energy consumption in UAVs followed by a general categorization of the factors affecting energy consumptions of UAVs.

Keywords: Unmanned Aerial Vehicles · UAV routing
Energy consumption of UAVs

1 Introduction

Transportation problems and their associated solution strategies has long been a study of interest for both academia and industry [1–3]. In recent years, unmanned aerial vehicles (UAVs) have become the subject of immense interest and have developed into a mature technology applied in areas such as defense, search and rescue, agriculture, manufacturing, and environmental surveillance [4–9]. Without any required alterations to the existing infrastructure such as deployment station on the wall or guiding lines on the floor, UAVs are capable of covering flexible wider areas in the field than ground-based equipment [10].

However, this advantage comes at a price. To efficiently utilize this flexible resource, it is necessary to establish a coordination and monitoring system for the UAV or fleet of UAVs to determine their outdoor route and schedule in a safe, collision-free, and time-efficient manner, that takes their operating environment into account [7, 11, 12]. Following recent advances in UAV technology, Amazon [13], DHL [14], Federal

J. Świątek et al. (Eds.): ISAT 2018, AISC 853, pp. 173–184, 2019.
https://doi.org/10.1007/978-3-319-99996-8_16

Express [15], and other large companies with an interest in package delivery have begun investigating the viability of incorporating UAV-based delivery into their commercial services. It seems very likely that future multi-modal transportation networks will include UAVs as they are less expensive to maintain than traditional delivery vehicles such as trucks, can lower labor costs by performing tasks autonomously [16, 17], and are fast and able to bypass congested roads. This gives rise to a new problem category: the UAV routing problem (UAVRP). To support varied applications of UAV routing in practice, this paper presents several contributions for energy consumption of UAVs.

1.1 Important Factors to Consider in Deriving Energy Consumption of UAVs

In UAV routing, the majority of studies either assume unlimited fuel capacities [18] or do not consider the fuel in their approach at all. The authors have only been able to identify a few studies which consider fuel constraints in UAV routing [19, 20]. To achieve a realistic and efficient routing, understanding the factors that determine the energy consumption is critical in deriving energy consumption models.

In vessel routing fuel consumption is typically considered to be a function of speed [21] and are heavily non-linear [22]. In the existing research of UAV routing linear approximations for consumption are used [16]. However, we know from industry that this is not reasonable for UAVs, as the weight of the payload in combination with speed and weather conditions are critical.

In the following sections of this contribution, we discuss the main factors as identified in the literature: weather conditions, flying speed, and payload. The aim is to define what UAV routing problems should take into account and how this differs from traditional routing problems.

1.2 Impact of Weather

In outdoor routing for UAVs, one must deal with the stochastics of weather conditions that influence energy consumption of UAVs [23–26]. These elements have some characteristics that potentially can strongly influence the solution strategy for the UAV routing problem. Two main factors for weather's influence on UAV routing are listed below.

a. Wind: The major environmental factor that affects the UAV is wind in the form of wind direction and speed. Wind may benefit the energy consumption or give increased resistance to the movement in other cases [27].
b. Temperature: Temperature conditions can affect the UAV's battery performance as it is linked to battery drain and capacity [16].

Ignoring the impact of weather will not provide more realistic solutions as flying with the wind could reduce energy consumption and cold temperatures may adversely affect battery performance [28]. Most existing research in UAV routing does not consider weather factors and, therefore, ignores the impact of weather on the performance [16, 20, 29, 30]. Furthermore, as weather changes over time in a stochastic manner [31], one must assume that a particular route will have different fuel consumption at different times.

1.3 UAV Flying Speed and Payload

The relative flying speed of the UAV is a critical factor in determining the fuel consumption. Wind speed and direction are linked with the flying speed because, depending on the wind direction, it may affect the flying status of UAV either positively or negatively. The flying status of UAV can be any of the following:

a. hovering,
b. horizontal moving or cruising or level flight, and
c. vertical moving: vertical take-off/landing/altitude change.

Hence, the flying status of the UAV should be considered as well as the flying speed in calculating the energy consumption [27], and relevant models are proposed in Sect. 3 in relation to these flight statuses.

UAVs typically carry some form of payload such as camera equipment or parcels. The impact of different weights of payloads can be significant enough that they should be considered when deriving the energy consumption models [27, 28]. From the airline industry, it is known that fuel/energy consumption depends on certain factors. For example, maximum flight distance or flight time of UAV could be constrained by take-off gross weight, empty weight, thrust to weight ratio [32], fuel weight, and payload [33]. From the airline industry, one can find comparable models for flight such as available fuel models for multirotor helicopters [34] that indicate that linear approximation of the energy consumption is not applicable for large variations of the payload carried [28].

2 Energy Consumption Models for UAVs

Based on the main factors influencing the energy consumption, different energy models can be proposed based on the context of the UAV routing. Theoretical understanding of flight identifies the primary design parameters for achieving the minimum lift for takeoff of a flying object. These include power, weight, width, air density, drag coefficient, and surface area of the flying object. Beyond the primary design parameters, there are numerous other critical secondary design issues regarding, for example, balance, control, and shape that must also be correct to achieve flight [35]. The implication is that each type of UAV in a particular configuration of these parameters has a unique behavior regarding fuel consumption.

These parameters must be considered when calculating a UAV's energy consumption for a particular route under a particular set of circumstances as the flight time of a UAV is defined by these parameters and its energy storage capacity. An energy consumption model helps balance these parameters by providing a function of energy consumed by the UAV. Such models are critical during UAV routing to compare the energy consumed by alternative routes. The aim of this study is not only to provide a global fuel consumption model for UAVs but also to identify the link and influence of the main factors on the consumption.

When the UAV is flying at a constant speed in a horizontal moving state, we have an example of Newton's first law of motion. In this flying state, all the forces cancel each other to produce no net force and so the UAV continues moving in a straight line [36–40]. The upward lift on the UAV equals the downward force of gravity; the forward thrust of the propeller or rotors is matched by the backward drag on the UAV (Fig. 1).

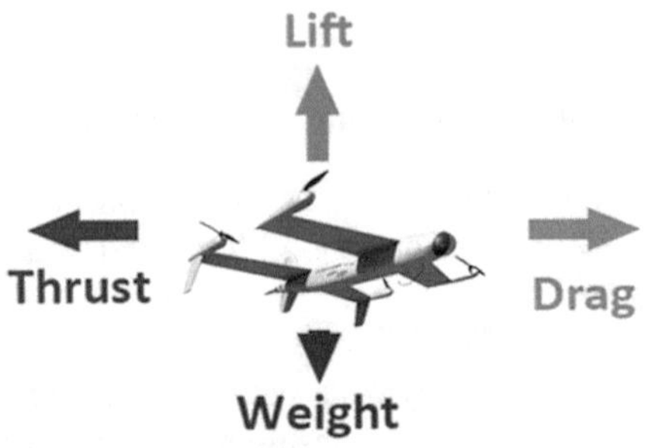

Fig. 1. Different forces act on UAV.

From Newton's second formula we can derive the following equation.

$$F = V\left(\frac{\mathrm{dm}}{dt}\right) \tag{1}$$

Because in horizontal moving the weight of the UAV is equal and opposite to the lifting force, this lifting force is the reaction to diverting the air downward. This lifting force is the reaction to diverting the air downward. So, the weight of the UAV is equal to the speed of the air being thrown down multiplied by the mass of air per unit of time that is being affected by the UAV [38, 41].

$$W = F_L = V\left(\frac{dm}{dt}\right) \tag{2}$$

W is the weight of the UAV, V is the downward speed of the air, and $\frac{d\mathrm{m}}{dt}$ is the mass of the air being thrown down per unit of time.

Let b denote the width of UAV. We can now calculate the mass of air per time unit as the density of the air multiplied by the speed of the UAV multiplied by the area influenced by the UAV [37, 38, 41].

$$\frac{dm}{dt} = \frac{1}{2}Db^2v \tag{3}$$

Substituting this into our main lift Eq. (2) gives:

$$W = V\frac{1}{2}Db^2v \tag{4}$$

Where W is the weight of the UAV, V is the speed of the air being thrown down, D is the density of the air, v is the relative speed of the UAV through the air, and ½ b^2 is the effective area affected by the UAV body.

2.1 Power Consumption in Horizontal Moving

Power is required to lift the UAV into the air and some power is needed to overcome the parasitic drag that is impeding its forward movement through the air [38, 39, 42, 43]. Let us first focus on the power required to overcome parasitic drag. The parasitic drag can be modeled as [37]:

$$F_P = \frac{1}{2}C_DADv^2 \tag{5}$$

Where F_P is the parasitic drag, C_D is the aerodynamic drag coefficient, A is the front facing area, D is the density of the air, and v is the UAV's relative speed through the air. The general equation for power is [41]:

$$P = Fv \tag{6}$$

Hence, the power needed to overcome the parasitic drag is:

$$P_P = \frac{1}{2}C_DADv^3 \tag{7}$$

The UAV needs power both to overcome parasitic drag and for lifting the UAV [44].

$$P_T = P_P + P_L \tag{8}$$

The power needed to overcome parasitic drag is the greatest at high speeds while the power needed for lift is the greatest at the low speeds. Between these two extremes, the power requirement is the lowest, that is, there is an optimal cruising speed (Fig. 2). Because the power requirement is greater at slower speeds, it does not make sense to travel at a speed lower than where this power requirement is the lowest, unless the UAV does not wish to arrive early at a destination [45, 46].

The purpose of lift is to transfer energy from the UAV to surrounding air to lift the UAV. This energy is the kinetic energy that the air is given as it is thrown downward [37, 41, 43]. For individual objects, this is calculated as:

$$E = \frac{1}{2}mV^2 \tag{9}$$

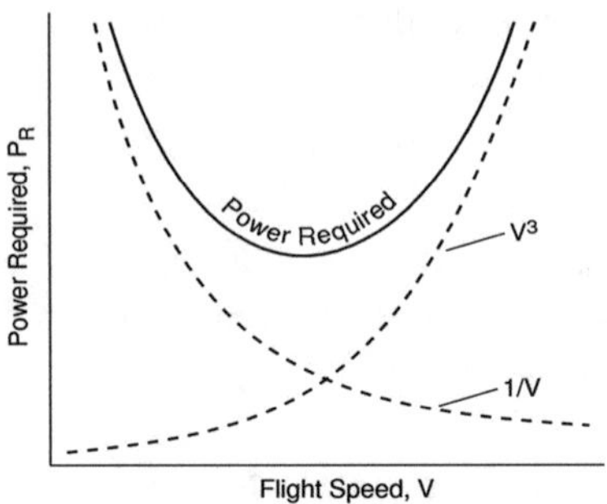

Fig. 2. Power vs Flight Speed

The power required for lift is the amount of energy given to the air per unit of time and substituting (8) we have:

$$P_L = \frac{dE}{dt} = \frac{d\left(\frac{1}{2}mV^2\right)}{dt} = \frac{1}{2}\left(\frac{dm}{dt}\right)V^2 \tag{10}$$

From substituting (7) and (8) we can have the power required to lift as [41]:

$$P_L = \frac{W^2}{Db^2v} \tag{11}$$

Where P_L is the power needed for lift, W is the total weight of the UAV, D is the density of the air, b is the width of UAV, and v is the relative speed of the UAV through the air. Recalling our total power Eq. (8), the power needed for flight is as follows.

$$P_T = \frac{1}{2}C_DADv^3 + \frac{W^2}{Db^2v} \tag{12}$$

Where P_T is the power needed for flight in watts, C_D is the aerodynamic drag coefficient, A is the front facing area in m^2, W is the total weight of the UAV in kg, D is the density of the air in $kg/m^{3,}$ b is the width of UAV in meters, and v is the relative speed of the UAV in m/s considering the wind speed and direction.

The speed cube is in the numerator of parasitic power term, and speed is in the denominator of the power for lift term.

By taking the derivative of the total power equation with respect to speed then setting the result equal to zero, we can find the speed for minimum power [37, 38, 41, 42, 47].

$$v_{\min} = \left(\frac{2W^2}{3C_DAb^2D^2}\right)^{0.25} \tag{13}$$

We can now take the calculated minimum power speed and substitute it into the total power equation to calculate the minimum power needed for flight [37, 41, 42, 47].

$$P_{\min} = \frac{4}{3}\left(\frac{W^2}{Db^2 v_{\min}}\right) \tag{14}$$

The minimum power speed is not the normal cruising speed of the UAV, but rather, it would be the bare minimum speed to use.

2.2 Optimum Flying Speed

Optimum speed is the speed that gives the least amount of drag (Fig. 3). The speed that has the least amount of drag on the aircraft is the optimum cruising speed [38, 41, 43, 47]. The total drag is the parasitic drag that was calculated earlier plus the drag generated in throwing the air down.

$$F_T = F_P + F_I \tag{15}$$

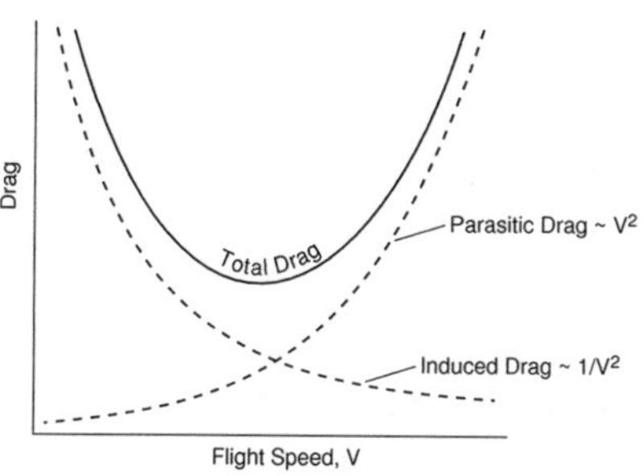

Fig. 3. Drag vs Flight Speed

The drag generated in throwing the air down, or induced drag, is simply a way to account for the fact that more force is needed to move the UAV through the air [38, 44, 46]. The induced drag is calculated by starting with the power lift equation and applying the standard power Eq. (11). The induced drag is then just the power lift equation divide by speed [38, 41, 47]. We can now add the parasitic drag and the induced drag.

$$F_T = \frac{1}{2} C_D A D v^2 + \frac{W^2}{Db^2 v^2} \tag{16}$$

Like before when we found the minimum power speed, we determine the cruising speed by taking the derivative of the total force equation with respect to speed then setting the result equal to zero we can find the speed for the minimum drag force [37, 41, 42, 47].

$$v_{\text{optimum}} = \left(\frac{2W^2}{C_D A b^2 D^2} \right)^{0.25} \tag{17}$$

Where v_{optimum} is the optimum cruising speed, W is the weight of the flying object, D is the density of the air, A is the frontal area, C_D is the drag coefficient, and b is the width of the UAV. With this relatively simple equation, we can input data about a UAV and the density of the air to calculate the correct speed to fly.

2.3 UAV Energy Consumption in High Speeds

From the literature according to [41], in steady level flight, the thrust of the UAV is equal to the drag of the UAV, and the lift is equal to the total weight of UAV, in which case the propulsive thrust power can then be given as follows.

$$T = W * \frac{C_D}{C_L} \tag{18}$$

From Power Eq. (10), we can derive that

$$P_P = Tv \tag{19}$$

C_D is the Drag coefficient, and C_L is Lift coefficient of the UAV.

$$P_P = \frac{C_D}{C_L} * Wv \tag{20}$$

From (12), the total power in higher speeds is;

$$P_T = \frac{C_D}{C_L} * Wv + \frac{W^2}{Db^2 v} \tag{21}$$

2.4 Energy Consumption in Hovering, Vertical Takeoff and Landing

Studies have used the following equation in calculating the energy consumption of UAVs, which is derived using power consumed by a multirotor helicopter, and they have proved that the power it consumes is approximately linearly proportional to the weight of its battery and payload under practical assumptions [16, 34].

$$p^* = \frac{T^{\frac{3}{2}}}{\sqrt{2D\varsigma}} \tag{22}$$

Also, this study has assumed that the power consumed during takeoff and landing is, on average, approximately equivalent to the power consumed during hover. Power $p*$ in watts and the thrust T in Newtons. Air density of air D in kg/m^3, and the facing area ς of the UAV is in m^2, where the thrust T = W g, given the UAV total weight W in kg, and gravity g in N.

As air density is dependent on temperature, different temperature conditions will lead to different air densities and thus will affect the power consumption of UAVs.

3 Relationship of Factors Affecting UAV Energy Consumption

Figure 4 presents an overview of the relationships between different factors which are linked with energy consumption of UAVs. Among these factors, speed of UAV and wind direction has a correlation as speed of the UAV is affected by the wind speed and direction. Based on the existing research, smaller power consumptions were observed when flying into headwind [38, 48], which is due to the increasing thrust by translational lift, when the UAV moves from hovering to forward flight [27]. When flying into a headwind, translational lift increases due to the relative airflow increases, resulting in less power consumption to hover the UAV [49]. When the wind speed exceeds a certain limit, the aerodynamic drag may outweigh the benefit of translational lift [27].

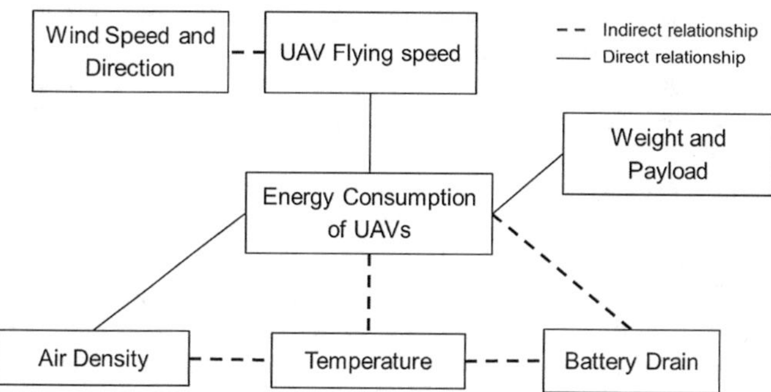

Fig. 4. Factors that affect energy consumption of UAVs

Moreover, temperature and air density have a relationship and this is linked with the battery drain. Air density influences the lifting capacity of aircraft and varies with temperature [50]. On the other hand, studies have shown that in cold conditions at or below zero degrees, shorter flying times and increased risk of UAV malfunction are observed [51]. In contrast to all other factors, weight and payload act as an individual factor which influences the energy consumption of UAVs in general.

4 Conclusion

This paper focuses mainly on deriving the energy consumption of UAVs, which is highly non-linear and dependent on weather, speed, and payload. This makes UAV routing differ significantly from all other types of routing we traditionally deal with, as

UAVs are expected to travel in certain high altitudes, and are, therefore, significantly susceptible to wind and weather conditions.

We have presented equations to calculate the total power consumption of UAVs in different flight scenarios including horizontal moving, vertical moving, and hovering based on the existing literature. In the future, we will further analyze these models by experimenting with industrial data and different models of available UAVs.

References

1. Sitek, P., Wikarek, J.: Capacitated vehicle routing problem with pick-up and alternative delivery (CVRPPAD): model and implementation using hybrid approach. Ann. Oper. Res. (2017). https://doi.org/10.1007/s10479-017-2722-x
2. Sitek, P.: A hybrid approach to the two-echelon capacitated vehicle routing problem (2E-CVRP) BT - recent advances in automation, robotics and measuring techniques. In: Szewczyk, R., Zieliński, C., Kaliczyńska, M. (eds.), pp. 251–263. Springer International Publishing, Cham (2014)
3. Nielsen, I., Dang, Q.V., Bocewicz, G., Banaszak, Z.: A methodology for implementation of mobile robot in adaptive manufacturing environments. J. Intell. Manuf. **28**, 1171–1188 (2017). https://doi.org/10.1007/s10845-015-1072-2
4. Yakici, E.: Solving location and routing problem for UAVs. Comput. Ind. Eng. **102**, 294–301 (2016). https://doi.org/10.1016/j.cie.2016.10.029
5. Bolton, G.E., Katok, E.: Learning-by-doing in the newsvendor problem a laboratory investigation of the role of experience and feedback. Manuf. Serv. Oper. Manag. **10**, 519–538 (2004). https://doi.org/10.1287/msom.1060.0190
6. Avellar, G.S.C., Pereira, G.A.S., Pimenta, L.C.A., Iscold, P.: Multi-UAV routing for area coverage and remote sensing with minimum time. Sensors (Switzerland) **15**, 27783–27803 (2015). https://doi.org/10.3390/s151127783
7. Khosiawan, Y., Nielsen, I.: A system of UAV application in indoor environment. Prod. Manuf. Res. **4**, 2–22 (2016). https://doi.org/10.1080/21693277.2016.1195304
8. Barrientos, A., Colorado, J., del Cerro, J., et al.: Aerial remote sensing in agriculture: a practical approach to area coverage and path planning for fleets of mini aerial robots. J. F. Robot. **28**, 667–689 (2011). https://doi.org/10.1002/rob
9. Khosiawan, Y., Park, Y., Moon, I., et al.: Task scheduling system for UAV operations in indoor environment. Neural Comput. Appl. **9**, 1–29 (2018). https://doi.org/10.1007/s00521-018-3373-9
10. Zhang, M., Su, C., Liu, Y., et al.: Unmanned aerial vehicle route planning in the presence of a threat environment based on a virtual globe platform. ISPRS Int. J. Geo Inf. **5**, 184 (2016). https://doi.org/10.3390/ijgi5100184
11. Xiang, J., Liu, Y., Luo, Z.: Flight safety measurements of UAVs in congested airspace. Chin. J. Aeronaut. **29**, 1355–1366 (2016). https://doi.org/10.1016/j.cja.2016.08.017
12. Khosiawan, Y., Khalfay, A., Nielsen, I.: Scheduling unmanned aerial vehicle and automated guided vehicle operations in an indoor manufacturing environment using differential evolution-fused particle swarm optimization. Int. J. Adv. Robot. Syst. **15**, 1–15 (2018). https://doi.org/10.1177/1729881417754145
13. Popper, B.: Drones could make Amazon's dream of free delivery profitable - The Verge (2016). http://www.theverge.com/33

14. Bonn: DHL | Press Release | English. In: DHL (2017). http://www.dhl.com/en/press/releases/releases_2014/group/dhl_parcelcopter_launches_initial_operations_for_research_purposes.html. Accessed 11 Apr 2017
15. Wang, X., Poikonen, S., Golden, B.: The Vehicle Routing Problem with Drones : A Worst-Case Analysis Outline Introduction to VRP Introduction to VRPD, pp. 1–22 (2016)
16. Dorling, K., Heinrichs, J., Messier, G.G., Magierowski, S.: Vehicle routing problems for drone delivery. IEEE Trans. Syst. Man Cybern. Syst. **47**, 1–16 (2016). https://doi.org/10.1109/tsmc.2016.2582745
17. Aasen, H., Gnyp, M.L.: Spectral comparison of low-weight and UAV- based hyperspectral frame cameras with portable spectroradiometers measurements. In: Proceedings of Work UAV-Based Remote Sensing Methods for Monitoring Vegetation, pp. 1–6 (2014). https://doi.org/10.5880/tr32db.kga94.2
18. Frazzoli, E., Bullo, F.: Decentralized algorithms for vehicle routing in a stochastic time-varying environment. In: 2004 43rd IEEE Conference on Decision Control (IEEE Cat No04CH37601), vol. 4, pp. 3357–3363 (2004). https://doi.org/10.1109/cdc.2004.1429220
19. Sundar, K., Venkatachalam, S., Rathinam, S.: An Exact Algorithm for a Fuel-Constrained Autonomous Vehicle Path Planning Problem (2016)
20. Sundar, K., Rathinam, S.: Algorithms for routing an unmanned aerial vehicle in the presence of refueling depots. IEEE Trans. Autom. Sci. Eng. **11**, 287–294 (2014). https://doi.org/10.1109/TASE.2013.2279544
21. Zhang, J., Zhao, Y., Xue, W., Li, J.: Vehicle routing problem with fuel consumption and carbon emission. Int. J. Prod. Econ. **170**, 234–242 (2015). https://doi.org/10.1016/j.ijpe.2015.09.031
22. Feng, Y., Zhang, R., Jia, G.: Vehicle routing problems with fuel consumption and stochastic travel speeds (2017). https://doi.org/10.1155/2017/6329203
23. Kinney, G.W., Hill, R.R., Moore, J.T.: Devising a quick-running heuristic for an unmanned aerial vehicle (UAV) routing system. J. Oper. Res. Soc. **56**, 776–786 (2005). https://doi.org/10.1057/palgrave.jors.2601867
24. Yu, V.F., Lin, S.-W.: Solving the location-routing problem with simultaneous pickup and delivery by simulated annealing. Int. J. Prod. Res. **54**, 1–24 (2015). https://doi.org/10.1016/j.asoc.2014.06.024
25. Qian, Z., Wang, J., Wang, G.: Route Planning of UAV Based on Improved Ant Colony Algorithm, pp. 1421–1426 (2015)
26. Sarıçiçek, İ., Akkuş, Y.: Unmanned aerial vehicle hub-location and routing for monitoring geographic borders. Appl. Math. Model. **39**, 3939–3953 (2015). https://doi.org/10.1016/j.apm.2014.12.010
27. Tseng, C-.M., Chau, C-.K., Elbassioni, K., Khonji, M.: Autonomous Recharging and Flight Mission Planning for Battery-operated Autonomous Drones, pp. 1–10 (2017)
28. Dorling, K., Heinrichs, J., Messier, G.G., Magierowski, S.: Vehicle routing problems for drone delivery. IEEE Trans. Syst. Man Cybern. Syst. **47**, 70–85 (2017). https://doi.org/10.1109/TSMC.2016.2582745
29. Guerriero, F., Surace, R., Loscri, V., Natalizio, E.: A multi-objective approach for unmanned aerial vehicle routing problem with soft time windows constraints. Appl. Math. Model. **38**, 839–852 (2014). https://doi.org/10.1016/j.apm.2013.07.002
30. Habib, D., Jamal, H., Khan, S.A.: Employing multiple unmanned aerial vehicles for co-operative path planning. Int. J. Adv. Robot. Syst. **10**, 1–9 (2013). https://doi.org/10.5772/56286
31. Wu, J., Zhang, D., Pei, D.: Autonomous route planning for UAV when threats are uncertain. In: 2014 IEEE Chinese Guidance, Navigation and Control Conference (CGNCC 2014), pp. 19–22 (2015). https://doi.org/10.1109/cgncc.2014.7007214

32. Shetty, V.K., Sudit, M., Nagi, R.: Priority-based assignment and routing of a fleet of unmanned combat aerial vehicles. Comput. Oper. Res. **35**, 1813–1828 (2008). https://doi.org/10.1016/j.cor.2006.09.013
33. Zhang, J., Jia, L., Niu, S., et al.: A space-time network-based modeling framework for dynamic unmanned aerial vehicle routing in traffic incident monitoring applications. Sensors (Switzerland) **15**, 13874–13898 (2015). https://doi.org/10.3390/s150613874
34. Leishman, D.S., (Eng.. PDFRASJG) Principles of Helicopter Aerodynamics (2006). https://doi.org/10.1002/1521-3773(20010316)40:6%3c9823::aid-anie9823%3e3.3.co;2-c
35. Joo, H., Hwang, H.: Surrogate aerodynamic model for initial sizing of solar high-altitude long-endurance UAV. J. Aerosp. Eng. **30**, 04017064 (2017). https://doi.org/10.1061/(ASCE)AS.1943-5525.0000777
36. Nancy, H.: Bernoulli and Newton. In: NASA Off (2015). https://www.grc.nasa.gov/WWW/K-12/airplane/bernnew.html. Accessed 4 Oct 2017
37. David, E.: Deriving the Power for Flight Equations (2003). http://www.dinosaurtheory.com/flight_eq.html. Accessed 4 Oct 2017
38. Tennekes, H.: The Simple Science of Flight, 2nd edn. The MIT Press Cambridge, London (2009)
39. National Academies of Sciences and Medicine: Commercial Aircraft Propulsion and Energy Systems Research: Reducing Global Carbon Emissions. National Academies Press (2016)
40. Farokhi, S.: Aircraft Propulsion. Wiley, Hoboken (2014)
41. Greitzer, E.M., Spakovszky, Z.S., Waitz, I.A.: 16.Unified: Thermodynamics and Propulsion Prof. Z. S. Spakovszky (2008). http://web.mit.edu/16.unified/www/FALL/thermodynamics/notes/notes.html. Accessed 4 Oct 2017
42. Trips, D.: Aerodynamic Design and Optimization of a Long Range Mini-UAV (2010)
43. Hill, P.G., Peterson, C.R.: Mechanics and thermodynamics of propulsion, 764 p. Addison-Wesley Publ. Co., Reading (1992)
44. Francis, N.H.: Learning and Using Airplane Lift/Drag (2014)
45. Nigam, N., Bieniawski, S., Kroo, I., Vian, J.: Control of multiple UAVs for persistent surveillance: algorithm and flight test results. IEEE Trans. Control Syst. Technol. **20**, 1236–1251 (2012). https://doi.org/10.1109/TCST.2011.2167331
46. Kunz, P.J.: Aerodynamics and Design for Ultra-Low Reynolds Number Flight (2003)
47. Edlund, U., Nilsson, K.: Optimum Design Cruise Speed for an Efficient Short Haul Airliner, pp. 960–966 (1984)
48. Moyano Cano, J.: Quadrotor UAV for wind profile characterization (2013)
49. Administration USD of TFA: Helicopter Flying Handbook. US Department of Transportation Federal Aviation Administration, vol. 5, pp. 22–117 (2012). https://doi.org/10.1088/1751-8113/44/8/085201
50. Thøgersen, M.L.: WindPRO/ENERGY Modelling of the Variation of Air Density with Altitude through Pressure, Humidity and Temperature (2000)
51. Cessford, J.R., Barwood, M.J.: The Effects of Hot and Cold Environments on Drone Component Performance and Drone Pilot Performance (2015)

Agile Approach in Crisis Management – A Case Study of the Anti-outbreak Activities Preventing an Epidemic Crisis

Jan Betta[1(✉)], Stanisław Drosio[2], Dorota Kuchta[1], Stanisław Stanek[3], and Agnieszka Skomra[1]

[1] Faculty of Computer Science and Management, Wrocław University of Science and Technology, Wrocław, Poland
dorota.kuchta@pwr.edu.pl

[2] Faculty of Informatics, Katowice University of Economics, Katowice, Poland

[3] Faculty of Management, General Tadeusz Kościuszko Academy of Land Forces, Wrocław, Poland

Abstract. The paper presents a case study which illustrates a possible application of Agile approach to crisis management. The proposal of such a merger was done in another paper, here its implementation to epidemic crisis management is described. It is shown that the nature of activities during the epidemic crisis suits the Agile philosophy fairly well and that the application of Agile frameworks to epidemic crisis management may substantially increase its efficiency, mainly due to the communication patterns which are required by the Agile approach.

Keywords: Crisis management · Epidemic crisis · Agile management

1 Introduction

In an earlier work [3] a merger of crisis management and Agile management (especially the Scrum framework) was proposed. The aim of this paper is to present a first trial of a validation of this proposal, by means of a single case study method [7]. The application of the merger "Crisis management – Scrum framework", which was described in detail in [3], was proposed to one of Polish Crisis Management Centers. As one of the authors of the present paper works as a consultant for this Crisis Management Center, it was possible to conduct a theoretical common reflection, together with practitioners from the Center, on how such a merger might look like in a concrete case and whether it would be useful. The aim of the present paper is to present the main results of this refection and to propose a rudimentary scheme of a possible application of the merger in practice, together with theoretical foundations of crisis management and Agile management.

J. Świątek et al. (Eds.): ISAT 2018, AISC 853, pp. 185–195, 2019.
https://doi.org/10.1007/978-3-319-99996-8_17

2 Crisis and Crisis Management

Research focused specifically on crises was initiated in the 1960s and 1970s in such sciences as psychology and sociology, giving birth to the science of disaster response [8]. A broad overview of relevant research findings was recently provided in [4]. The 2 concept of crisis is, for a number of reasons, an ambiguous one, hence its definition has raised lots of controversy and has received extensive treatment in literature. Etymologically, the word itself can be traced back to the Greek *krino*, that may be interpreted as e.g. a turning point, a fork in a development path, or a sudden unforeseen situation requiring prompt decision and action. In Chinese, the crisis (*Weji*) symbol is composed of two simpler symbols: *We* – for threat, and *Ji* – for opportunity. A linguistic-lexical analysis therefore shows crisis as having both negative and positive potential.

The most widely cited definition of the term crisis was proposed in [12]. It describes crisis as a low-probability high-impact event that threatens the viability of the organization and is characterized by ambiguity of cause, effect, and means of resolution, as well as by a belief that decisions must be made swiftly. As is the case with illness, crisis is not a merely accidental occurrence, but an incident that is usually preceded by certain symptoms that can be identified by insightful and experienced managers.

In [4] the concept of crisis management is positioned in relation to the internal and external perspective:

- the internal perspective involves the coordination of complex technical and relational systems and design of organizational structures to prevent the occurrence, reduce the impact, and learn from a crisis;
- the external perspective involves shaping perceptions and coordinating with stakeholders to prevent, solve, and grow from a crisis.

The features which are common for crisis, directly resulting from the definitions above, are: abnormal/unusual situation, instability, loss of control, specificity, changes, serious consequences, disruption of the balance, disaster. Moreover, there are also other ones, such as: increasing citizen participation [6], stakeholders making decisions under stress, experience improvement, engagement and realism, quick decision-making in critical conditions, complexity of information [5], engagement of security forces [4], early warning, external and internal influences [2], total disruptive event, panic, lack of morale, misinformation, loss of knowledge, loss of leadership, cancelling recruitments, loss of reputation [17], instability and discontinuity [1].

On the other hand, crisis management can be characterised by the following features:

- Managing a crisis involves the participation of various stakeholders. Main phases are: mitigation, preparedness, response and recovery. Successful crisis management requires full integration of all of the involved parties [6];
- Crisis management involves quick decision-making in critical conditions, acting in an urgent decision-making situation; their goal is to minimise the potential negative consequences. The human factor is frequently a main source of errors in the decision-making process. Decision-making, communication, leadership and coordination are critical skills to be used in crisis management [16];

- Business crisis management is a system which tries to summarise the law of crisis occurrence and development, avoid and reduce the harm of the crisis, strengthen crisis warning, crisis decision-making and crisis handling. The main part of crisis management consists in crisis monitoring. Many signs of crisis appear before the crisis [2].

Let us now pass to the second element of the merger discussed in this paper, the Agile project management.

3 Agile Approach

The Agile approach to project management, invented and applied originally to IT projects, breaks away with the classic approach to project management and the sequential implementation of phases in favour of enhancing the flexibility of operations. The basic principle is 'fast and flexible response to customer needs' so as to be able to provide the customer with a product that fulfils his or her expectations [18]. This approach is characterised primarily by openness to changes and accepting them as an integral part of the project. It focuses on building relationships between project team members based on trust and commitment, which contributes to the reduction of monitoring and documentation [18].

The agile approach is based on the iterative work model, consisting of the division of the project into smaller parts, which remain constant over time (iterations) and culminate in the delivery to the client of ready-to-release pieces of software. The division of the project into iterations allows for greater openness to changes and undertaking of challenges on the part of the project team, since any potential failure is limited to a single iteration. The essence of the approach is the self-organising team, which itself makes decisions about the way the project is carried out, accounting for the need to adapt to changes [18].

The so-called Agile Manifesto [11] is considered to be the credo of the Agile approach to project management. Elaborated in 2001 by a group of programmers, it presents the demands placed on the modern approach to software development. However, its creation was not dictated by the elaboration of a new methodology; rather, the manifesto is only an indication of certain features it was intended to characterize. Thus, the Agile Manifesto presents the following values [11]:

- "Individuals and interactions over processes and tools;
- Working software over comprehensive documentation;
- Customer collaboration over contract negotiation;
- Responding to change over following a plan."

As mentioned above, the Agile approach was first applied to IT projects, but nowadays it is also applied to other project types, for example R&D projects [9]. The Agile approach is important in cases when even the project goal can change or should be made more precise in the course of project implementation. In the Agile philosophy this is not considered as a problem: the project team and the project stakeholders welcome changes if they are justified from the point of view of stakeholders satisfaction.

There are several approaches or frameworks within the Agile philosophy. The most important one is the Scrum framework [18]. Scrum assumes five events (Sprint Planning Meeting, Daily Scrum, Sprint, Sprint Review, Sprint Retrospective). Scrum Team consists from three Scrum roles (Scrum Master, Product Owner, Development Team) 4 and from tree artefacts (Product Backlog, Sprint Backlog, Increment). Figure 1 shows the diagram of the Scrum framework:

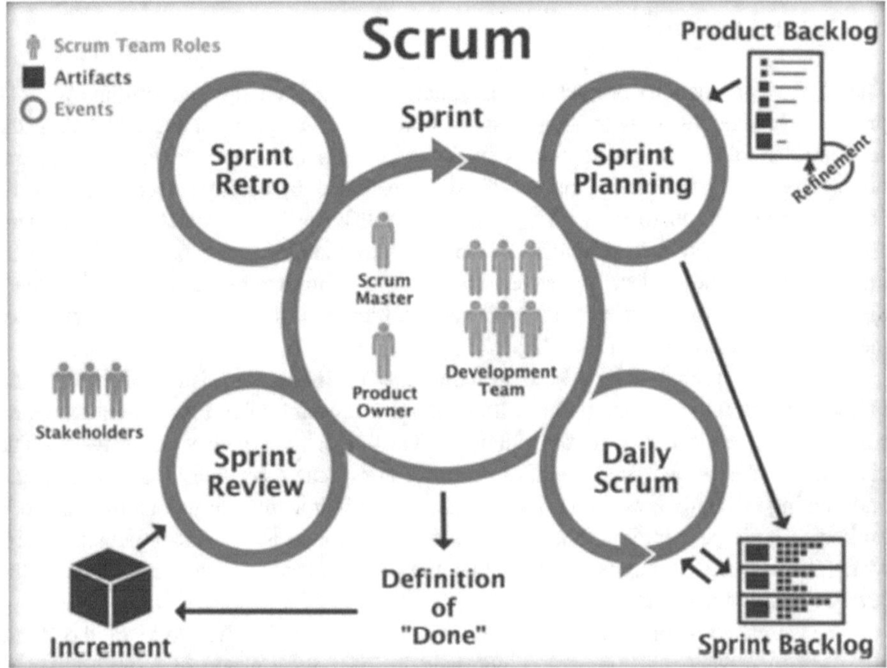

Fig. 1. Scrum framework (Source: Wysocki 2009)

Here we will describe only selected elements from Fig. 1, those which are needed in the remainder of the paper.

Scrum Master is a person who supports the project team in a correct application of Scrum principles. Product Owner is a person representing the knowledge and understanding of the needs of the business realized through product design. She or He is responsible for the Product Backlog Items list, called Product Backlog. Product Backlog can be described as a list of elements or tasks to be accomplished. Each item of the Product Backlog has two quantitative attributes or weights: the importance of the item (the higher the importance, the sooner the item should be accomplished) and an estimation of the effort needed to accomplish the item. The question how both attributes are determined is complex. Here it suffices to say that the importance assigned to an item depends above all on the influence of the successful accomplishment of the item on the fulfilment of project goals and on stakeholders satisfaction. The effort necessary to accomplish the item can be measured in various units, the unit can be selected by the project team for each case. It can be for example man-hours.

During the Sprint Planning, the Development Team selects the Product Backlog Items to be implemented in the next Sprint. A Sprint may be a period 2-6 weeks. The idea of Agile project management is to elaborate detailed plans only for the next Sprint, the rest of the project is not planned in detail. The total available effort of the next Spring is known and is called the Sprint capacity. The Sprint capacity depends on the number of project team members, their availability and experience. During the Sprint 5 Planning, the members of the project should select such a subset of Product Backlog items which maximizes the total importance of the items selected which a total effort required to accomplish the selected items not exceeding the Sprint capacity. Formally, this can be formulated as a knapsack problem [10, 14, 15]. Each Sprint can be considered as small project and has its own goal describing what is to be implemented within it.

Every day a short meeting, called Daily Scrum, is organized. It is a meeting during which each team member answers three questions:

- What has been done since the last meeting to achieve the goal of the Sprint?
- What will be done before the next meeting to achieve the goal of the Sprint?
- What obstacles stand in the way of achieving the goal of the Sprint?

Daily Scrum should be very short, its duration can be measured in minutes. Even its recommended form – a standing meeting – supports the idea of not losing time, of concentrating on the essentials necessary to achieve the goals of the Sprint within the planned time and to contribute as much as possible to the fulfillment of the goals of the whole project and to the stakeholders satisfaction.

4 Agile Approach Applied to Crisis Management

So far, the Agile approach has been applied to project management, but in [3] its application to crisis management was proposed. This stemmed from the observation that crisis management and Agile project management may have a lot in common. A full analysis of the similarities between crisis management and agile management can be found in [3]. Table 1 gives some of its elements.

Table 1 together with [3] justify the application of Agile management to crisis management. Here we will propose a case study of how Agile crisis management might evolve in practice. The case study concerns epidemic crises and their treatment in a Polish voivodeship.

Table 1. Analyse comparative Crisis vs. Agile (Source: Betta, Skomra 2017)

Main crisis features	Agile items
Abnormal/unusual situation	**Customer collaboration** over contract negotiation **Responding to change** over following a plan
Instability	**Individuals and interactions** over processes and tools
Loss of control	**Responding to change** over following a plan
Specificity	**Customer collaboration** over contract negotiation
Changes	**Working software** over comprehensive documentation
Serious consequences	**Working software** over comprehensive documentation
Disruption of the balance	**Responding to change** over following a plan
Disaster	**Working software** over comprehensive documentation **Customer collaboration** over contract negotiation

5 Case Study: Agile Approach in the Anti-outbreak Activities Stage of the Epidemic Crisis

We consider here a Polish Voivodeship Crisis Management Center. The term "project team" will be used to design the group of persons from the Center assigned to the management of the crisis situation in question.

We refer here more specifically to epidemic crises, linked to highly contagious diseases. The activities in the Center referring to epidemics are divided into four groups:

A. Activities before the outbreak of a highly contagious disease;
B. Activities in the stage when the occurrence of a highly contagious disease is suspected;
C. Activities in the stage of an epidemic outbreak;
D. Activities in the stage when an epidemic turns out to be impossible to control.

Activities from group A are carried out on a continuous basis, thus they cannot be regarded as a project. However, each of the remaining groups of activities are undertaken only if there occurs a relevant necessity and can be considered to be a project. Each of those projects can – and, in the opinion of the authors of the present paper, should be managed using the Agile approach. It is so because each of them regard a very dynamic situation, where quick reaction to continuous changes, a highly flexible attitude and efficient continuous communication are necessary.

Here we will concentrate on group B: "Activities in the stage when the occurrence of a highly contagious disease is suspected". If we are dealing with the suspicion of the occurrence of a highly contagious disease, we have to start a project which has a threefold objective (or rather an objective composed of three goals):

(a) to control the behavior and mood of the population (so that no panic outbreaks, people are obedient to the instructions given and no unnecessarily negative rumors spread out);
(b) to stave off the disease;

(c) to determine that with a high probability it is not possible to stop the epidemic and thus group C of activities should be started, which would constitute another project.

The exact goal will be clarified in the course of the implementation of group B activities, according the development of the situation. Subgoals b. and c. are contradictory, it will have to be clarified which one will be pursued ultimately. All the subgoals are formulated fuzzily, it will be clear what has to be done only in the course of events.

Crisis Product Owner will be an expert in epidemic crises who will have the final word in all the current decisions. The Crisis Product Backlog will be here composed of the activities which now are elements of group B in the considered Crisis Management Center. They will be presented and analyzed below. Because of the very dynamic nature of the project realized by means of activities from group B, we propose to introduce very short Sprints, possibly of one day duration (or even shorter, it will be possible to change the Sprint duration if necessary, contrary to the classical Scrum approach), and to merge Sprint Planning Meeting with Daily Scrum into a meeting which we propose to call Crisis Scrum Meeting. Thus, every day or even at shorter intervals, the project team, which in the considered Crisis Management Centre would be composed of medical doctors, epidemiologists, psychologists and relevant decision makers, would meet for a brief meeting, where they would share their opinions on the hitherto results, conclusions and future actions. Crisis Scrum Master will be a member of the project team responsible for convoking the Crisis Scrum Meetings and for their due course. He or she does not have to be an expert in crisis management, but has to possess soft competences and leadership abilities so that the project team trust her or him and are obedient to her or him, as there will not much time for useless discussions.

The Crisis Scrum Meeting should not be confused with ad hock meetings or other communication forms which in the considered situation would take place fairly often, involving selected members of the project team. The Crisis Scrum Meeting is a meeting which has to take place whatever the situation is, in order to enforce communication within the entire project team. In a situation when a contagious disease is suspected, the members of the project team will face various situations, meet and talk to various persons (frightened local residents, ill persons before or after the diagnosis and their relives, doctors etc.), share various emotions, always seeing only a portion of the entire picture. The Crisis Scrum Meeting goal is to enforce a communication encompassing the views and experiences of the whole team and all the important stakeholders, in order to give due weights to individual reports and enable everyone to see the whole picture of the situation. The team should be "forced", in short intervals, to quickly reflect (and, above all, to discuss, as each member may have different pieces of information and a different intuitions) on the current goal they are pursuing (is it still possible to stave off the disease or are we rather turning to the point when the epidemic is unavoidable), on the questions what the most important current tasks are, where the most burning locations are in which direct actions should be undertaken and how the work can be shared within the team.

During the Crisis Scrum Meeting weights should be given to the elements of the Crisis Product Backlog (listed below), expressing both the current importance of the items and the effort needed to accomplish them. The Crisis Scrum Master should ask all the members, once they have decided what the current goal is, to give weights to the Product Backlog elements and use a simple software, which may be based for example on Solver in the widely used Excel application and on the knapsack model, to determine which tasks have to be accomplished immediately. This step is necessary, because the capacity of the crisis management team (measured in available man-hours) is always limited and a choice of actions to be carried out immediately has to be accomplished. Moreover, because of the possibly high level of emotions, the choice should be made as objectively as possible, where the mathematical model and a software funding within seconds an optimal or close to optimal solution might be helpful. The team may be then given a few minutes to raise objections to the list of actions to be implemented immediately determined by means of the computer. After this short time the Crisis Product Owner makes the final decision and everybody sets out to execute the currently selected tasks.

The Crisis Product Backlog will be composed of the activities which now belong, in the Crisis Management Center in question, to the above mentioned group B. Here are the most important ones, together with a short description relating to the Agile approach. It should be noted that the description of the activities in itself is not precise, thus the Crisis Scrum Meeting should also serve to make them sufficiently precise to be able to be carried out (Table 2).

Table 2. Crisis Product Backlog in the anti-outbreak activities stage of the epidemic crisis and the Agile approach (an extract) (Source: own elaboration)

Activity	Factors influencing the importance evaluation (measured e.g. in the scale from 0 (unimportant) to 5 (extremely important))	Factors influencing the effort evaluation (expressed in man-hours)
Asking the local government for additional forces (medical doctors, transport means etc.)	The more the disease is spreading out, the more important is additional help	The effort needed depends on the variety of means asked for (e.g. medical doctors will have to be asked in another administrative unit than transport means)
Conducting epidemiological investigations (about source of disease explosion, incubation period, number and location of infected persons etc.)	This activity is especially important in the first stage of the project, when the choice of the weights of subgoals b. and c. should be made and medical forces with the necessary means sent to specific locations	The effort needed depends on the number of locations where disease symptoms have been reported and the number of persons potentially affected in all the identified locations

(*continued*)

Table 2. (*continued*)

Activity	Factors influencing the importance evaluation (measured e.g. in the scale from 0 (unimportant) to 5 (extremely important))	Factors influencing the effort evaluation (expressed in man-hours)
Setting up isolating places for potentially infected persons	Like the previous activity	Like the previous activity
Verifying the supply of relevant medications and injections	This activity is especially important at the beginning and each time when during the Crisis Scrum Meeting an abrupt increase in the number of persons infected is reported	The effort needed depends on the number of locations where an essential increase of infected persons has been reported
Sending psychologists to places where panic is about to break out	The importance of this activity depends on the mood among population reported in the Crisis Scrum Meeting	The effort needed depends both on the number of places where there is a threat of panic and the population size in each of those places
Applying to self-government authorities for issuing decisions on matters in which the State Sanitary Inspector has no competence	The importance of this activity depends on the importance of the decisions that are applied for	The effort needed depends on the relations of the project team with the relevant selfgovernment authority

During the Crisis Scrum Meeting it should be remembered not to plan out the whole capacity. About 20% of the team capacity should be kept as a buffer for unexpected events which are unavoidable in crises.

6 Conclusions

In this paper it was shown how the Agile approach, or more exactly the Scrum framework, can be applied to crisis management. A case study was used: a part of epidemic crisis management process as it is defined in one of Polish Voivodeship Crisis Management Center.

It was not possible to apply the approach in practice. First of all, fortunately no epidemic was threatening the region in question at the moment of the research. But also, in order to apply the approach in practice, a series of simulations would have to be run beforehand, as epidemic crisis is a too serious situation to allow for unverified procedures. However, as one of the authors of the present paper works as a consultant in the Crisis Management Center, we have been able to get to know the opinion of practitioners, which was on the whole positive.

Especially the Crisis Scrum Meeting was highly evaluated, as a forced possibility to meet at short intervals, even in hectic crisis situation, in order to communicate and form a general opinion and feeling of the situation, taking into account the views end experiences of the whole team.

Of course, further research and more case studies in the form of exhaustive simulations are needed to develop a well formed Agile approach to crisis management. But generally the Agile approach, and especially the Scrum framework, seem to have much in common with the needs of crisis management: flexibility, cooperation, communication, trust are in both more important than earlier developed plans. In the management of many crises, like epidemic crises, humans should be at the center, and as it happens, the Agile approach puts humans at the center. Whether these humans are stakeholders of an IT project in the original Scrum framework or human beings in danger in a crisis situation, they are humans and the satisfaction of their current needs must constitute the main goal of the project. It seems that a merger of Agile management and crisis management may help to ensure this in the situations when these needs are really basic – like the needs to rescue life and health. That is why it seems important to continue the research proposed in [3] and in this paper.

References

1. Benabena, F.: A formal framework for crisis management describing information flows and functional structure. Procedia Eng. **159**, 353–356 (2016)
2. Betta, J.: Resistance-conflict-crisis – factors of the triad risk (in Polish). J. Sci. (2013). General Tadeusz Kosciuszko Military Academy of Land Forces, Wrocław
3. Betta, J., Skomra, A.: Agile crisis management. In: Conference Material, IV. Medial International Scientific Conference of the Series "Decisions in Situations of Endangerment", The General T. Kościuszko Military University of Land Forces, Wrocław (2017)
4. Bundy, J., Pfarrer, M.D., Short, C.E., Coombs, W.T.: Crises and crisis management: integration, interpretation, and research development. J. Manag. **43**(6), 1661–1692 (2017)
5. Cruz-Mil, O., et al.: Reassurance or reason for concern: security forces as a crisis management strategy. Tour. Manag. **5**, 114–125 (2016)
6. da Silva Avanzi, D., et al.: A framework for interoperability assessment in crisis management. J. Ind. Inf. Integr. **5**, 26–38 (2017)
7. Dyer, W., Wilkins, A.: Better stories, not Better constructs, to generate better theory: a rejoinder to Eisenhardt. Acad. Manag. Rev. **16**, 613–619 (1991)
8. Jaques, T.: Issue management as a post-crisis discipline: identifying and responding to issue impacts beyond the crisis. J. Public Aff. **9**(1), 35–44 (2009)
9. Kuchta, D., Skowron, D.: Traditional versus agile scheduling and implementation of R&D projects: a case study. In: Vopava, J., et al. (eds.) Proceedings of AC 2017 International Conferences, pp. 622–630. MAC Prague, Prague (2017)
10. Kuchta, D., Skowron, D.: Scheduling of high uncertainty projects. In: Grosicki, R., et al. (eds.) Decisions in Situations of Endangerment: Interdisciplinarity of the Decision Making Process, pp. 50–61. Publishing House of the General Tadeusz Kościuszko Military Academy of Land Forces (2017)
11. Manifesto for Agile Software Development. http://agilemanifesto.org/
12. Pearson, C.M., Clair, J.A.: Refraining crisis management. Acad. Manag. Rev. **23**(1), 59–76 (1998)

13. Stanek, S., Drosio, S.: A hybrid decision support system for disaster/crisis management. In: 16th IFIP WG8.3 International Conference on Decision Support Systems, pp. 279–290. IOS Press, Amsterdam (2012)
14. Sysło, M., Deo, N., Kowalik, J.S.: Discrete Optimization Algorithms: with Pascal Programs. Dover Books on Computer Science, New York (2006)
15. Szőke, Á.: Conceptual scheduling model and optimized release scheduling for agile environments. Inf. Softw. Technol. **53**(6), 574–591 (2011)
16. Tena-Chollet, F., et al.: Training decision-makers: Existing strategies for natural and technological crisis management and specifications of an improved simulation-based tool. Saf. Sci. **97**, 144–153 (2016)
17. Vardarlıer, P.: Strategic approach to human resources management during crisis. Procedia-Soc. Behav. Sci. **235**, 463–472 (2016)
18. Wysocki, R.K.: Effective Project Management: Traditional, Agile, Extreme. Wiley Publishing, Indianapolis (2009)

Multiple Criteria Optimization for Emergency Power Supply System Management Under Uncertainty

Grzegorz Filcek[1(✉)], Maciej Hojda[1], and Joanna Gąbka[2]

[1] Faculty of Computer Science and Management, Wroclaw University of Science and Technology, 27 Wyb. Wyspianskiego Street, 50-370 Wroclaw, Poland
grzegorz.filcek@pwr.edu.pl
[2] Faculty of Mechanical Engineering, Wroclaw University of Science and Technology, 27 Wyb. Wyspianskiego Street, 50-370 Wroclaw, Poland

Abstract. The paper deals with a problem of an emergency power supply in the case of a blackout. It needs to be decided which of the power consuming devices should stay active and under what modes of operation. For this purpose a decision making problem with multiple criteria such as cost, systems operation time and priority usage, was formulated. It was assumed that the information about the recovery time is given by an expert in the form of certainty distributions. Then the results were provided under the assumption that the planned execution time is not shorter than the estimated recovery time with a given certainty threshold. This methodology is illustrated with a computational example.

Keywords: Emergency power system · Uncertain variables
Multiple criteria optimization

1 Introduction

Consistent delivery of electrical energy is essential in contemporary living and working environments. Interruptions, caused by power outages, prevent access to modern technological advances, thus reducing the quality of life and preventing a majority of work-related activities [1]. Power outages are commonplace across the world and affect a great number of people [2]. Causes of blackouts include natural disasters such as storms, hurricanes and earthquakes. In the USA, in 2017 alone, four major power outages were caused by storms [3]. Other reasons for power grid failures consist of grid strain and machine wear [4].

Blackouts can vary in size, from highly localized to nationwide. In India, a record number of six hundred million people were left without access to electricity in an aftereffect of a power grid failure in 2012 [5], while simultaneously making a headline in the Washington Post article titled "India blackout, on second day, leaves 600 million without power" [6]. Power outages, due to their prolonged and repeating nature, are destructive to an even greater effect in developing countries, where they have a negative impact on education, health, economic growth and security [7].

J. Świątek et al. (Eds.): ISAT 2018, AISC 853, pp. 196–206, 2019.
https://doi.org/10.1007/978-3-319-99996-8_18

Typically, a power facilities try to ensure that energy demand and supply are balanced while simultaneously guaranteeing that power generators are under their full capacity. Growing power consumption puts an additional strain on the grid, increasing the risk of overload. There are intelligent systems developed, that are designed to prevent large-scale emergencies in power systems [8–10], unfortunately, they do not cover all the situations, when power loss appears. Use of alternative power sources such as solar and wind are the cause of additional liabilities due to their unpredictable character [11]. In consequence, power outages are more frequent nowadays, which leads to the situations in which it is justified and even necessary to invest into an emergency power distribution system. The system would ensure continuous power delivery to high priority receivers by reducing the load caused by low priority ones. High priority receivers are those, for which stable performance is essential to prevent the loss of human life or property damage. This includes facilities such as hospitals, mines, air traffic control centers, manufacturing systems, data centers and scientific laboratories.

The paper is divided into 3 sections. It starts with this introduction, followed by the problem formulation. The closing section describes the numerical experiment.

1.1 Emergency Power Supply System Outline

The emergency power supply system consists of independent sources of electrical energy that support important electrical devices upon loss of normal power supply. The system may include standby generators, batteries and other apparatus. Central part of the system is a management unit which optimizes power usage by enabling or disabling power receivers and routing the energy to them. The system's main objective is to ensure a high level of operations during the recovery time in the case of a blackout [12, 13].

Recovery time is typically unknown and is dependent on multiple factors such as the cause of the blackout and the workforce available for repairs. Due to different character of circumstances leading to a blackout it is often difficult to estimate, using historic data, how long will the power supply recovery take. In such cases, the estimation is often relegated to an expert who provides uncertain information about the recovery time. This information may be modeled with the use of uncertain variables [14, 15].

1.2 Uncertain Variables

Uncertain variables were introduced and developed by Zdzislaw Bubnicki in series of publications [14, 15].

The definition of an uncertain variable $\bar{x}$ refers to two soft properties (such properties $\varphi(x)$ for which the logic value of $v[\varphi(x)]$ belongs to a non-crisp interval of [0, 1]): "$\bar{x} \cong x$" which means "$\bar{x}$ is approximately equal to x" or "$\bar{x}$ is the approximate value of x", and "$\bar{x} \tilde{\in} D_x$" which means "$\bar{x}$ approximately belongs to the set D_x" or "the approximate value of $\bar{x}$ belongs to D_x". The uncertain variable $\bar{x}$ is defined by a set of values X, the function $h(x) = v(\bar{x} \tilde{=} x)$ called certainty distribution (i.e. the certainty index that $\bar{x} \tilde{=} x$, given by an expert), and the following definitions for $D_x, D_1, D_2 \subseteq X$:

$$v(\bar{x} \tilde{\in} D_x) = \begin{cases} \max\limits_{x \in D_x} h(x) & \text{for} \quad D_x \neq \varnothing \\ 0 & \text{for} \quad D_x = \varnothing \quad \text{(empty set)}, \end{cases} \tag{1}$$

$$v(\bar{x} \tilde{\notin} D_x) = 1 - v(\bar{x} \tilde{\in} D_x), \tag{2}$$

$$v(\bar{x} \tilde{\in} D_1 \vee \bar{x} \tilde{\in} D_2) = \max\{v(\bar{x} \tilde{\in} D_1), v(\bar{x} \tilde{\in} D_2)\}, \tag{3}$$

$$v(\bar{x} \tilde{\in} D_1 \wedge \bar{x} \tilde{\in} D_2) = \begin{cases} \min\{v(\bar{x} \tilde{\in} D_1), v(\bar{x} \tilde{\in} D_2)\} & \text{for} \quad D_1 \cap D_2 \neq \varnothing \\ 0 & \text{for} \quad D_1 \cap D_2 = \varnothing. \end{cases} \tag{4}$$

It is assumed that $\max\limits_{x \in X} h(x) = 1$.

1.3 Management of Emergency Power Supply System Under Uncertainty

Management of the emergency power supply system under uncertainty consists of a series of decisions made in order to protect the high priority devices from failure due to the loss of power. The system must first decide which power receivers can remain enabled, then it must decide how to allocate the power to these receivers. The decision must take into consideration all the given constraints regarding power production capabilities of the emergency system. It is assumed that to provide power to a device means providing at least a minimal level of power that ensures the correct functioning of that device. However, increasing the level of power can affect the quality of the work performed by the device. The decisions can be evaluated using the following quality criteria: *the priority at which the receivers run, the time of system operation* before the resources become exhausted (or insufficient to sustain the minimal levels of power), and *the cost of using power resources*.

In the case when the estimated time of the power supply recovery is not known, it is only possible to ensure that the system will work at minimum power levels – providing power only to essential devices. If the information about the recovery time is given, it can be used to make power allocation decisions which can put the system (the receivers) into higher power modes, thus increasing the quality/usability of the system and reducing the relative cost of using power resources.

Recovery time is assumed to be provided in the form of a certainty distribution (certainty index that $\bar{x}$ is approximately equal to the amount of time at which the primary power source is restored). The emergency power supply system can run at higher or lower levels of power consumption as long as its estimated operation time will not invalidate a given certainty threshold ($v(\bar{x} \tilde{\in} D_x) \geq \alpha$, $D_x = \{x \in X : x \leq t\}$, where t is estimated recovery time and α is the certainty threshold). The full decision-making problem formulation is given in the next section.

In fields where human life and property are likely to receive damage, the majority of experts tend to formulate their knowledge in a conservative way. This minimizes their liability in case their estimations provide to be too optimistic. In our case, this almost guarantees that the provided value of time until power failure is resolved will be more than sufficient. This is likely to affect the results of the proposed emergency power supply system by decreasing the number of enabled devices or generating higher costs. To alleviate this problem, the decision maker can influence the decision making process by adjusting the certainty threshold to a lower value. Setting the expected certainty below the maximum value can result in decrease of the execution costs and/or increase of the number of enabled devices. However, it increases the risk that the repair time will exceed the emergency system uptime. Selected value of the certainty level should be based on the type of the application, where more power critical situations such as providing power in a hospital yields a different value than a less power critical situations such as providing power to a private household.

2 Problem Statement

Decision making problem formulated in this paper consists of mode selection for a number of power consuming devices (controlled electrical circuits) in a situation of temporary failure of the primary power source. This selection has to ensure that most important devices continue their work until the primary power source is restored. To meet this goal, it is necessary to manage the costs of power consumption and efficiently use the expert's knowledge regarding the power failure duration.

On one hand, it is assumed that the receivers can work in a finite number of modes, where each mode is characterized by different values of power consumption and work priorities. For example, a ventilation system has several settings of fan speed, a lighting system can provide light selectively (not all lights are on at the same time), a production system can work at different speeds. On the other hand, power suppliers, which are accumulators, aggregates and batteries, use a given amount of non-renewable resource (often shared between several devices). They can also work with different efficiency which influences the amount of used resource and provided energy. Consumption of resources and use of power supplies generates a cost.

Power failure duration (time till the primary power source is restored) is provided by a subjective expert, who gives information about the certainty levels for different possible durations. Expert can base his information on experience, on knowledge of the cause of the power failure, on the general judgment of the situation, and on the conditions and priority of the faulty power line or device. Furthermore, the expert can include objective, historic data, if any is available. It is assumed, that certainty distribution given by the expert has a triangular shape (see Fig. 1).

The goal of decision making is to provide a power supply plan, that is to decide which devices are to remain enabled and in what operation mode, so that the given certainty level can be met. Certainty level is the level of satisfaction of a soft property that the planned execution time is no shorter than the estimated duration of the power failure. Certainty level is calculated based on the knowledge provided by an expert and is in the [0, 1] interval.

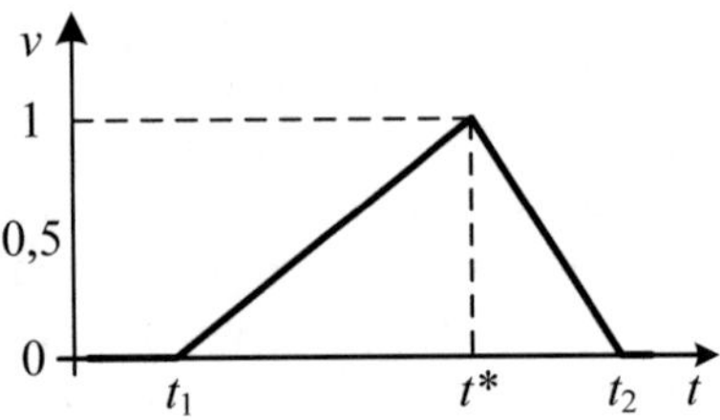

Fig. 1. Triangle certainty distribution

2.1 Notation

The basic notation used in the model is gathered in the following Tables.

Table 1. Parameters referring to power receivers

$L \in \mathrm{N}$	Number of power receivers
$M \in \mathrm{N}$	Maximum number of power receivers modes
$p_{l,m} \in \mathrm{N}_0$	Priority of the power receiver l working in the mode m (It is assumed, that the lower is the variable value, the higher is the priority), element of the matrix $\mathbf{p} = [p_{l,m}]_{\substack{l=1,2,\ldots,L,\\ m=1,2,\ldots,M.}}$
$w_{l,m} \in \mathrm{R}_+$	Energy consumption in time units of the power receiver l working in the mode m (It is assumed that the higher is the mode number the bigger is energy consumption), element of the matrix $\mathbf{w} = [w_{l,m}]_{\substack{l=1,2,\ldots,L,\\ m=1,2,\ldots,M.}}$
$r_{l,m} \in \{0,1\}$	Indicator of availability of the working power receiver l in the mode m, element of the matrix $\mathbf{r} = [r_{l,m}]_{\substack{l=1,2,\ldots,L,\\ m=1,2,\ldots,M.}}$

Auxiliary Variables Definition

$$EC_l(\mathbf{x}) = \sum_{m=1}^{M} w_{l,m} x_{l,m} \tag{5}$$

$$ES_k(\mathbf{y}) = \sum_{z=1}^{Z} s_{z,k}^{\max} y_{z,k}. \tag{6}$$

$$T_k(\mathbf{y}, \mathbf{u}) = \sum_{z \in \{1,2,\ldots,Z\}: h_{z,k} > 0} \frac{\bar{D}_z u_{z,k} h_{z,k}}{f_{z,k}(s_{z,k}^{\max} y_{z,k})}. \tag{7}$$

$$C_k(\mathbf{y}, \mathbf{u}) = \bar{c}_k T_k(\mathbf{y}, \mathbf{u}). \tag{8}$$

$$CS_z(\mathbf{y}, \mathbf{u}) = \sum_{k=1}^{K} \bar{D}_z u_{z,k} \tilde{c}_z, \tag{9}$$

Table 2. Parameters referring to power sources and power resources

$K \in \mathsf{N}$	Number of power sources
$Z \in \mathsf{N}$	Number of resources types
$s_{z,k}^{\min} \in \mathsf{R}_+$	Minimum value of power produced by the power source k, element of the matrix $\mathbf{s}^{\min} = [s_{z,k}^{\min}]_{Z} = 1, 2, \ldots, Z, zk1, 2, \ldots, K.$
$s_{z,k}^{\max} \in \mathsf{R}_+$	Maximum value of power produced by the power source k, element of the matrix $\mathbf{s}^{\max} = [s_{z,k}^{\max}]_{Z} = 1, 2, \ldots Z, k = 1, 2, \ldots, K.$
$h_{z,k} \in \{0, 1\}$	Indicator of possibility of use of the resource z by the power source k, element of the matrix $\mathbf{h} = [h_{z,k}]_{Z} = 1, 2, \ldots, Z, k = 1, 2, \ldots, K.$
$f_{z,k}(s) \in \mathsf{R}_+$	Function of consumption of the resource z by the power source k producing power s, element of the matrix $\mathbf{f}(\mathbf{s}) = [f_{z,k}(s)]_{Z} = 1, 2, \ldots, Z, k = 1, 2, \ldots, K.$
$\bar{c}_k \in \mathsf{R}_+$	Cost of work for the power source k in time unit, element of the matrix $\bar{\mathbf{c}} = [\bar{c}_k]_{k=1,2,\ldots,K}$
$\tilde{c}_z \in \mathsf{R}_+$	Cost of unit of the resource z, element of the matrix $\tilde{\mathbf{c}} = [\tilde{c}_z]_{z=1,2,\ldots,Z}$
$\bar{D}_z \in \mathsf{R}_+$	Available amount of the resource z, element of the matrix $\bar{\mathbf{D}} = [\bar{D}_z]_{z=1,2,\ldots,Z}$

Table 3. Parameters referring to uncertain parameter

$t_1 \in \mathsf{R}_+$	Minimum time needed for the primary power supply restoration, as claimed by the expert
$t_2 \in \mathsf{R}_+$	Maximum time needed for the primary power supply restoration, as claimed by the expert
$t^* \in \mathsf{R}_+$	Time needed for the primary power supply restoration, as claimed by the expert to be the most certain

Table 4. Other parameters

$\alpha \in [0,1]$	Minimum certainty threshold, at which the system may work during the primary power supply failure
$\beta \in \mathsf{R}_+$	Maximum cost of using the emergency system
$\rho \in [0,1]$	Weight of the cost
$\omega \in [0,1]$	Weight of the priority
$\lambda \in [0,1]$	Weight of the time.

$$C_{\max} = \sum_{z=1}^{Z} \bar{D}_z \big(\max_{k \in \{1,2,\ldots,K\}} \big(\frac{\bar{c}_k}{f_{z,k}(s_{z,k}^{\min})} \big) h_{z,k} + \tilde{c}_z \big), \tag{10}$$

$$P_{\max} = \sum_{l=1}^{L} \max_{m \in \{1,2,\ldots,M\}} \{ p_{l,m} r_{l,m} \}, \tag{11}$$

Table 5. Decision variables

$x_{l,m} \in \{0,1\}$	Decision describing if the power receiver l works in the mode $m(x_{l,m} = 1(0)$- receiver l works in mode m (otherwise)), element of the matrix $\mathbf{x} = [x_{l,m}]_{l = 1,2,\ldots,L, \; m = 1,2,\ldots,M}$.
$y_{z,k} \in [0;\, 1]$	Decision describing the utilization level of the power source k with the use of the resource z, element of the matrix $\mathbf{y} = [y_{z,k}]_{z = 1,2,\ldots,Z, \; k = 1,2,\ldots,K}$.
$u_{z,k} \in [0;\, 1]$	Decision describing the allocation of the resource z to the power source k, element of the matrix $\mathbf{u} = [u_{z,k}]_{z = 1,2,\ldots,Z, \; k = 1,2,\ldots,K}$.
$b_{z,k} \in \{0,1\}$	Auxiliary variable indicating if the power source k is running with the use of the resource z, element of the matrix $\mathbf{b} = [b_{z,k}]_{z = 1,2,\ldots,Z, \; k = 1,2,\ldots,K}$.

Table 6. Auxiliary variables and main criteria

$EC_l(\mathbf{x})$	Total energy consumption in time unit by the receiver l	$P_{\max}$	The best priority usage
$ES_k(\mathbf{y})$	Power available at the power source k in time unit	$T_{\max}$	The longest time of system operation
$T_k(\mathbf{y},\mathbf{u})$	Time of power availability at the power source k	$T(\mathbf{y},\mathbf{u})$	Total planned time of system operation
$C_k(\mathbf{y},\mathbf{u})$	Cost of use the power source k in time unit	$C(\mathbf{y},\mathbf{u})$	Total cost
$CS_z(\mathbf{y},\mathbf{u})$	Cost of use of the resource z	$P(\mathbf{x})$	Priority usage
$C_{\max}$	Maximum cost	$Q(\mathbf{x},\mathbf{y},\mathbf{u})$	Objective function

$$T_{\max} = \max_{k \in \{1,2,\ldots,K\}} \sum_{z=1}^{Z} \frac{\bar{D}_z h_{z,k}}{f_{z,k}(s_{z,k}^{\min})}. \tag{12}$$

Evaluation Criteria Definition

$$C(\mathbf{y},\mathbf{u}) = \sum_{k=1}^{K} C_k(\mathbf{y},\mathbf{u}) + \sum_{z=1}^{Z} CS_z(\mathbf{y},\mathbf{u}), \tag{13}$$

$$T(\mathbf{y},\mathbf{u}) = \min_{k \in D_b}\{T_k(\mathbf{y},\mathbf{u})\}, \quad D_b = \{k \in \{1,2,\ldots,K\} : \sum_{z=1}^{Z} b_{z,k} > 0\}. \tag{14}$$

$$P(\mathbf{x}) = \sum_{l=1}^{L}\sum_{m=1}^{M} x_{l,m} p_{l,m}. \tag{15}$$

Table 7. Constraints

Description	Definition
1. Each power source is considered to be running with the use of some resource if it has a positive value of this resource allocated and a positive utilization set	(17)–(20).
2. Each power receiver may work in at most one mode	(21)
3. The resources allocated to the power sources cannot exceed the available amounts	(22)
4. Each power receiver, which has priority with value 0 at any mode is enabled	(23)
5. Certainty level cannot be smaller than the given threshold α, i.e. $v(\tilde{x} \leq T(\mathbf{y}, \mathbf{u})) \geq \alpha$. For the triangular distribution this constraint has the form of (23)	(24)
6. The total cost cannot exceed the given threshold β	(25)
7. The power receivers may work only in the available modes	(26)
8. The amount of energy produced by the power sources must be sufficient to satisfy the demand of the power receivers	(27)
9. Each enabled power source must produce minimal power	(28)
10. A resource may be allocated only to power sources designed to use it	(29)

Objective Function Definition

The following objective function is parameterized by weights ρ, ω, and λ, which correspond to the importance of each criterion and satisfy the relation $\rho + \omega + \lambda = 1$.

$$Q(\mathbf{x}, \mathbf{y}, \mathbf{u}) = \rho \frac{C(\mathbf{y}, \mathbf{u})}{C_{\max}} + \omega(1 - \frac{P(\mathbf{x}) + 1}{P_{\max} + 1}) + \lambda(1 - \frac{T(\mathbf{y}, \mathbf{u})}{T_{\max}}). \tag{16}$$

Constraints Definition

$$\forall_{z \in \{1,2,\ldots,Z\}} \forall_{k \in \{1,2,\ldots,K\}} (u_{z,k} + 1 - b_{z,k} > 0), \tag{17}$$

$$\forall_{z \in \{1,2,\ldots,Z\}} \forall_{k \in \{1,2,\ldots,K\}} (u_{z,k} * (1 - b_{z,k}) = 0), \tag{18}$$

$$\forall_{z \in \{1,2,\ldots,Z\}} \forall_{k \in \{1,2,\ldots,K\}} (y_{z,k} + (1 - b_{z,k}) > 0), \tag{19}$$

$$\forall_{z \in \{1,2,\ldots,Z\}} \forall_{k \in \{1,2,\ldots,K\}} (y_{z,k} * (1 - b_{z,k}) = 0), \tag{20}$$

$$\forall_{l \in \{1,2,\ldots,L\}} \sum_{m=1}^{M} x_{l,m} \leq 1, \tag{21}$$

$$\forall_{z \in \{1,2,\ldots,Z\}} \sum_{k=1}^{K} u_{z,k} \leq 1, \tag{22}$$

$$\forall_{l \in \{1,2,\ldots,L\}} (1 - \sum_{m=1}^{M} x_{l,m}) \leq \min_{m \in \{1,2,\ldots,M\}} \{p_{l,m}\}, \tag{23}$$

$$T(\mathbf{y},\mathbf{u}) \geq \alpha(t^* - t_1) + t_1, \tag{24}$$

$$C(\mathbf{y},\mathbf{u}) \leq \beta, \tag{25}$$

$$\forall_{l=1,2,\ldots,L} \forall_{m=1,2,\ldots,M} \, x_{l,m} \leq r_{l,m}, \tag{26}$$

$$\sum_{l=1}^{L} EC_l(\mathbf{x}) \leq \sum_{k=1}^{K} ES_k(\mathbf{y},\mathbf{u}), \tag{27}$$

$$\forall_{z\in\{1,2,\ldots,Z\}} \forall_{k\in\{1,2,\ldots,K\}} s_{z,k}^{\max} y_{z,k} \geq s_{z,k}^{\min} b_{z,k}, \tag{28}$$

$$\forall_{z\in\{1,2,\ldots,Z\}} \forall_{k\in\{1,2,\ldots,K\}} u_{z,k} \leq h_{z,k}. \tag{29}$$

2.2 Problem Formulation

For the given data: L, K, M, Z, $\mathbf{p}$, $\mathbf{w}$, $\mathbf{r}$, $\mathbf{s}^{\min}$, $\mathbf{s}^{\min}$, $\mathbf{f}(s)$, $\bar{\mathbf{c}}$, $\bar{\mathbf{D}}$, $\tilde{\mathbf{c}}$, t_1, t_2, t^*, α, β, find $\mathbf{x}^*, \mathbf{y}^*, \mathbf{u}^*$ minimizing $Q(\mathbf{x},\mathbf{y},\mathbf{u})$, feasible with respect to (17)-(29), i.e.

$$(\mathbf{x}^*,\mathbf{y}^*,\mathbf{u}^*) = \arg\min Q(\mathbf{x},\mathbf{y},\mathbf{u}). \tag{30}$$

2.3 Solution Procedure

The problem is defined as a mixed integer nonlinear programming (MINLP), where functions $\mathbf{f}(s)$ are linear, variables $\mathbf{y}$ and $\mathbf{u}$ are bounded to be values of less than 100 integers divided by 100 (e.g. for $y_{z,k}$, condition $n = 100y_{z,k}$ must be satisfied for n $\{0,1,\ldots,100\}$). To solve the problem, LINGO solver has been used.

3 Numerical Example

The numerical example presents the use of the model to plan emergency power system usage in case of a primary power supply failure in a small manufacturing company. The company has three main electric circuits that power the production line ($L = 3$). Circuits may work in different modes ($M = 3$) and in each mode a different set of power receivers is enabled, thus the power consumption is different. The first circuit, responsible for ventilation, may run in two modes. In the first mode, where some fans are disabled, the power consumption is $w_{1,1} = 15$ kW and the priority is $p_{1,1} = 0$ (the highest), and in the second mode it is $w_{1,2} = 20$ kW with the priority $p_{1,2} = 4$. This circuit must run in at least one mode. The second circuit, responsible for providing lighting, works in three modes with the following power consumption: $w_{2,1} = 14$, $w_{2,2} = 20$, and $w_{2,3} = 25$ kW, and the respective priorities $p_{2,1} = 1$, $p_{2,2} = 2$, and $p_{2,3} = 5$. The third circuit, responsible for powering production machines, works in only one mode with priority $p_{3,1} = 0$ and consumes $w_{3,1} = 18$ kW (non-zero indicators are $r_{1,1}$, $r_{1,2}$, $r_{2,1}$, $r_{2,2}$, $r_{2,3}$, and $r_{3,1}$). The emergency system consists of three aggregators ($K = 3$), that may produce different amounts of power and use different fuel ($Z = 2$). All devices have linear functions $f_{z,k}(s) = a_{z,k}\ s$ of resource consumption, with coefficient $a_{z,k}$. The first with the maximum power of $s_{1,1}^{\max} = 12$ kW consumes $a_{1,1} = 0.3$ l/kWh of diesel, the second one, with the maximum power of $s_{2,2}^{\max} = 10$ kW consumes $a_{2,2} = 0.41$ l/kWh of petrol, and the third one, with $s_{2,3}^{\max} = 22$ kW, consumes $a_{2,3} = 0.42$ l/kWh of petrol. The minimum power they can provide is $s_{1,1}^{\min} = 6$, $s_{2,2}^{\min} = 6$ and $s_{2,3}^{\min} = 10$ kW. The cost of usage for each aggregate is respectively $\bar{c}_1 = 100$, $\bar{c}_2 = 130$, and $\bar{c}_3 = 200$ PLN (polish zloty) per hour. Fuel available for these power sources is $\bar{D}_1 = 1350$ l of diesel and $\bar{D}_2 = 1400$ l of petrol. The assumed prices of diesel is $\tilde{c}_1 = 5.08$, and petrol $\tilde{c}_2 = 5.12$ PLN per liter. It is further assumed, that the emergency power system is equipped with an UPS subsystem with a battery, which is capable of powering the system at full power consumption for short period of time, which is sufficient to make and execute the decisions concerning the system after a power failure (approx. 15 min). The decision situation is as follows. The system has lost the primary power supply for an unknown period of time. The energy supplier expert has estimated, that the restoration may last from $t_1 = 1$ to $t_2 = 10$ h, but he claims that $t^* = 5$ h is, in his opinion, the most certain time. The manufacturing company owner decides, that it is enough for him if the system will work effectively and for the lowest possible cost for the certainty threshold of at least $\alpha = 0.6$ or as long as possible if the cost will not be greater than $\beta = 6000$ PLN. He also set the weight for the priority use to $\omega = 0.7$ and $\lambda = 0.3$, $\rho = 0$ (for cost minimization $\lambda = 0$, $\rho = 0.3$) to the remaining criteria. With the use of the solver, he gets the following results.

For a case when time of operation was maximized, he gets a solution in which the system works with the lights turned off and with all the fans enabled for about $T(\mathbf{y}^*, \mathbf{u}^*) = 11.38$ h and generates cost of $C(\mathbf{y}^*,\mathbf{u}^*) = 5982.01$ PLN. For a case when the cost is minimized, he gets a solution where the planned time of operation is $T(\mathbf{y}^*, \mathbf{u}^*) = 3.4$ h, but total cost is only $C(\mathbf{y}^*,\mathbf{u}^*) = 1783.49$ PLN. He also checks the solution for a situation when the system operates for a time at least as long as the one which expert provided to be the most certain for power restoration. The results he obtained are as follows: Cost $C(\mathbf{y}^*,\mathbf{u}^*) = 2836.78$ PLN and time $T(\mathbf{y}^*,\mathbf{u}^*) = 5.02$ h. The decisions concerning the modes of circuits operation were the same in every case ($x_{1,2} = x_{3,1} = 1$). After situation analysis, the company owner decided to apply the third solution ($x_{1,2} = x_{3,1} = 1$ $y_{1,1} = 1$, $y_{2,2} = 0.68$, $y_{2,3} = 0.9$, $u_{1,1} = 0.02$, $u_{2,2} = 0.01$, $u_{2,3} = 0.03$, other variable values are zeros). Computation time for a computer equipped with Intel Core i7-6500 CPU, 8 GB RAM, and Windows 10, was not more than 13 s for each experiment.

4 Conclusions

The formalized approach to emergency power management under blackouts was presented. The proposed approach is based on an backup subsystem capable of providing power to power consuming devices over a limited period of time. There was proposed a method of deciding which devices can remain enabled and in what mode of operation to ensure that the continuous operation is possible until the blackout is resolved.

Further work will concern itself with extending the model to include scheduling start times of power sources and the development of an efficient solution algorithm. A decision support system for emergency power systems will then be developed.

References

1. Matthewman, S.D., Byrd, H.: Blackouts: a sociology of electrical power failure. Social Space, pp. 1–25 (2014)
2. Chen, Q., Yin, X., You, D., Hou, H., Tong, G., Wang, B., Liu, H.: Review on blackout process in China Southern area main power grid in 2008 snow disaster. In: IEEE Power and Energy Society General Meeting (2009)
3. Major Power Outage Events. https://poweroutage.us/about/majorevents. Accessed 11 Dec 2017
4. Eleschová, Ž., Beláň, A.: Blackout in the power system. In: Murgaš, J. (ed.) AT&P Journal Plus, pp. 58–60 (2008)
5. Hundreds of millions without power in India, 31 August 2012. http://www.bbc.com/news/world-asia-india-19060279. Accessed 10 Apr 2018
6. Blackout for 19 states, more than 600 million Indians, 31 July 2012. Ndtv.com. Accessed 10 Apr 2018
7. Hachimenum, A.: Impact of power outages on developing countries: evidence from rural households in Niger Delta, Nigeria. J. Energy Technol. Policy **5**(3), 27–38 (2015)
8. Negnevitsky, M., Tomin, N., Panasetsky, D., Kurbatsky, V.: Intelligent approach for preventing large-scale emergencies in electric power systems. In: IEEE International Conference on Electric Power Engineering PowerTech 2013, Grenoble, France, 16–20 2013
9. Negnevitsky, M., Voropai, N., Kurbatsky, V., Tomin, N., Panasetsky, D.: Development of an intelligent system for preventing large-scale emergencies in power systems. In: IEEE/PES General Meeting, Vancouver, BC, Canada, 21–25 2013
10. Baldick, R., et al.: Initial review of methods for cascading failure analysis in electric power transmission systems IEEE PES CAMS task force on understanding, prediction, mitigation and restoration of cascading failures. In: 2008 IEEE Power and Energy Society General Meeting - Conversion and Delivery of Electrical Energy in the 21st Century, Pittsburgh, PA, pp. 1–8 (2008)
11. Chertkov, M., et al.: Predicting failures in power grids: the case of static overloads. IEEE Trans. Smart Grid **2**(1), 162–172 (2011)
12. Makarov, Y.V., Reshetov, V.I., Stroev, A., Voropai, I.: Blackout prevention in the United States, Europe and Russia. Proc. IEEE **93**(11), 1942–1955 (2005)
13. Wang, X., Shao, W., Vittal, V.: Adaptive corrective control strategies for preventing power system blackouts. In: 15th Power Systems Computation Conference, PSCC 2005 Power Systems Computation Conference (PSCC) (2005)
14. Bubnicki, Z.: Uncertain variables and their applications in knowledge-based decision systems: New results and perspectives. Int. J. Intell. Syst. **23**(5), 574–587 (2008)
15. Bubnicki, Z.: Analysis and Decision Making in Uncertain Systems. Springer, London (2004)

Overcoming Challenges in Hybrid Simulation Design and Experiment

Jacek Zabawa(✉) and Bożena Mielczarek

Faculty of Computer Science and Management, Wrocław University of Science and Technology, ul. Ignacego Łukasiewicza 5, 50-371 Wrocław, Poland
jacek.zabawa@pwr.edu.pl

Abstract. The purpose of this paper is to present the concept of modules and interfaces for a hybrid simulation model that forecasts demand for healthcare services on the regional level. The interface, developed with the Visual Basic for Application programming tools for spreadsheets, enables comprehensive planning of simulation experiment for the combined model that operates based on two different simulation paradigms: continuous and discrete-event. This paper presents the capabilities of the developed tools and discusses the results of the conducted experiments. The cross-sectional age-gender specific demographic parameters describing population of two subregions of Lower Silesia were calculated based on historical data retrieved from Central Statistical Office databases. We demonstrated the validity of the developed interface. The model correctly responded to the seasonal increased intensity of patients arrivals to healthcare system.

Keywords: Healthcare services · Simulation modeling
Continuous simulation · Discrete-event simulation · Hybrid simulation

1 Introduction

This paper builds up on our previous study that focuses on the use of combined simulation methods to support healthcare demand predictions [14, 15]. The hybrid model simulates the consequences of the demographic changes, the variability in the incidence rates that result from the population ageing, and the seasonal fluctuations in epidemic trends on the future demand for healthcare services. This in turn may help the healthcare managers to adjust the resources needed to cover the future healthcare needs expressed by the population inhabiting the region.

This is still the on-going project that aims to develop a fully operative hybrid model that combines two simulation approaches: continuous and discrete-event. Hitherto, we were able to solve the "drainage problem" in the aging chain demographic simulation [18] and we proposed a method to eliminate the differences between historical data and simulation results when projecting the population evolution within the predefined time range. This was achieved by designing the hierarchical blocks and increasing the number of elementary cohorts up to 210 elementary one-year male/female items. In our research we were faced however with another challenge. Simulation experiments have revealed that due to the very large number of results (millions of records) it was necessary to

J. Świątek et al. (Eds.): ISAT 2018, AISC 853, pp. 207–217, 2019.
https://doi.org/10.1007/978-3-319-99996-8_19

develop a set of analytical tools for simulation experiment planning. It was also essential to construct the appropriate data sheets for the fast and accurate input/output data analysis, especially when the more advanced sampling methods are applied.

The overall aim of this paper is to present the approach for credible experimental design and output data analysis to be applied in the hybrid simulation model.

2 Theoretical Background - Premises for Hybrid Solution

Literature survey proves that simulation is widely and successfully used in healthcare decision making [7, 10]. Simulation methods may be divided into different categories based on various criteria, whereas practice in the area of health care applications indicates that the most common criterion [2, 9, 12, 13, 16] is related to time perception. According to this criterion, the simulation methods are divided into:

- Monte Carlo techniques which, generally, ignore the passage of time,
- continuous modeling that considers the cause-and-effect relationships, feedback loops, and fixed-interval time steps,
- discrete-event modeling that registers changes caused by individual objects moving through the system and random-interval time steps closely related to state changes occurring in the system,
- agent-based system, the sub-method of discrete modeling with the ability to focus on the behaviors and interactions between particular objects.

The type of the problem determines the simulation approach best fitted to model the issue. For example, when modeling the factors affecting the epidemic health condition [8] or the susceptibility to a given type of disease a continuous approach is preferred. However, to model a performance of a health care facility one usually selects discrete-event approach [7]. Factors that cannot be identified with certainty lead to stochastic simulation techniques [3], i.e. Monte Carlo or discrete-event.

When modelling the performance of health care systems the specific concepts appear: *cohort modelling* is useful to represent the flow of individuals between age-gender groups; *temporal factors* such as hour of day, day of week, month, season, calendar year are helpful to describe the patients arrival rates to facilities; *geographic characteristics* such as the distance to the facility may be used to determine the reaction time needed to effectively provide emergency service. Each of these concepts is usually more closely connected with only one simulation approach. For example, the temporal changes are more easily managed using discrete-event simulation, while cohort modelling is more typical for continuous modelling.

Due to the heterogeneity of approaches used in the simulation of health care systems, hybrid concepts [17] have been developed to combine different methods in one master model [1, 4, 6]. In our study we applied three approaches: Monte Carlo in the context of repetitive experiments and sampling, continuous simulation to model demographic evolutions, and discrete-event method to generate objects representing patients arriving to a healthcare facility with service requests. One of the benefits of such a solution, observed also in our study, is the ability to consider large scale problems, i.e. many millions of patients arriving to health care facilities, [5].

3 Description of the Hybrid Model

3.1 Basic Assumptions

The hybrid model consists of two submodels: continuous model created in accordance with the system dynamics approach and discrete model built in accordance with the discrete-event approach. The continuous model performs demographic simulations for the years 2010–2030 based on historical data for the period 2010–2015 and official governmental forecasts describing the expected population changes. The discrete model uses demographic data from the continuous model, empirical data on hospital admissions drawn from National Health Fund regional branch and the elaborated parameters that describe seasonality trends occurring in patients arrivals.

3.2 Population Model

The first model (see Fig. 1) is essential for predicting population aspects: the population size, the number of births and deaths, migration and growing up processes. The model was built in Extendsim [11] and the detailed description of this model may be found in [18]. In order to increase the clarity of the text the brief recapitulation is given below.

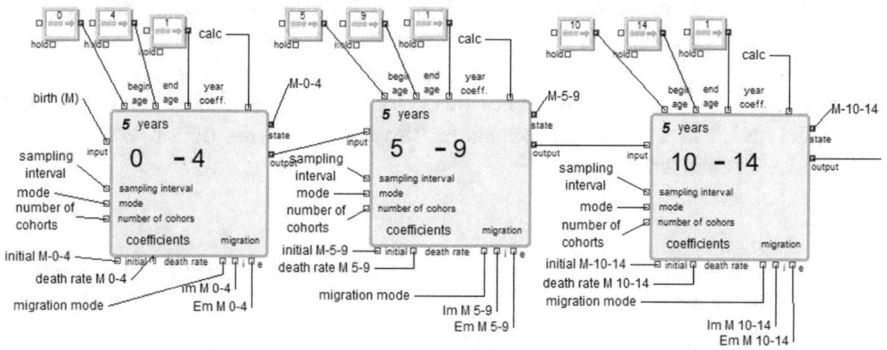

Fig. 1. An excerpt from the first model: continuous simulation approach (young males part).

There are two gender chains, female (F) and male (M), and each chain consists of 18 major cohorts. All major cohorts, except the oldest, consist of 5 elementary cohorts Each chain has two special-type cohorts: marginal left (F_0_4 and M_0_4) and marginal right (F_85+ and M_85+). The youngest cohort (marginal left) interacts with a stream of births (inflow), while the oldest cohort (marginal right) contains a large number of (20) elementary cohorts representing the entire population of the oldest people. Each cohort also interacts with one or two streams such as a growing up stream, a stream of deaths (outflow) and a stream of migration (balance of inflow and outflow). All cohorts with birth, growing up and death streams are situated inside positive or negative feedbacks loops, while migration streams are defined as the proportion of the size of the given cohort or an independent parameter, for example the absolute values

Record #	Time	Year	Month	Day	M_0_4	M_5_9	M_10_14	M_15_19
1	38718	2006	1	1	24829,00	27000,00	32474,00	40307,00
2	39083	2007	1	1	26209,23	26586,40	31343,04	38697,24
3	39448	2008	1	1	27679,91	26153,67	30301,23	37109,93
4	39814	2009	1	1	29228,08	25875,55	29258,94	35582,25
5	40179	2010	1	1	30616,83	25943,23	28375,10	34176,45
6	40544	2011	1	1	31798,83	26442,83	27608,44	32868,80
7	40909	2012	1	1	32876,84	27291,57	27023,86	31678,34
8	41275	2013	1	1	33604,80	28390,78	26672,63	30574,47
9	41640	2014	1	1	34145,97	29601,05	26552,72	29584,94
10	42005	2015	1	1	34456,81	30790,82	26734,86	28727,90
11	42370	2016	1	1	34568,74	31898,90	27214,89	28021,25
12	42736	2017	1	1	33919,78	32856,81	27949,26	27502,24
13	43101	2018	1	1	33169,34	33625,22	28865,35	27202,85
14	43466	2019	1	1	32409,78	34149,16	29878,34	27147,72
15	43831	2020	1	1	31722,49	34373,33	30905,51	27344,77
16	44197	2021	1	1	31145,87	34295,85	31873,39	27781,11
17	44562	2022	1	1	30577,00	33972,28	32715,96	28418,26
18	44927	2023	1	1	29994,10	33489,09	33376,59	29202,37
19	45292	2024	1	1	29383,32	32926,70	33814,56	30069,53
20	45658	2025	1	1	28749,97	32336,44	34014,04	30952,37
21	46023	2026	1	1	28406,20	31740,36	33988,02	31786,65
22	46388	2027	1	1	28072,10	31143,13	33771,70	32512,71
23	46753	2028	1	1	27750,70	30559,82	33411,53	33084,47
24	47119	2029	1	1	27430,52	30017,54	32952,83	33472,82
25	47484	2030	1	1	27110,63	29532,20	32433,16	33667,92

Fig. 2. An excerpt from the results of the continuous simulation model.

of individuals. At the end of demographic simulation we receive multicolumn table that contains the predictions of cohorts sizes in subsequent years (2010–2030), (see Fig. 2).

Our research is based on the situation in a Polish administrative region called the Wrocław Region (WR). The demographic forecasts are usually prepared by the government scientific institutions and take into account the various combinations of population parameters, such as fertility rate (similar to birth rate), mortality (death) rate, life expectancy, rate or number of migrations, which are described using the qualitative categories expressed by: very high, high, average, low, very low.

In our study, we chose Wrocław area population and one of the population forecast option (see Table 1) developed by Polish Government Population Council for 2014–2050 [19] called in our other publications "Scenario 3". This option was randomly selected for examination.

Table 1. The description of demographic parameters. One of the official population forecast, selected for our study.

Variant	Fertility rate	Mortality rate	Life expectancy	Migrations
"No 3"	High	Medium	Medium	Medium

In the further part of the paper we present the elements of the model that considers all the assumptions described above. Our main goal was to develop, implement and validate the operation of the proposed IT solution. It is clear that such a computer tool depends strongly on the structure of input data and during the verification/validation process it is necessary to consider different sets of input data, also coming from our previous research. Such an approach ensures the effectiveness of the whole scientific process. This paper however focuses only on some specific IT solutions and does not aim at the discussion of the results for the complete set of the population forecast options.

3.3 Arrival Model

The second model was built in accordance with the discrete-event approach in Extendsim, too. It contains 36 hierarchical blocks, i.e. the same number as the number of main cohorts. The hierarchical blocks allow us to quickly build large models because each of their identical structure. Hierarchical blocks can be controlled by different parameters and can also represent multiplied stream sources. All outputs of hierarchical blocks are connected to a single output stream (see Fig. 3). In every source cohort the specific object (i.e. service request) has been assigned an attribute value "cohort number" that enables us to recognize the source cohort of that object. Each object may be linked to the particular parameter of random distribution, such as the service time or the code of the disease.

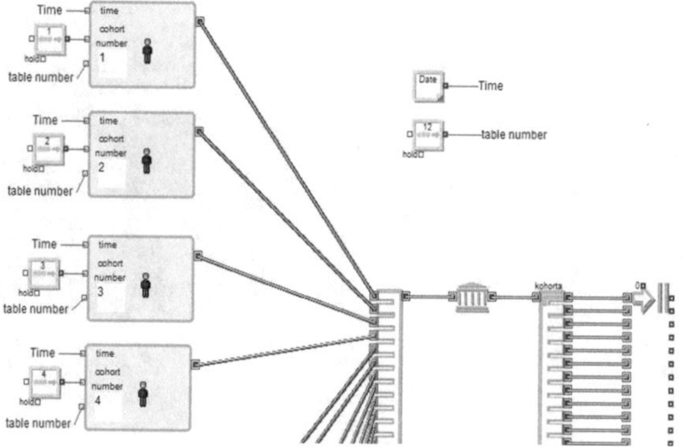

Fig. 3. An excerpt from the structure of the discrete-event model – "hierarchical blocks" level.

The data describing cohorts are read from the tables, for example the parameters of random exponential distributions that define time between subsequent requests. The simulated size of the population in a given cohort is calculated based on the intensity of the requests (historical, monthly) and the size of the population (historical, yearly).

3.4 Integration of the Models

One of the challenges to be overcome when trying to integrate two different simulation approaches in one master model is the issue of mutual communication between two sub-models. The monthly intensity of patients arrivals is associated not only with historical monthly data from 2010 but also with the sizes of population cohorts in the end of the simulated year. Therefore, theoretically, each year should be simulated twice. First, the number of arriving patients should be generated according to the parameters describing every cohort and second, the simulation should be repeated using the coefficient calculated previously.

The next challenge to be overcome when performing hybrid simulation is ensuring the compliance between the deterministic continuous simulation and the stochastic discrete approach. In case of deterministic simulation repetitions are unnecessary.

Hence it seems that the best option is to prepare the demographic forecasts by the continuous model and store the results in the external table which is at the same time the “input” table for discrete model.

3.5 Spreadsheet Interface

We have developed the spreadsheet interface (MS Excel) to enable the fast and accurate modification of the parameters necessary to define the seasonality of patients arrivals. The user first selects a range of months for which the modified seasonality will be applied. In the next step the cohorts for the modifications are selected. Some cohorts may be excluded from the modification. For example, one can select months from February to April and only men’s cohorts (see Fig. 4). It is also possible to select only cohorts with the highest number of arrivals (see Fig. 5).

	coefficient				M_0_4	M_5_9	M_10_14	M_15_19
X	2	start; end	2	5	1	1	0	1
40179	31	2010	1	1	0,016434	0,011109	0,010292	0,011789
40210	28	2010	2	1	0,049167	0,030479	0,012055	0,025861
40238	31	2010	3	1	0,060818	0,044756	0,017633	0,028886
40269	30	2010	4	1	0,051995	0,033046	0,01605	0,026032
40299	31	2010	5	1	0,054218	0,044275	0,019036	0,034823
40330	30	2010	6	1	0,026503	0,02318	0,018353	0,01804
40360	31	2010	7	1	0,027277	0,018689	0,012739	0,018982
40391	31	2010	8	1	0,020475	0,01171	0,012307	0,017041
40422	30	2010	9	1	0,022731	0,014879	0,018317	0,014072
40452	31	2010	10	1	0,024415	0,014959	0,019108	0,0147
40483	30	2010	11	1	0,022563	0,012192	0,015654	0,013929
40513	31	2010	12	1	0,026031	0,014277	0,013135	0,012474
40544	31	2011	13	1	0,026031	0,014277	0,013135	0,012474

Fig. 4. An excerpt from the MS Excel interface. The aim is to indicate the largest stream intensity values (the smallest mean in random an exponential distribution) then decrease its intensity (coefficient X = 2) in months from 2 to 5 (“start; end”) for cohorts M_0_4, M_5_9 and M_15_19.

Due to the very large number of resulting output records, i.e. millions of records that contain information about the time arrival and cohort’s number, we have also developed an analytical tool in MS Excel spreadsheet to easily observe patients arriving in particular months.

Year	Month	Number of cases (arrivals) M_0_4	M_5_9	M_10_14	M_15_19	M_20_4	M_25_29
2010	1	488	277	286	413	347	489
2010	2	730	380	335	453	386	514
2010	3	903	558	490	506	550	574
2010	4	772	412	446	456	459	469
2010	5	805	552	529	610	551	683
2010	6	787	578	510	632	550	650
2010	7	810	466	354	665	512	784
2010	8	608	292	342	597	428	627
2010	9	675	371	509	493	506	548
2010	10	725	373	531	515	550	680
2010	11	670	304	435	488	456	546
2010	12	773	356	365	437	391	514

Fig. 5. An excerpt from the historical number of arrivals (2010). The months with the highest frequencies are highlighted.

4 Simulation Experiment

4.1 Basic Assumptions

Simulation experiments were conducted according to the demographic scenario described earlier (see Table 1). We decided to study the effects of changes in the values of the seasonality indicators on the intensity of patients arriving to healthcare facilities. It seems that the seasonality is caused by the variabilities in morbidity trends separately for different cohorts during the year. We will conduct the research on the impact of hypothetical changes in the seasonal morbidity trends on the intensity of simulated arrivals to the healthcare system.

4.2 Results and Discussion

We propose coefficient C as the independent variable and the historical intensity from 2010 as the reference intensity. By multiplying the reference value by the value of parameter C we would like to increase the number of arrivals in a specified month for a specified cohort.

The formula for calculations is as follows (1):

$$\begin{aligned} &\textit{Parameter of exponential random distribution (time between arrivals) [hours]} \\ &= (1 / \textit{historic number of arrivals in a given month in a given cohort}) \\ &* (\textit{given cohort size in } 2010\ \textit{year/cohort size}) \\ &* (\textit{number of hours in this month in } 2010 / \textit{number of} \\ &\quad \textit{hours in a given month in a given year}) \end{aligned} \tag{1}$$

In our experiment the coefficient C is selected as the independent variable. For each cohort we found a month with the highest number of historical (2010) arrivals (see Fig. 5): for both gender the month of July is described by the highest numbers of arriving patients. This intensity is particularly often observed for the cohorts of the age groups from 25 to 50 years. Historical data reveal also that the highest number of women older than 60 years registers in healthcare facilities in March. Several experiments were performed in order to check the conformity of the model with the historical data.

The coefficient C is multiplied by the number of arrivals in each of the highlighted month. The intensity of arrivals throughout the period from 2010 to 2030 relative to historical intensity (year 2010) was multiplied by a constant value in the range of 0.1 to 2 with step 0.1. This means that, for example, the coefficient C = 2 in the March 2010 causes that in simulated March 2010 we have 1620 arrivals in a cohort M_0_4 instead of 903 (but on average 1806). Figure 6 shows the simulated arrivals with the increased intensity (C = 2).

		Number of cases (simulated)					
year	month	M_0_4	M_5_9	M_10_14	M_15_19	M_20_24	M_25_29
2015	1	404	234	299	519	405	489
2010	2	593	321	353	563	509	514
2010	3	1620	483	538	626	660	574
2010	4	703	333	475	538	554	469
2010	5	680	477	598	770	1354	683
2010	6	678	974	511	771	656	650
2010	7	712	393	379	1676	564	784
2010	8	535	237	339	696	517	627
2010	9	579	319	541	565	651	548
2010	10	634	280	1107	633	637	680
2010	11	533	257	504	610	611	546
2010	12	700	337	391	550	486	514

Fig. 6. An excerpt from the simulated number of arrivals (2015). The value of factor C was increased (C = 2). The months with the highest frequencies are highlighted. It should be noted that the values in the table are affected not only by seasonality but also by changes in the population size as a result of continuous system simulation.

The parameters of interarrival time distributions were calculated based on the historical number of arrivals in the year 2010, separately for each calendar month and each age-gender cohort. The sizes of cohorts are extracted from historical data or – beyond the range of historical data – from the demographic simulation model.

The relationship between the total number of simulated arrivals (2010–2030) and the changing value of coefficient C is demonstrated in Fig. 7. The interesting phenomenon was observed. As expected, the higher coefficient C leads to the higher number of arrivals (in a statistical sense), however a noticeable irregularity can be seen when the small value of coefficient C is applied, i.e. C = 0.1. Smaller values of C reduce to almost zero the significance of the previously leading month. In Fig. 8 the histogram of the simulated frequency distribution of interarrival times resulting from the changes in coefficient C, is presented. The observed interarrival times are consistent with historical data (almost 200,000 arrivals in one year), the histogram shape corresponds to the exponential distribution and the basic statistics (as variance, not shown in the paper) are very close to each other within the tested range of coefficient C values (the experiment assumes that the Poisson process parameters change only at the turn of the month).

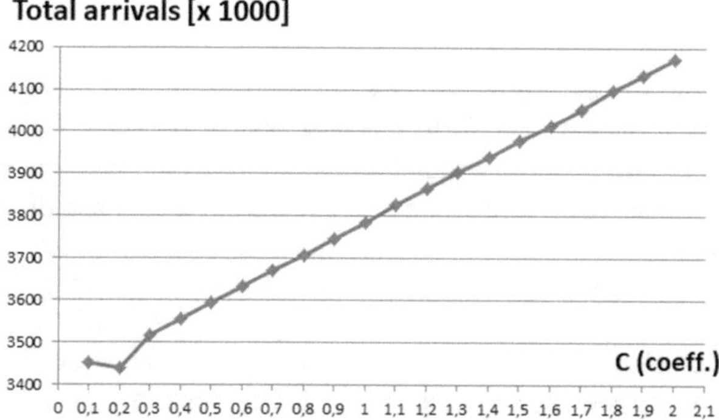

Fig. 7. The number of arrivals in the function of coefficient C. The almost linear relationship.

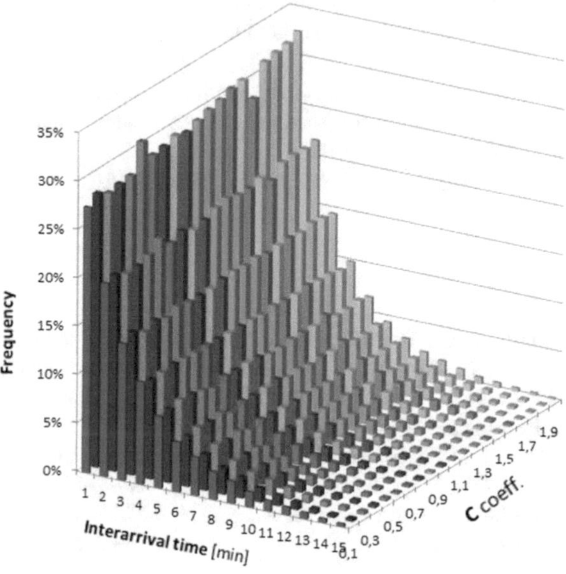

Fig. 8. The histogram of the frequency distribution of the simulated intervals in the function of the coefficient C.

The output values generated by the model are consistent with the historical data. The growth of C coefficient increases the intensity of the simulated arrival stream and the overall number of simulated arrivals to healthcare facilities. It can also be observed that a histogram of the interarrival time distribution preserves its original exponential pattern, however the parameters follow step-wise changes according to monthly seasonality and the trend of population size generated by the continuous model. The slight differences observed in the simulated values are the result of the fact that changes were introduced only in one month.

5 Summary

Our contribution is summarized as follows. We have developed the hybrid simulation model composed of two sub-models which were elaborated using different simulation paradigms. Both models, i.e. the population model based on the continuous approach and the arrivals model built with discrete-event methodology, are created on one IT platform (Extendsim). The overall aim of the simulation was to forecast future demand for healthcare services, taking into account the probable demographic changes. We were faced with the challenge of overcoming a large number of data, resulting from the experiments, when planning and conducting the simulation. The MS Excel interface (in VBA language) was developed to overcome these difficulties.

We performed the series of experiments to check the consistency of results with the assumption that the seasonality of incidences overlaps on population trends. We managed to demonstrate the correct response of the model to the modifications made on the independent variable. The modified values of the parameter C influenced the intensity of arrivals however the seasonal character of the arrivals was maintained.

Acknowledgements. This project was financed by the grant *Simulation modelling of the demand for healthcare services* from the National Science Centre, Poland, and was awarded based on the decision 2015/17/B/HS4/00306.

References

1. Balaban, M.: Return to work behavior of people with disabilities: a multi-method approach. In: Tolk, A., Diallo, S., Ryzhov, I., Yilmaz, L., Buckley, S., Miller, J. (eds.) Winter Simulation Conference 2014, pp. 1561–1572. Institute of Electrical and Electronics Engineers Inc., Piscataway (2014)
2. Brailsford, S., Harper, P., Patel, B., Pitt, M.: An analysis of the academic literature on simulation and modelling in health care. J. Simul. **3**(3), 130–140 (2009)
3. Cardoso, T., Oliveira, M., Barbosa-Póvoa, A., Nickel, S.: Modeling the demand for long-term care services under uncertain information. Health Care Manag. Sci. **15**(4), 385–412 (2012)
4. Crowe, S., Gallivan, S., Vasilakis, C.: Informing the management of pediatric heart transplant waiting lists: complementary use of simulation and analytic modeling. In: Yilmaz, L., Chan, W., Moon, I., Roeder, T., Macal, C., Rossetti, M. (eds.) Winter Simulation Conference 2015, pp. 1654–1665. Institute of Electrical and Electronics Engineers Inc., Piscataway (2015)
5. Djanatliev, A., German, R.: Towards a guide to domain-specific hybrid simulation. In: Yilmaz, L., Chan, W., Moon, I., Roeder, T., Macal, C., Rossetti, M. (eds.) Winter Simulation Conference 2015, pp. 1609–1620. Institute of Electrical and Electronics Engineers Inc., Piscataway (2015)
6. Gao, A., Osgood, N., An, W., Dyck, R.: A tripartite hybrid model architecture for investigating health and cost impacts and intervention tradeoffs for diabetic end-stage renal disease. In: Tolk, A., Diallo, S., Ryzhov, I., Yilmaz, L., Buckley, S., Miller, J. (eds.) Winter Simulation Conference 2014, pp. 1676–1687. Institute of Electrical and Electronics Engineers Inc., Piscataway (2014)

7. Gul, M., Guneri, A.: A comprehensive review of emergency department simulation applications for normal and disaster conditions. Comput. Ind. Eng. **83**, 327–344 (2015)
8. Homer, J., Hirsch, G.: System dynamics modeling for public health: background and opportunities. Am. J. Public Health **96**(3), 452–458 (2006)
9. Kasaie, P., Kelton, W., Vaghefi, A., Naini, S.: Toward optimal resource allocation for control of epidemics: an agent-based simulation approach. In: Johansson, B., Jain, S., Montoya-Torres, J., Hugan, J., Yücesan, E. (eds.) Winter Simulation Conference 2010, pp. 2237–2248. Institute of Electrical and Electronics Engineers Inc., Piscataway (2010)
10. Katsaliaki, K., Mustafee, N.: Applications of simulation within the healthcare context. J. Oper. Res. Soc. **62**(8), 1431–1451 (2011)
11. Krahl, D.: Extendsim: a history of innovation. In: Laroque, C., Himmelspach, R., Pasupathy, R., Rose, O., Uhrmacher, A. (eds.) Winter Simulation Conference 2012. Institute of Electrical and Electronics Engineers Inc., Piscataway (2012)
12. Marshall, D., Burgos-Liz, L., IJzerman, M., Crown, W., Padula, W., Wong, P., Pasupathy, K., Higashi, M., Osgood, N.: Selecting a dynamic simulation modeling method for health care delivery research – part 2: report of the ISPOR dynamic simulation modeling emerging good practices task force. Value Health **18**(2), 147–160 (2015)
13. Mielczarek, B., Uziałko-Mydlikowska, J.: Application of computer simulation modeling in the health care sector: a survey. Simul. Trans. Soc. Model. Simul. Int. **88**(2), 197–216 (2012)
14. Mielczarek, B., Zabawa, J.: Simulation model for studying impact of demographic, temporal, and geographic factors on hospital demand. In: Chan, W., D'Ambrogio, A., Zacharewicz, G., Mustafee, N., Wainer, G., Page, E. (eds.) Winter Simulation Conference 2017, pp. 4498–4500. Institute of Electrical and Electronics Engineers Inc., Piscataway (2017)
15. Mielczarek, B., Zabawa, J.: Healthcare demand simulation model. In: Nolle, L., Burger, A., Tholen, C., Werner, J., Wellhausen, J. (eds.) 32nd European Conference on Modelling and Simulation 2018, pp. 53–59. European Council for Modelling and Simulation (2018)
16. Sobolev, B.G., Sanchez, V., Vasilakis, C.: Systematic review of the use of computer simulation modeling of patient flow in surgical care. J. Med. Syst. **35**(1), 1–16 (2011)
17. Viana, J.: Reflections on Two Approaches to hybrid simulation in healthcare. In: Tolk, A., Diallo, S., Ryzhov, I., Yilmaz, L., Buckley, S., Miller, J. (eds.) Winter Simulation Conference 2014, pp. 1585–1596. Institute of Electrical and Electronics Engineers Inc., Piscataway (2014)
18. Zabawa J., Mielczarek B., Hajłasz M.: Simulation approach to forecasting population ageing on regional level. In: Wilimowska, Z., Borzemski, L., Świątek, J. (eds.) ISAT 2017, Advances in Intelligent Systems and Computing 657, Part III, pp. 184–196. Springer (2017)
19. Rządowa Rada Ludnościowa. http://bip.stat.gov.pl/organizacja-statystyki-publicznej/rzadowa-rada-ludnosciowa/

Medium-Term Electric Energy Demand Forecasting Using Generalized Regression Neural Network

Paweł Pełka and Grzegorz Dudek(✉)

Department of Electrical Engineering, Czestochowa University of Technology, Al. Armii Krajowej 17, 42-200 Czestochowa, Poland
dudek@el.pcz.czest.pl

Abstract. Medium-term electric energy demand forecasting is becoming an essential tool for energy management, maintenance scheduling, power system planning and operation. In this work we propose Generalized Regression Neural Network as a model for monthly electricity demand forecasting. This is a memory-based, fast learned and easy tuned type of neural network which is able to generate forecasts for many subsequent time-points in the same time. Time series preprocessing applied in this study filters out a trend and unifies input and output variables. Output variables are encoded using coding variables describing the process. The coding variables are determined on historical data or predicted. In application examples the proposed model is applied to forecasting monthly energy demand for four European countries. The model performance is compared to performance of alternative models such as ARIMA, exponential smoothing, Nadaraya-Watson regression and neuro-fuzzy system. The results show high accuracy of the model and its competitiveness to other forecasting models.

Keywords: Generalized Regression Neural Network
Medium-term load forecasting · Pattern-based forecasting

1 Introduction

Medium-term load forecasting (MTLF) provides useful information for energy management, maintenance scheduling, power system planning and operation. It includes forecasts from one month to several years. In competitive markets, where energy is traded, the accurate forecast of monthly, quarterly and yearly energy demands can provide an advantage in negotiations and concluding contracts for medium term generation, transmission and distribution.

The mid-term electric load as a function of time has a complex nonlinear behavior. It expresses a trend following the economic and technological development of a country, yearly seasonality corresponding to climatic factors and weather variations and random component disturbing the time series.

In literature MTLF methods can be categorized into two general groups [1]. The first one includes the conditional modeling approach and focuses on economic analysis, management and long term planning energy load and energy policies. As input

J. Świątek et al. (Eds.): ISAT 2018, AISC 853, pp. 218–227, 2019.
https://doi.org/10.1007/978-3-319-99996-8_20

information are considered: historical load data, weather factors, economic indicators and electrical infrastructure measures. A MTLF model of this type can be found in [2], where macroeconomic indicators, such as the consumer price index, average salary earning and currency exchange rate are taken into account as inputs.

The second group includes the autonomous modeling approach, which requires a smaller set of inputs: primarily past loads and weather variables. This approach is more suited for stable economies. The forecasting methods applied in this approach are classical methods such as ARIMA or linear regression [3], and computational intelligence methods, such as neural networks [4].

Neural networks have many attractive features, such as: universal approximation property, learning capabilities, massive parallelism, robustness in the presence of noise, and fault tolerance. They are often use to modeling of complex, nonlinear problems such as MTLF [1, 2]. In this work we propose MTLF model based on Generalized Regression Neural Network (GRNN). This is a memory-based, fast learned and easy tuned type of neural network which is able to generate forecasts for many subsequent time-points in the same time. Time series preprocessing applied in this study filters out the trend and unifies input and output variables. Output variables are encoded using coding variables describing the process. ARIMA and exponential smoothing models are applied for prediction of coding variables.

The rest of this paper is organized as follows. In Sect. 2 we define a forecasting model based on GRNN describing network architecture and learning, and data preprocessing methods. In Sect. 3 we test the model on real load data. We compare results of the proposed methods to other MTLF methods. Finally, Sect. 4 concludes the paper.

2 Forecasting Model Based on GRNN

2.1 GRNN

GRNN is a type of supervised neural network with radial basis activation functions. It was introduced by Specht in 1991 [5] as a memory-based network that provides estimates of continuous variables. In comparison of other NN types, where data are propagated forward and backward many times until an acceptable error is found, in GRNN data only needs to propagate forward once. Thus, the training of GRNN is very fast. Other advantages of GRNN are: easy tuning, highly parallel structure and smooth approximation of a target function even with sparse data in a multidimensional space.

The GRNN architecture in Fig. 1 is shown. The network is composed of four layers: input, pattern (radial basis layer), summation and output. The input layer distributes inputs x_j without processing to the next layer. In the pattern layer nonlinear transformation is applied to the inputs. Each neuron of this layer uses a radial basis function which is commonly taken to be Gaussian:

$$G_i(\mathbf{x}) = \exp\left(-\frac{\|\mathbf{x} - \mathbf{x}_i\|^2}{s_i^2}\right) \tag{1}$$

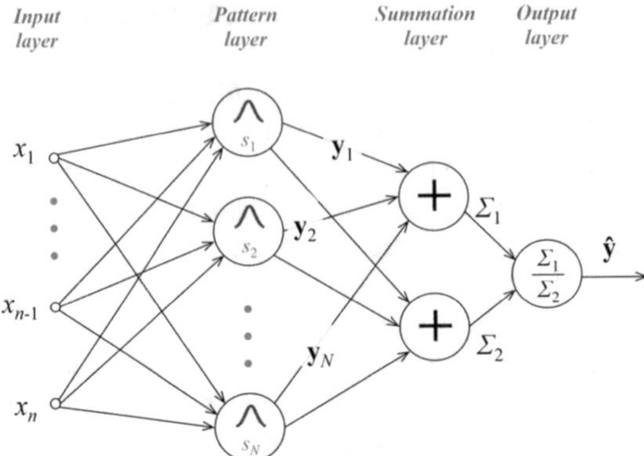

Fig. 1. GRNN architecture.

where: $\mathbf{x}_i$ is the i-th learning sample which is a center vector of the Gaussian function, s_i is a smoothing parameter and $\|.\|$ is a Euclidean norm.

Each neuron represents individual training vector. Its output expresses the similarity between the input vector $\mathbf{x}$ and the i-th training vector. So the pattern layer maps the n-dimensional input space into N-dimensional space of similarity, where N is the number of training vectors.

The summation layer contains two neurons. The first one calculates the sum Σ_1 of the target patterns $\mathbf{y}_i$ weighted by the neuron outputs, whiles the second one calculates the arithmetic sum Σ_2 of the pattern layer outputs.

The GRNN output calculated by the output layer neuron expresses the weighted sum of the target patterns $\mathbf{y}_i$:

$$\widehat{\mathbf{y}} = g(\mathbf{x}) = \frac{\sum_{i=1}^{N} G_i(\mathbf{x})\mathbf{y}_i}{\sum_{i=1}^{N} G_i(\mathbf{x})} \tag{2}$$

Note that the lower distance between $\mathbf{x}$ and $\mathbf{x}_i$ entails the higher i-th neuron output and consequently the higher contribution of the target pattern $\mathbf{y}_i$ to the sum (2).

A smoothing parameter s is the only parameter to estimate. It determines the smoothness of the fitted function and generalization performance of the model. When s becomes larger, the neuron output increases (weights for $\mathbf{y}_i$ in (2) are bigger), with the result that the fitted function becomes smoother. Smoothing parameter s can be the same for all neurons or individually adjusted for each neuron. Finding the optimal smoothing parameter value is a key issue in GRNN learning. In [6] for adjusting s, the same for all neurons, simple enumerative method was used. In [7] for searching N-dimensional space of smoothing parameters a differential evolution algorithm was applied. In this study we assume the same s for all neurons calculated as $s = 0.02{\cdot}l{\cdot}\text{median}(\mathbf{x})$, where median($\mathbf{x}$) is the median of pairwise distances between learning x-patterns and l is tuned by enumerating.

2.2 Time Series Preprocessing

Vector **x** called an input pattern represents predictors, and vector **y** called an output pattern represents the forecasted time series fragment. The input pattern is an n-component vector representing a time series fragment preceding the forecasted fragment. Let us denote the forecasted fragment by $Y_i = \{E_{i+1}\ E_{i+2}\ \ldots\ E_{i+m}\}$, and the preceding fragment by $X_i = \{E_{i-n+1}\ E_{i-n+2}\ \ldots\ E_i\}$, where E_k is the monthly energy consumption and k is the time index. An input pattern $\mathbf{x}_i = [x_{i,1}\ x_{i,2}\ \ldots\ x_{i,n}]^T$ represents the fragment X_i. Components of this vector are preprocessed points of the sequence X_i. Different preprocessing methods are considered [8]:

$$x_{i,t} = E_{i-n+t} \tag{3}$$

$$x_{i,t} = \frac{E_{i-n+t}}{\bar{E}_i} \tag{4}$$

$$x_{i,t} = E_{i-n+t} - \bar{E}_i \tag{5}$$

$$x_{i,t} = \frac{E_{i-n+t} - \bar{E}_i}{D_i} \tag{6}$$

where $t = 1, 2,\ldots, n$, $\bar{E}_i$ is the mean value of the sequence X_i, and $D_i = \sqrt{\sum_{j=1}^{n} (E_{i-n+j} - \bar{E}_i)^2}$ is a measure of their dispersion.

Pattern components defined using (3) are the same as elements of the sequence X_i. Pattern components defined using (4) are the points of the sequence X_i divided by the mean value of this sequence. Patterns (5) are composed of the differences between points and the mean sequence value. Pattern (6) is the normalized vector $[E_{i-n+1}\ E_{i-n+2}\ \ldots\ E_i]^T$. All patterns defined using (6) have the unity length, mean value equal to zero and the same variance.

Similarly to input patterns, output patterns $\mathbf{y}_i = [y_{i,1}\ y_{i,2}\ \ldots\ y_{i,m}]^T$ representing the forecasted sequence Y_i, are defined as follows:

$$y_{i,t} = E_{i+t} \tag{7}$$

$$y_{i,t} = \frac{E_{i+t}}{\bar{E}_i} \tag{8}$$

$$y_{i,t} = E_{i+t} - \bar{E}_i \tag{9}$$

$$y_{i,t} = \frac{E_{i+t} - \bar{E}_i}{D_i} \tag{10}$$

To calculate the forecast of the monthly energy consumption E_{i+t} on the basis of the forecasted y-pattern generated by the GRNN model we use transformed Eqs. (7)–(10).

For example, in the case of (10) the forecasted energy consumption for the horizon t is calculated as follows:

$$\widehat{E}_{i+t} = \widehat{y}_{i,t} D_i + \bar{E}_i \tag{11}$$

where $\widehat{y}_{i,t}$ is the t-th component of the pattern $\widehat{\mathbf{y}}$ predicted by GRNN (2).

In the above formulas (7)–(11), the coding variables $\bar{E}_i$ and D_i are determined in three ways [9]:

C1. In the first approach they are calculated from the sequence X_i. So, $\bar{E}_i$ and D_i for Y_i are the same as for X_i. This enables us to calculate the forecast substituting in (11) coding variables for Y_i, which are unknown at the moment of forecasting, by known coding variables determined for X_i.

C2. In the second approach $\bar{E}_i$ and D_i in (7)–(10) are determined from the sequence Y_i. Note, that in this case coding variables are not available for the forecasted sequence Y_i at the time of making the forecast. Thus, they should be forecasted. We use ARIMA and exponential smoothing (ETS) for this purpose. The forecasted coding variables are inserted into (11) to calculate the forecasted energy consumption.

C3. In the third approach, which is used only for one-step ahead forecasts (variant B in the experimental part of the work), the coding variables $\bar{E}_i$ and D_i are determined from the annual period including time series fragments $\{E_{i-n+2}, E_{i-n+3}, \ldots, E_{i+1}\}$. In this case when using (11) the coding variables cannot be calculated from time series elements because the value of E_{i+1} is not known. Thus, $\bar{E}_i$ and D_i should be predicted. Just like in the case of C2, we use for this ARIMA and ETS.

3 Application Examples

In this section the proposed GRNN model is applied to model and forecast the electricity load demand for four European countries: Poland (PL), Germany (DE), Spain (ES) and France (FR). The data including monthly electricity demand time series were obtained from the ENTSO-E repository (www.entsoe.eu). Data for PL cover the period from 1998 to 2014 and data for the other countries cover the period from 1991 to 2014. The forecasts are made for data from 2014, using data from previous years to GRNN learning. The forecasts were prepared in two variants:

A. for all 12 months 2014 simultaneously (GRNN generates output pattern $\mathbf{y}$ representing the sequence $Y_i = \{E_{i+1}\ E_{i+2}\ \ldots\ E_{i+12}\}$),

B. individually for 12 consecutive months of 2014 (12 GRNN models are created each of which generates a forecast for one month from the period January 2014–December 2014).

In variant A the training set contains pairs ($\mathbf{x}_i$, $\mathbf{y}_i$), which are historical for the forecasted sequence. The y-pattern having 12 components ($m = 12$) represents 12 months from January to December. The x-pattern represents n months directly preceding the forecasted sequence. In variant B y-pattern having only one component ($m = 1$) represents one month of the year. The x-pattern represents n months directly preceding the forecasted month. In variant A the y-patterns are encoded using C1 or C2 approach, whilst in variant B they are encoded using C1 or C3 approach. In variants C2 and C3 the coding variables are predicted using ARIMA and ETS. In Fig. 2 results of forecasting the coding variables in variant C2 are shown.

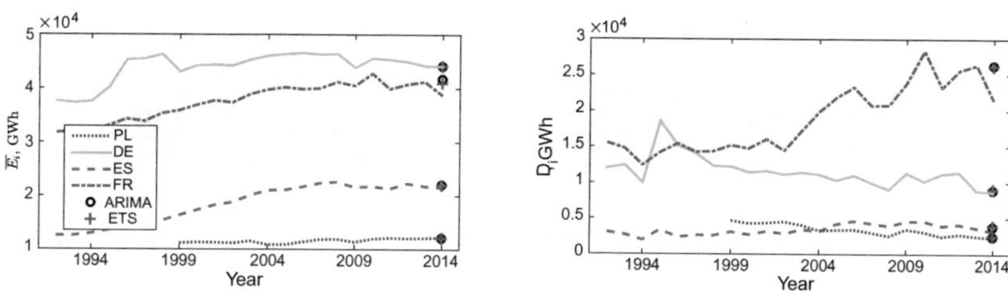

Fig. 2. Forecasts of coding variables in variant C2.

There are two parameters to estimate in GRNN model: the input pattern length n and the smoothing parameter s which is tuned by enumerating variable l (see Sect. 2.1). The model parameters were selected using grid search in leave-one-out procedure, where n was searched in the range from 3 to 24, and l was searched in the range from 1 to 10.

Figures 3 and 4 show forecast errors for 2014 depending on the model variant and definition of patterns. In most cases the best results were achieved for C1 variant, which does not need additional forecasting of the coding variables. Only for DE data in B variant a little better results were obtained when using C3-ETS. In five out of eight considered variants the lowest errors were achieved when patterns were defined by normalization (6)–(10). In two cases definitions (5)–(9) gave better results, and in one case, for FR data and variant A, the model without time series preprocessing turned out to be the most accurate.

The real and forecasted monthly demand are presented in Figs. 5 and 6, and errors for each month of 2014 in Figs. 7 and 8. Forecast errors for validation and test samples for best variants of pattern definitions in Tables 1 and 2 are presented. In these tables the results of comparative models are also shown: ARIMA, ETS, Nadaraya-Watson estimator (N-WE) [8] and neuro-fuzzy system (N-FS) [9]. Best results are shown in bold. As you can see from these tables the proposed GRNN model looks quite good against the comparative models. In all cases it outperformed the classical models such as ARIMA and ETS and was competitive in accuracy with state-of-the-art models.

Variant B which generates one-step ahead forecasts, usually provides better results than variant A. An exception is DE, where higher errors in variant B are observed. It is difficult to draw conclusions from Figs. 7 and 8, where errors for successive months are very diverse and there is no regularity here.

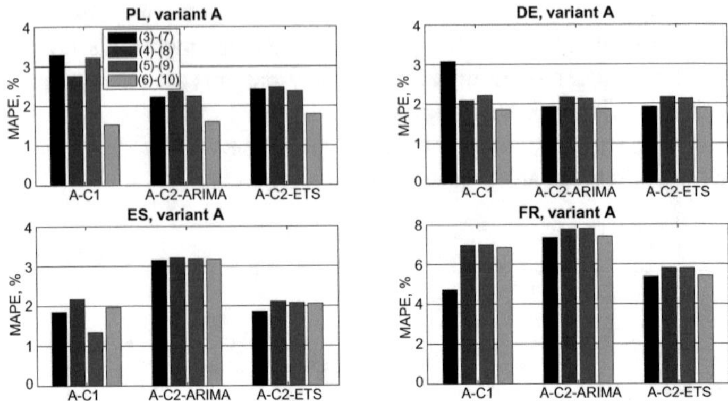

Fig. 3. Errors for different variants of coding variables determination and pattern definitions, variant A.

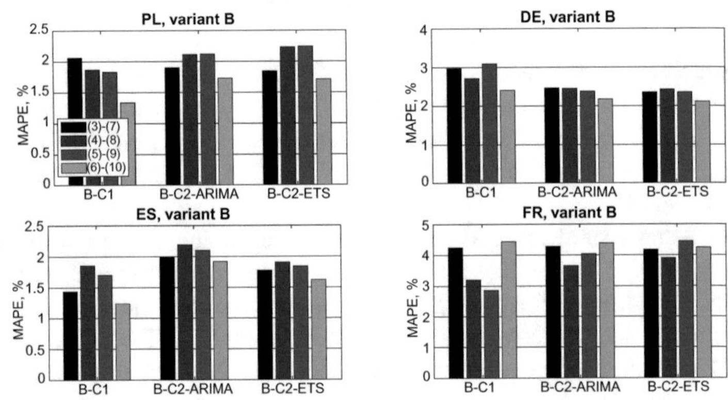

Fig. 4. Errors for different variants of coding variables determination and pattern definitions, variant B.

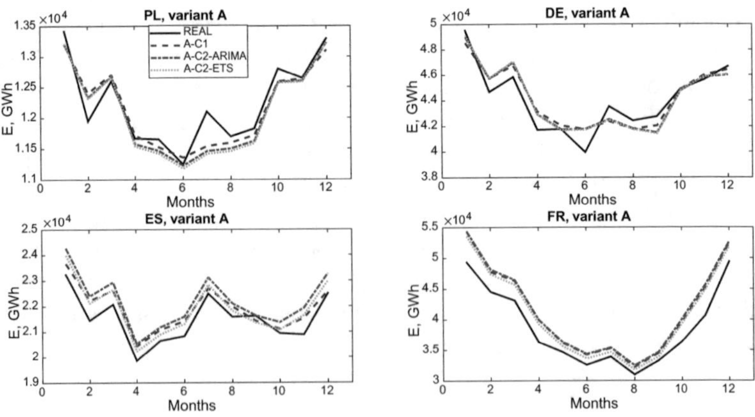

Fig. 5. Real and forecasted monthly demand for 2014, variant A.

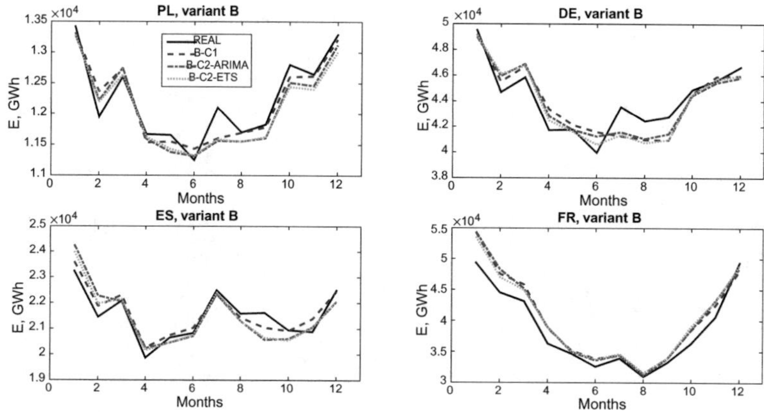

Fig. 6. Real and forecasted monthly demand for 2014, variant B.

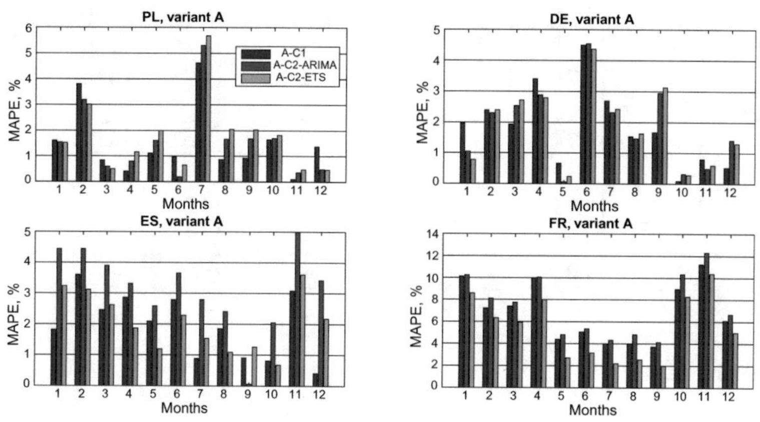

Fig. 7. Errors for consecutive months of 2014, variant A.

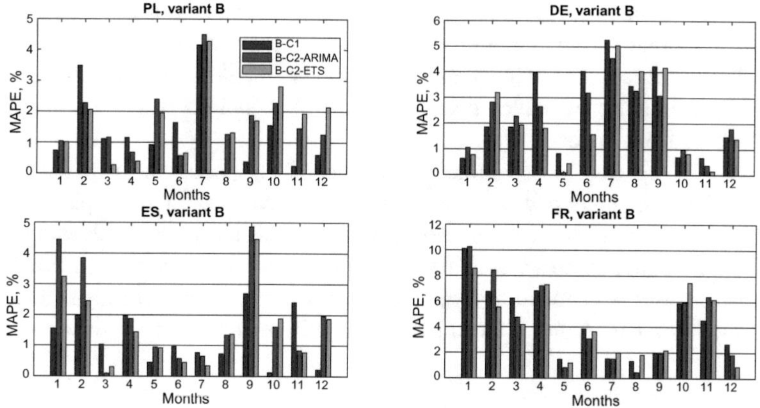

Fig. 8. Errors for consecutive months of 2014, variant B.

Table 1. Forecast errors, variant A.

	PL		DE		ES		FR	
	$MAPE_{val}$	$MAPE_{tst}$	$MAPE_{val}$	$MAPE_{tst}$	$MAPE_{val}$	$MAPE_{tst}$	$MAPE_{val}$	$MAPE_{tst}$
A-C1	2.95	**1.52**	3.25	1.85	2.90	**1.34**	3.20	4.73
A-C2-ARIMA	1.61	1.60	2.03	1.86	2.47	3.16	2.50	7.36
A-C2-ES	1.61	1.78	2.03	1.89	2.47	1.86	2.50	5.37
ARIMA	–	3.25	–	4.36	–	1.93	–	10.76
ETS	–	6.42	–	2.82	–	2.36	–	6.77
N-WE	–	1.53	–	**1.80**	–	1.49	–	4.71
N-FS	–	1.57	–	4.94	–	1.67	–	**3.34**

Table 2. Forecast errors, variant B.

	PL		DE		ES		FR	
	$MAPE_{val}$	$MAPE_{tst}$	$MAPE_{val}$	$MAPE_{tst}$	$MAPE_{val}$	$MAPE_{tst}$	$MAPE_{val}$	$MAPE_{tst}$
B-C1	2.04	1.34	2.52	2.41	2.44	1.24	3.05	2.86
B-C3-ARIMA	1.97	1.73	2.33	2.18	2.10	1.92	3.01	4.04
B-C3-ES	1.97	1.71	2.33	**2.11**	2.10	1.62	2.80	3.90
ARIMA	–	1.75	–	2.33	–	1.43	–	4.10
ETS	–	2.28	–	2.64	–	2.85	–	3.66
N-WE	–	1.30	–	2.47	–	1.16	–	**2.83**
N-FS	–	**1.06**	–	2.87	–	**0.95**	–	5.85

4 Conclusion

In this work we present GRNN model for medium-term load forecasting. In this approach the forecast is derived from the neighborhood of the query pattern using locally weighted regression. The model works on preprocessed time series sequences to filter out a trend and unify input and output patterns. Four methods of preprocessing are considered. Output variables are encoded using coding variables calculated from historical data or forecasted using classical methods: ARIMA or ETS. In most cases forecasting the coding variables does not improve model accuracy compared to calculating them from history.

The model has only two parameters: the smoothing parameter of radial activation functions and the input pattern length. They are searched in a simple grid search procedure. Fast one pass learning and easy tuning are the biggest advantages of the GRNN. In the light of the experimental study, it can be concluded that GRNN has been proven to be useful in medium-term load forecasting. It outperformed the classical models such as ARIMA and ETS and was competitive in accuracy with state-of-the-art models.

References

1. Ghiassi, M., Zimbra, D.K., Saidane, H.: Medium term system load forecasting with a dynamic artificial neural network model. Electr. Power Syst. Res. **76**, 302–316 (2006)
2. Gavrilas, M., Ciutea, I., Tanasa, C.: Medium-term load forecasting with artificial neural network models. In: 16th International Conference and Exhibition on Electricity Distribution, IET, Amsterdam (2001)
3. Hor, C.L., Watson, S., Majithia, S.: Analyzing the impact of weather variables on monthly electricity demand. IEEE Trans. Power Syst. **20**, 2078–2085 (2005)
4. González-Romera, E., Jaramillo-Morán, M., Carmona-Fernández, D.: Monthly electric energy demand forecasting based on trend extraction. IEEE Trans. Power Syst. **21**, 1946–1953 (2006)
5. Specht, D.F.: A general regression neural network. IEEE Trans. Neural Netw. **2**(6), 568–576 (1991)
6. Dudek, G.: Neural networks for pattern-based short-term load forecasting: a comparative study. Neurocomputing **2015**, 64–74 (2016)
7. Dudek, G.: Generalized regression neural network for forecasting time series with multiple seasonal cycles. In: Filev D., et al. (eds.) Intelligent Systems 2014, Advances in Intelligent Systems and Computing, vol. 323, pp. 839–846 (2015)
8. Dudek, G., Pełka, P.: Medium-term electric energy demand forecasting using Nadaraya-Watson estimator. In: Rusek, S., Gono, R. (eds.) Proceedings of 18th International Scientific Conference on Electric Power Engineering (EPE), pp. 300–305. IEEE, New York (2017)
9. Pełka, P., Dudek, G.: Neuro-fuzzy system for medium-term electric energy demand forecasting. In: Borzemski, L., Świątek, J., Wilimowska, Z. (eds.) Information Systems Architecture and Technology: Proceedings of 38th International Conference on Information Systems Architecture and Technology – ISAT 2017, Advances in Intelligent Systems and Computing, vol. 655, pp. 38–47. Springer, Cham (2018)

Factors Affecting Energy Consumption of Unmanned Aerial Vehicles: An Analysis of How Energy Consumption Changes in Relation to UAV Routing

Amila Thibbotuwawa[1(✉)], Peter Nielsen[1], Banaszak Zbigniew[1], and Grzegorz Bocewicz[2]

[1] Department of Materials and Production, Aalborg University, Aalborg, Denmark
{amila,peter}@mp.aau.dk, Z.Banaszak@wz.pw.edu.pl
[2] Faculty of Electronics and Computer Science, Koszalin University of Technology, Koszalin, Poland
bocewicz@ie.tu.koszalin.pl

Abstract. Unmanned Aerial Vehicles (UAV) routing is transitioning from an emerging topic to a growing research area and one critical aspect of it is the energy consumption of UAVs. This transition induces a need to identify factors, which affects the energy consumption of UAVs and thereby the routing. This paper presents an analysis of different parameters that influence the energy consumption of the UAV Routing Problem. This is achieved by analyzing an example scenario of a single UAV multiple delivery mission, and based on the analysis, relationships between UAV energy consumption and the influencing parameters are shown.

Keywords: Unmanned Aerial Vehicles · UAV routing
Energy consumption of UAVs

1 Introduction

UAVs have developed into a mature technology applied in areas such as defense, search and rescue, agriculture, manufacturing and environmental surveillance [1–8]. A UAV can replace manned aerial vehicles in unsafe and uninhabitable situations. UAVs are opening new possibilities to perform complex missions with some degree of autonomy without any required alterations to the existing infrastructure, e.g. deployment station on the wall or guiding lines on the floor and UAVs are capable of covering flexible wider areas in the field [9]. The development of UAVs is evolving, and there has been an increased interest to make UAVs operate increasingly autonomously perform missions [10]. A critical aspect of proper mission planning is considering the energy consumption by the UAV in the completion of the mission.

The content of the article is organized as follows. Section 2 describes the energy consumption models used for the study, focusing on various factors affecting the energy consumption of UAVs by analyzing how each factor influence the energy

J. Świątek et al. (Eds.): ISAT 2018, AISC 853, pp. 228–238, 2019.
https://doi.org/10.1007/978-3-319-99996-8_21

consumption respectively. Section 3 outlines the formal description of the problem scenario used in the study, accompanied with the representative illustrations followed by a description containing the effects of weather to the study. In the Sect. 4 the results of the analysis are presented and the Sect. 5 provides the conclusions.

2 Analysis of the Factors Affect UAV Energy Consumption

In this study consumption is calculated using the total power consumption calculation equation for cruise (1) and power consumption equation for take-off and landing (2) which detailed in previous work by the authors [11].

$$P_T = \frac{1}{2} C_D A D v^3 + \frac{W^2}{D b^2 v} \tag{1}$$

Where P_T is the power needed for flight in watts, C_D is the aerodynamic drag coefficient, A is the front facing area in m^2, W is the total weight of the UAV in kg, D is the density of the air in kg/m^3, b is the width of UAV in m, and v is the relative speed of the UAV in m/s through the air.

$$p^* = \frac{T^{\frac{3}{2}}}{\sqrt{2D\varsigma}} \tag{2}$$

Where, p^* is the power needed to vertical take-off and landing (VTOL) in watts and the thrust T in Newtons. Air density of air D in kg/m^3, and the facing area ς of the UAV is in m^2, where the thrust T = W g, given the UAV total weight W in kg, and gravity g in N. It is critical to note that this power consumption profile differs significantly from that of other autonomous vehicles such as AGVs and mobile robots [12–14]. In addition, it is critical to note that e.g. AGVs that run out of energy during execution of a mission can stop and wait, UAVs will potentially have a catastrophic failure leading to loss of the UAV and potential damage to infrastructure and humans.

Using these equations, we prepare an experiment to demonstrate the influence on various parameters during routing on UAV energy consumption. The characteristics are given in Table 1. Parameters were analysed by changing each parameter while keeping all other parameters as constants and the results are shown in Sect. 2.3. Afterwards, energy consumption of UAV was calculated against two changing parameters. The results are shown in Sect. 2.3.

2.1 UAV Power Consumption and UAV Flying Speed, UAV Power Consumption and Payload

Figure 1 shows how the energy consumption changes against the flying speed of the UAV and the payload carried by the UAV. The highly non-linear relationship for both flying speed and payload is clear. For lower flying speeds the energy consumption is clearly convex non-linear. However, for higher-flying speeds the energy consumptions tends towards a linear relationship with flying speed. Hence, linear approximations

Table 1. UAV specifications

Width	8.7 m
Front reference area	1.2 m^2
Top reference area	7.5 m^2
Empty weight	57.5 kg
Maximum takeoff weight	120 kg
UAV drag coefficient	0.546

could be made in higher-flying speeds. The non-linearity is comparably low in payload factor in comparison to the flying speed.

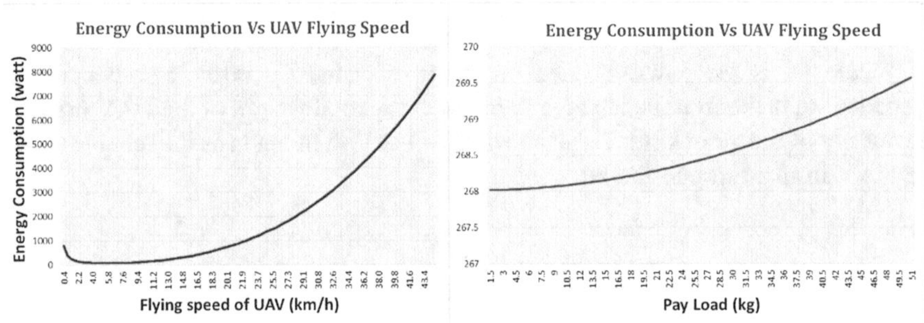

Fig. 1. UAV energy consumption vs UAV flying speed and vs pay load.

2.2 UAV Power Consumption and Air Density

Figure 3 shows how the UAV energy consumption changes against the air density. In shows, a linear relationship with higher air densities leads to higher energy consumption. The air density is directly linked with the atmospheric temperature and the ambient temperature is thus indirectly linked to the energy consumption [11] (Fig. 2).

2.3 UAV Power Consumption Against Two Changing Parameters

Figure 3 shows the non-linear energy consumption as a function of the flying speed and payload carried by the UAV. In this case we have changed both the payload and flying speed simultaneously and Fig. 3 shows the 3D illustration of the results. These illustrations were replicated for payload vs air density and flying speed vs air density with a similar behaviour. For reasons of brevity, these are omitted.

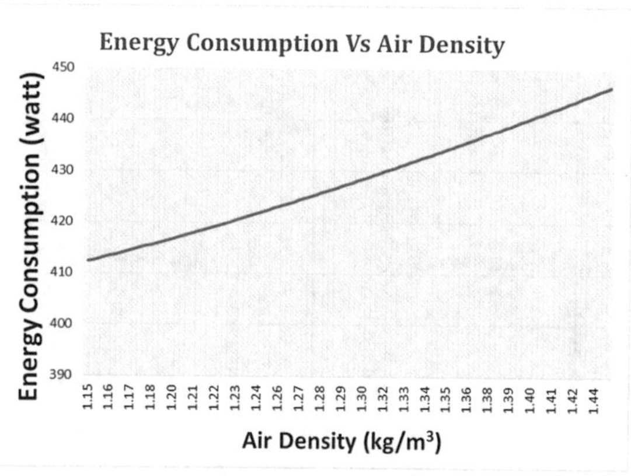

Fig. 2. Power consumption vs air density

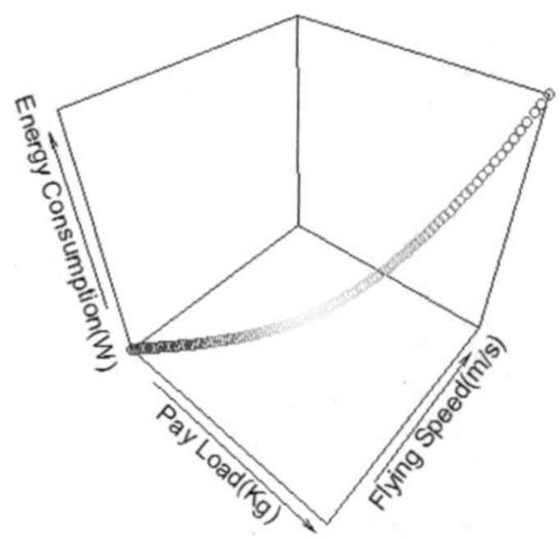

Fig. 3. Energy consumption against speed and payload

3 Scenario Analysis

While the non-linear energy consumption is straightforward to observe, the real challenge is how this behaviour influences the UAV routing problem. To analyse this we considered a single-UAV-routing scenario as illustrated in Figs. 4 and 5. Figure 5 shows the 2.5D illustration of the scenario. Here we considered an UAV starting from a base and then visiting three delivery locations before the UAV returns to the base. In each node, node 1 excluded, a delivery is completed and at node 4 the UAV finishes its assigned delivery tasks. The UAV returns to the depot using the same route after the final delivery. We define a route as the path that starts and ends at the depot, which is located vertically below the node 1. Each location i has a demand Di that represents the weight of the payload in kg delivered to location i. A time t is spent at each location to descend and ascend while a time T is spent for deliver the packages. On the return trip, the UAV flies back to the base through the nodes.

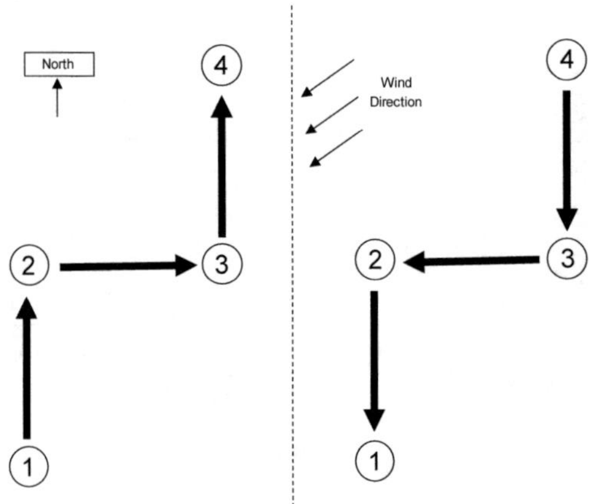

Fig. 4. Test scenario diagram

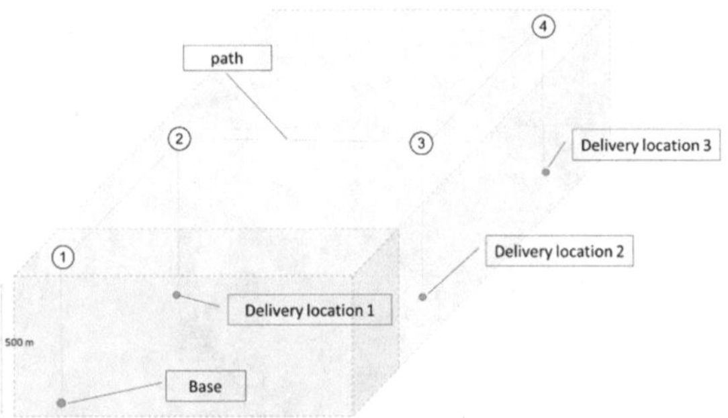

Fig. 5. Test scenario illustration

In this experiment, we consider that there are three deliveries in depot 2, 3 and 4 with a similar demand of q kg each. Thus, the UAV leaves the base carrying $3q$ kg payload and delivers q kg at each delivery location. The distance between 1–2, 2–3 and 3–4 is 500 m. The travel altitude is 500 m and we test the UAV speed from 25 km/h to 100 km/h and payloads, q, from 100 kg to 500 kg. We assume the air density is constant for the entire mission at 1.225 kg/m^3. Furthermore, the effect of weather is tested by changing the wind speed from 1 m/s to 20 m/s and the wind direction from 1–360°.

3.1 Weather Data

Weather data was taken based on the actual weather data of Denmark from January 1st of 2006 to December 31st of 2016. Figure 6 shows the distribution of weather data over the time of 10 years and based on the data we selected most occurred wind speed range and for the scenario, analysis and we did the analysis for all the possible wind directions. For the explanations presented in Sect. 2 we use the most frequently observed condition which is a wind speed of 8 m/s and wind direction of 220°.

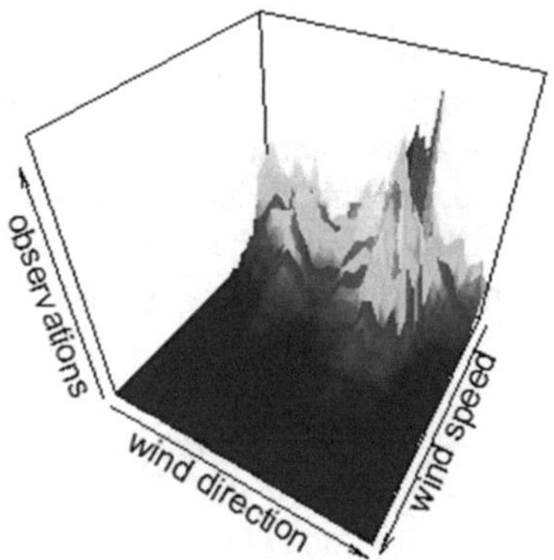

Fig. 6. Weather data

Wind Factor
To account for the effect of variations in wind speed and direction, when we calculate fuel consumption, let A_{ij} be the relative difference between the true traveling angle from target i to target j, θ_{ij}, and the wind direction, θ_w, that is $A_{ij} = |\theta_{ij} - \theta_w|$ [15, 16].

Wind Affecting Energy Consumption
Suppose the set of edges E is divided into two sets: E_t is the set of edges with tailwind, and E_h is the set of edges with headwind. When a UAV travels in an arc belonging to E_t $A_{ij} < 90$, the traveling will require a lower airspeed in order to maintain the scheduled groundspeed resulting in lower energy consumption. Vice versa, when a UAV travels arc belongs to E_h, $90 < A_{ij} < 180$, it requires a higher airspeed, in order to maintain the scheduled groundspeed, resulting in higher energy consumption. Let us consider v^w is the wind speed, the groundspeed of traveling from target i to target j with the wind speed adjustment can be expressed as [15, 16];

$$v_{ij}^g = v_{ij}^a + v^w \cos A_{ij} \tag{3}$$

3.2 Assumptions

We assume that UAVs fly between locations at a constant speed v in m/s, in a constant altitude. We assume a simple flight path, such that the UAV first ascends vertically to a desired altitude, and then travels in a straight path, and descends vertically in the location i, which has a demand Di.

We consider the impact of weather in this example as the wind affects energy consumption. Hence, we considered the wind speed and wind direction in calculating the relative flying speed of the UAV. We assume that the wind speed and direction is constant during the mission execution, that the demand at each location can be fully satisfied by the UAV and that the demand at a location is not higher than the carrying capacity. We assume that the UAV has sufficient energy capacity to complete the mission. For the calculations, we have considered an UAV with following specifications.

We assume for practical reasons that in a UAV trip all parameters will remain constant on a given arc. Payload and speed may change from one arc to another. Also we consider that UAV will travel at an constant speed of $v = v_{ij}$ on an arc (i, j) depending on the wind speed and wind direction with distance d_{ij} and a total load of $W = w + f_{ij}$, where w is the empty weight of the UAV and f_{ij} is the pay load carried by the UAV in the arc. The total amount of energy consumed on this arc can then be taken as:

$$P_{i\,to\,j} = P_t\left(\frac{d_{ij}}{v_{ij}^{g}}\right) \tag{4}$$

Which can be expanded using (1) is;

$$P_{ij} = \frac{1}{2}C_D AD(v_{ij}^{a})^3 + \frac{(w+f_{ij})^2}{Db^2 v_{ij}^{a}}\left(\frac{d_{ij}}{v_{ij}^{g}}\right) \tag{5}$$

4 Results of the Analysis

Experiments are carried out, with regards to the scenario explained for different flying speeds, different carrying payloads and changing wind speeds and directions mentioned in Sect. 3. We observe how the energy consumption changes with respect to inbound vs outbound travels and level flight vs take-off and landing.

The ratio of energy usage in the inbound vs outbound journeys and level flight vs take-off and landing against the different flying speeds are shown in the Fig. 7. It is visible that in higher speeds, the difference of energy consumption percentage between inbound and outward reduces. Furthermore, energy consumption in cruise vs take-off and landing shows a similar behaviour and in higher flying-speeds, the difference tends to be significantly less.

We observe, at lower flying speeds the UAV consumes more energy for take-off and landing compared to level flight. However, at higher speeds this changes, as energy consumption increases for level flight in higher speeds compared to VTOL.

The ratio of energy usage in the inbound vs outbound journeys and level flight vs take-off and landing against the different carrying payloads are shown in Fig. 8. We note that the relationship is clear between energy consumption and the weight of the payload carried.

With respect to payloads, we observe a similar behaviour as with the different flying speeds, at lower payload weights, the UAV consumes more energy for take-off and landing compared level flight. However, when the payload increases this change as energy consumption increases for level flight in higher speeds compared to VTOL.

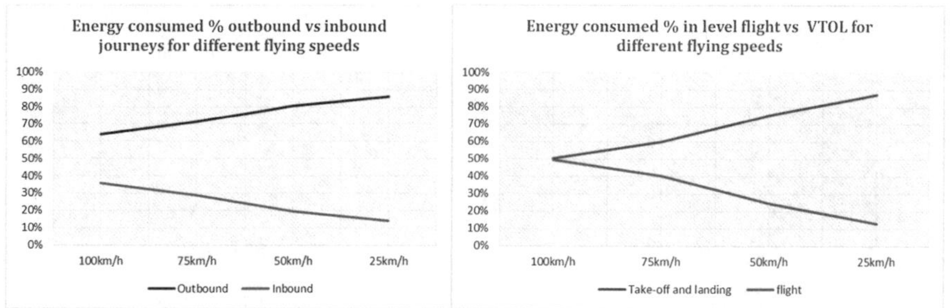

Fig. 7. Energy consumed percentage in outbound vs inbound trips and level flight vs VTOL for different flying speeds in most common weather conditions.

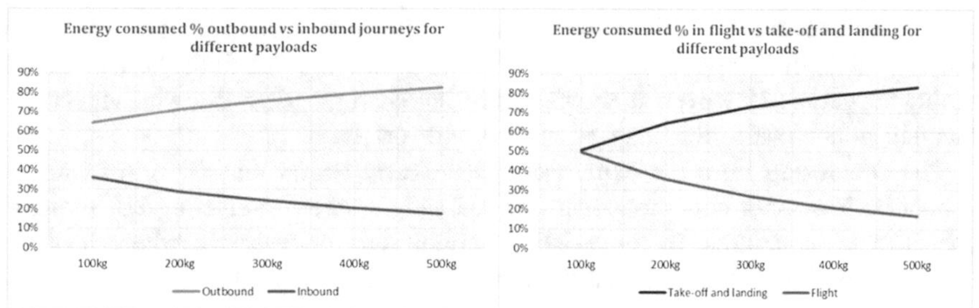

Fig. 8. Energy consumed percentage in outbound vs inbound trips and level flight vs VTOL for different carrying payloads in most common weather conditions.

To observe how the variations in wind speed affects the energy consumption we test the scenario with different wind speeds and show the results in Fig. 9. In addition, we test the energy consumption against varying wind directions and show the results in Fig. 10.

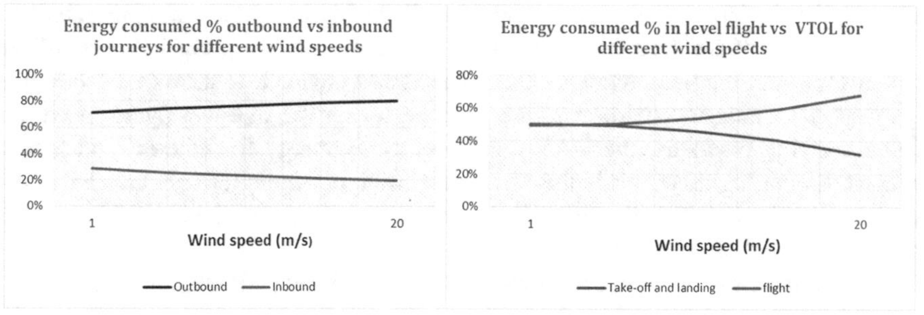

Fig. 9. Energy consumed percentage in outbound vs inbound trips and level flight vs VTOL for different wind speeds.

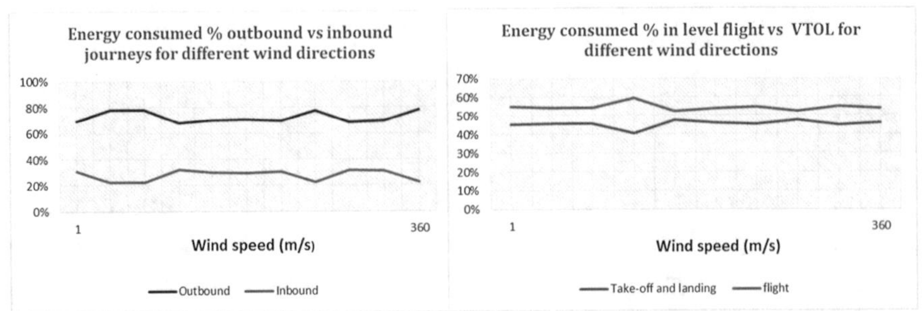

Fig. 10. Energy consumed percentage in outbound vs inbound trips and level flight vs VTOL for different wind directions.

We observe that both the wind speed and direction influences the energy consumption in both in bound vs out bound travels and in level flight vs take-off and landing scenarios. However, it shows significant changes when the wind direction is changing in contrast to the changes in the wind speeds.

The conclusion from the experiments are clearly shows that one must take into account both weather conditions and payload weight when conducting UAV routing. This is in stark contrast to the existing current state of research, where neither is typically considered. It seems that there is a significant gap in the current state.

5 Conclusion

This paper focuses on analyzing the energy consumption of UAVs, which is non-linear and dependent on weather, speed, direction, and payload. Wind has more of an effect on a UAV than it does on a passenger aircraft because of their lower cruising speed and weights. Furthermore, to differentiate from other time dependent routing problems [17, 18] the fuel/energy consumption also depends on the direction both w.r.t. wind direction, but also the changes in vertical level. The paper analyses an example scenario of a single UAV multiple delivery mission, and based on the analysis, relationships between UAV energy consumption and the influencing parameters are presented.

The results of the analysis shows that the energy consumption has a non-linear relationship with the parameters of flying speed and payload. It shows that in higher-flying speeds, linear approximations are possible. Moreover, in lower flying speeds and higher carrying payloads, the UAV consumes more energy for take-off and landing compared level flight, which is the vice versa in higher-flying speeds and lower carrying payloads. Results tells that changes wind direction can significantly affect the energy consumption of a UAV.

In the future, we will further analyze these models by experimenting with industrial data and different models of available UAVs.

References

1. Yakici, E.: Solving location and routing problem for UAVs. Comput. Ind. Eng. **102**, 294–301 (2016). https://doi.org/10.1016/j.cie.2016.10.029
2. Bolton, G.E., Katok, E.: Learning-by-doing in the newsvendor problem a laboratory investigation of the role of experience and feedback. Manuf. Serv. Oper. Manag. **10**, 519–538 (2004). https://doi.org/10.1287/msom.1060.0190
3. Avellar, G.S.C., Pereira, G.A.S., Pimenta, L.C.A., Iscold, P.: Multi-UAV routing for area coverage and remote sensing with minimum time. Sensors (Switzerland) **15**, 27783–27803 (2015). https://doi.org/10.3390/s151127783
4. Khosiawan, Y., Nielsen, I.: A system of UAV application in indoor environment. Prod. Manuf. Res. **4**, 2–22 (2016). https://doi.org/10.1080/21693277.2016.1195304
5. Barrientos, A., Colorado, J., del Cerro, J., et al.: Aerial remote sensing in agriculture: a practical approach to area coverage and path planning for fleets of mini aerial robots. J. Field Robot. 667–689 (2011). https://doi.org/10.1002/rob
6. Khosiawan, Y., Park, Y., Moon, I., et al.: Task scheduling system for UAV operations in indoor environment. Neural Comput. Appl. **9**, 1–29 (2018). https://doi.org/10.1007/s00521-018-3373-9
7. Khosiawan, Y., Khalfay, A., Nielsen, I.: Scheduling unmanned aerial vehicle and automated guided vehicle operations in an indoor manufacturing environment using differential evolution-fused particle swarm optimization. Int. J. Adv. Robot. Syst. **15**, 1–15 (2018). https://doi.org/10.1177/1729881417754145
8. Bae, H., Moon, I.: Multi-depot vehicle routing problem with time windows considering delivery and installation vehicles. Appl. Math. Model. **40**, 6536–6549 (2016)
9. Zhang, M., Su, C., Liu, Y., et al.: Unmanned aerial vehicle route planning in the presence of a threat environment based on a virtual globe platform. ISPRS Int. J. Geo Inf. **5**, 184 (2016). https://doi.org/10.3390/ijgi5100184
10. Tulum, K., Durak, U., Ider, S.K.: Situation aware UAV mission route planning. In: Proceedings of the IEEE Aerospace Conference (2009). https://doi.org/10.1109/AERO.2009.4839602
11. Thibbotuwawa, P., Peter, I., Zbigniew, B., Bocewicz, G.: Energy consumption in unmanned aerial vehicles: a review of energy consumption models and their relation to the UAV routing. In: 38th International Conference "Information Systems Architecture and Technology" (2018, to appear)
12. Dang, Q.V., Nielsen, I., Steger-Jensen, K., Madsen, O.: Scheduling a single mobile robot for part-feeding tasks of production lines. J. Intell. Manuf. **25**, 1271–1287 (2014). https://doi.org/10.1007/s10845-013-0729-y
13. Nielsen, I., Dang, Q.V., Bocewicz, G., Banaszak, Z.: A methodology for implementation of mobile robot in adaptive manufacturing environments. J. Intell. Manuf. **28**, 1171–1188 (2017). https://doi.org/10.1007/s10845-015-1072-2
14. Dang, Q.V., Nielsen, I.: Simultaneous scheduling of machines and mobile robots. Commun. Comput. Inf. Sci. **365**, 118–128 (2013). https://doi.org/10.1007/978-3-642-38061-7_12
15. Visoldilokpun, S.: UAV routing problem with limited risk. The University of Texas at Arlington (2008)

16. Biradar, A.S.: Wind estimation and effects of wind on waypoint navigation of UAVs. Masters thesis, Arizona State University (2014). https://repository.asu.edu/attachments/135075/content/Biradar_asu_0010N_13909.pdf
17. Huang, Y., Zhao, L., Van Woensel, T., Gross, J.P.: Time-dependent vehicle routing problem with path flexibility. Transp. Res. Part B Methodol. **95**, 169–195 (2017). https://doi.org/10.1016/j.trb.2016.10.013
18. Taş, D., Gendreau, M., Jabali, O., Laporte, G.: The traveling salesman problem with time-dependent service times. Eur. J. Oper. Res. **248**, 372–383 (2016). https://doi.org/10.1016/j.ejor.2015.07.048

Big Data Analysis, Knowledge Discovery and Knowledge Based Decision Support

Computer Based Methods and Tools for Armed Forces Structure Optimization

Andrzej Najgebauer(✉), Ryszard Antkiewicz, Dariusz Pierzchała, and Jarosław Rulka

Faculty of Cybernetics, Military University of Technology, Gen. Witolda Urbanowicza 2, 00-908 Warsaw, Poland
{andrzej.najgebauer, ryszard.antkiewicz, dariusz.pierzchala, jaroslaw.rulka}@wat.edu.pl

Abstract. The paper is devoted to a quantitative approach to support one of the most important problem solution in the area of defense planning process - armed forces (AF) structure optimization. The MUT team, taking part in Polish Strategic Defense Review has proposed a set of methods and tools to support the analyses for the evaluation of required capabilities of Polish Armed Forces in predicted security environment. The set of methods and tools presented in the paper is limited to AF structure optimization problem. The idea of optimization and particular components of the conflict model and methods of solving the problems are presented in the sequence of steps. The structure of AF is fixed for defined threat scenarios under the financial constraints or without such limitations. The measures of combat power of weapon systems for different participants of probable conflict and some important parameters like terrain or type of operation factors (multipliers) are defined and presented. The experimental results of the allocation process are based on the hypothetical conflict evaluation.

Keywords: Decision support · Structure optimization · Defense planning

1 Introduction

The systemic change in the country, including changes the Armed Forces the MOD faces new challenges. Both NATO's transformation, as well as the AF of a member country causes the need for a new approach to the operation of the AF. One of the important directions of the transformation is the issue of the development of the capability of the Armed Forces (AF) of the country and the identification of operational needs. This work was partially supported by the research co-financed by the National Centre for Research and Development and realized by Cybernetics Faculty at MUT: No DOBR/0069/R/ID1/2012/03, titled "System of Computer Based Support of Capability Development and Operational Needs Identification of Polish Armed Forces" [11]. Some results of the work were applied in the Polish Strategic Defense Review (SDR) for the evaluation of proposed by research teams future structures of Armed Forces.

J. Świątek et al. (Eds.): ISAT 2018, AISC 853, pp. 241–254, 2019.
https://doi.org/10.1007/978-3-319-99996-8_22

As the big challenge in defense planning process is the formulation of AF structure optimization model for the defined threat scenarios, where two important aspects – budgetary and operational can be considered simultaneously. As well as criteria the substantial constraints for the model were defined. Important aspect of the analysis there are combat power of reference modules (battalions and equivalent units) and combat assessment for different type of battle and also environmental conditions. Optimization of AF structures for fixed scenarios was considered for two important cases, the first, with cost of acquisition and exploitation constraints and the second without the limits. Some experimental results for hypothetical conflicts are the basis of presentation of the method.

Similar problem, however different formulation contains the work [6, 7, 9, 10, 11]. Our approach, in contrast to the referenced works, consider the cost of battle and relative-force ratios together in one problem formulation and solution and in that sense it seems to be unique.

2 Idea of the Methods of Armed Forces Structure Optimization

Combat power of weapon system can be defined as the effect created by combining maneuver, firepower, protection, and leadership, the dynamics of combat power, in combat against the enemy [5]. The effects of these elements with any other potential combat multipliers against the enemy, can be applied by the commander to achieve aims at minimal cost. It requires an assessment of many factors and by analyzing relative-force ratios defense planners can gain some insight into friendly forces capabilities needed to the operation, what type combat may be possible from both own and enemy perspectives and weakness of enemy.

The structure of AF can be fixed by solving for the most probable threat scenarios the allocation problem of required types of modules for different type of operations (battle) without of cost limitations or with the limitations of acquisition and exploitation of desired structure of Armed Forces. Proposed types of combat (battle) in the analysis are as follows:

	Deliberate defense	Deliberate attack	Hasty attack	Hasty defense
Deliberate defense		DD-DA	DD-HA	
Deliberate attack	DA-DD			DA-HD
Hasty attack	HA-DD			HA-HD
Hasty defense		HD-DA	HD-HA	

Combat power of warfare is determined for both sides in the common category measure. The structure and combat power of enemy should be determined in the scenario preparation process. The scenario contains the number of enemy divisions or brigades. Reference modules (battalions and equivalent units) of own forces, their combat power and cost of acquisition and exploitation are defined separately.

The optimization problem solving is based on the fixing of input data and applying of evolutionary approach in order to achieve the optimal allocation of reference modules according to 2 criteria: cost of acquisition and exploitation of modules for the estimated structure of troops, possible losses for the fixed type battle and terrain conditions. We can fix the importance each of the criteria. The terrain conditions we have considered are divided into five categories (OPEN, MIXED, ROUGH, URBAN, MOUNT). For different type of battle combat power of weapon systems for the conflict participants is modified by multipliers different for defenders and attackers.

For fixing the optimal structure of own forces for determined enemy we need to define the loss function based on force ratios with consideration of terrain and type of combat factors. Preparing scenario for predicted security environment, the analyst should determine the type and number of divisions or brigades of enemy and category of combat (battle) and category of terrain and sometimes (for the optimization problem under the cost constraints) the range of defense.

In addition to fixing a number of specific types of modules are determined also the pace attack of the enemy [km/day] and the defense in the days of fighting at a fixed depth of Defense. Method can also be used to calculate the cost of acquisition and the functioning of the different variants of the armed forces in the fixed perspective.

3 Mathematical Model of the Conflict

3.1 Reference Modules and Their Combat Potential and Cost

We should consider organization and military equipment and armament of the reference module in order to evaluate its relative combat power and cost of exploitation.

The mathematical model of a reference module is represented as follows: $RMd_{F(E)}(i) = \left(\bar{n}_{F(E)}(i) = (n^{w}_{F(E)}(i))_{w\in W_{WS}}, \overline{MCP}_{F(E)}(i),\ AMC_{F(E)}(i)\right)_{i=1,..,N^{F(E)}_{RMd}}$, where $N^{F(E)}_{RMd}$ - number of types of reference modules for side F – the friendly forces (for side E – the enemy forces) and

- $\overline{MCP}_{F(E)}(i)$ - vector of combat potential general and different categories of the reference module type i of side F(E);
- $AMC_{F(E)}(i)$ - annual cost of the reference module type i of side F(E).

Each weapon system could be evaluated by its individual combat power index cpi_w. There is no universal methodology for weapon systems combat power evaluation. Different methodologies are used in different countries. Always some parameters of weapon are taking into account, in order to evaluate its combat power index.

Let us define the following denotations:

- $W_{WS} = \{1,..,N_W\}$, W_{WS} - the set of number of types of weapon system, N_W - number of weapon system types (for all potential members of analyzed conflict)
- cpi_w, $w = 1,..,W$ - where cpi_w is a combat power index for weapon system w.

In order to formulate some success conditions of military operation, we define category of weapon systems as follows: tanks - W_T, infantry fighting vehicle (BMP) - W_{AV}, long-range anti-armor weapon (LR) - W_{LR}, short-range anti-armor weapon (SR) - W_{SR} short-range artillery - W_{SArt}, multiple launch rocket system - W_{MLRS}, tactical ballistic missiles - W_{TBM}, infantry weapon (mortars, small arms) - W_{inf}, attack helicopters - W_H, air defense weapons - W_{AD}. We assume that:

$$\begin{aligned} &W_{WS} = \bigcup_{n \in WC} W_n \\ &WC = \{T, AV, BMP, LR, SR, SArt, MLRS, TBM, inf, H, AD\} \cdot \\ &\forall n_1 \neq n_2 : W_{n_1} \cap W_{n_2} = \emptyset \end{aligned}$$

Now, we can define vector of combat power: $\overline{MCP}_{F(E)}(i) = \left(MCP_{F(E)}(i), (MCP^n_{F(E)}(i))_{n \in WC}\right)$, where:

- $MCP_{F(E)}(i) = \sum_{w=1}^{W} cpi_w \cdot n^w_{F(E)}(i)$ - the general relative combat power of the reference module type i of side A(B) is calculated as follows;
- $MCP^n_{F(E)}(i) = \sum_{w \in W_n} cpi_w \cdot n^w_{F(E)}(i),\ n \in WC$ - the relative combat potential of weapons category $n \in WC$

where $n^w_{A(B)}(i)$ is a number of weapon system type w in the reference module type i of side F(E).

Relative combat power of reference modules (general and for each weapon category) should be modified considering terrain conditions and type of combat. Types of combat were defined in Sect. 3. There are defined five type of terrain: OPEN, MIXED, ROUGH, URBAN and MOUNT [3]. Modifications is made by multiplication of relative potential power of reference modules by proper situational and type of combat multipliers – Table 1:

Table 1. Based on [4]

Defender										
Type of combat	Type of terrain	Weapon category								
		Tanks	AV	LR	SR	inf	SArt& MLRS	H	AD
Hasty defense	OPEN	0,88	0,616	0,756	0,63	0,63	0,88	1,2	0,7
	MIXED	0,8	0,56	0,7	0,7	0,7	0,8	1	1
	ROUGH	0,72	0,504	0,644	0,84	0,84	0,72	0,9	1,3
	URBAN	0,64	0,64	0,294	1,05	1,05	0,56	0,5	1,1
	MOUNT	0,64	0,448	0,588	1,12	1,12	0,64	1	2
Deliberate defense	OPEN	1,15	0,8085	1,242	1,035	1,035	1,155	1,2	0,7
	MIXED	1,05	0,7	1,15	1,15	1,15	1,05	1	1
	ROUGH	0,9	0,66	1,058	1,38	1,38	0,945	0,9	1,3
	URBAN	0,84	0,84	0,483	1,725	1,725	0,735	0,5	1,1
	MOUNT	0,84	0,588	0,966	1,84	1,84	0,84	1	2

Similar multipliers are defined for type of operation – top = Attacker.

$$\begin{aligned}&mlp(top, tot, n, tod),\ where:\\&top \in \{defender, attacker\},\\&tot \in ToT = \{OPEN, MIXED, ROUGH, URBAN, MOUNT\}\\&n \in WC\\&tod \in \{hasty\ defense,\ deliberate\ defense\}\end{aligned}$$

Thus, for given type of terrain and type of modified vector of combat power $\overline{mMCP_{F(E)}}(i)$ is calculated according to the following equations:

$$mMCP_{F(E)}(i, top, tot, n, tod) = \sum_{w=1}^{W} cpi_w \cdot n^w_{F(E)}(i) \cdot mlp(top, tot, I_{WC}(w), tod)$$

where: $I_{WC}(w) \in WC$ is category of weapon type w;

$$mMCP^n_{A(B)}(i) = \sum_{w \in W_n} cpi_w \cdot n^w_{A(B)}(i) \cdot mlp(top, tot, n, tod),\ n \in WC.$$

Annual reference module cost is evaluated as follows:

$$AMC_{F(E)}(i) = ACP_{F(E)}(i) + ACW_{F(E)}(i)$$

where: $ACP_{F(E)}(i)$ annual cost of staff and $ACW_{F(E)}(i)$ annual cost of weapon.

$ACW_{F(E)}(i) = \sum_{w=1}^{W} n^w_{F(E)}(i) \cdot co(w) \cdot \frac{1}{T_E(w)}$, where $co(w)$ is a cost of purchase and $T_E(w)$ time of exploitation of w-type weapon.

3.2 Combat Assessment

Combat assessment means evaluation of loss of staff, weapon and military equipment both sides of conflict and movement rates. Loss calculation could be made in different way, using mathematical models of combat (Lanchester equations, stochastic model of combat, combat simulation tools) or using loss functions, which relates losses of sides to ratio of combat potential of fighting sides. Such function is approximate based on historical data. In this paper we present a formulation of loss function based on historical data [3]. There are presented in the following tables:

Table 2. Loss for sides F and E for combat type: HA-HD

COF	0,25	0,33	0,5	1	2	3	4
F	70%	60%	50%	20%	15%	10%	10%
E	10%	10%	15%	20%	30%	50%	60%

We can approximate function of loss for side F and E taking into account data from Table 2 as follows:

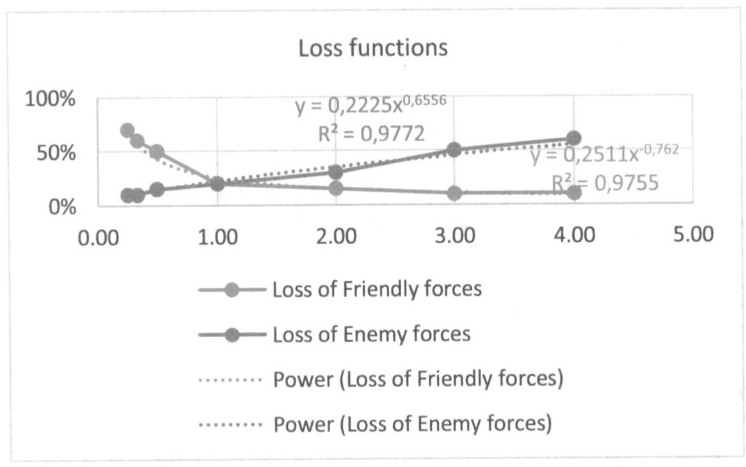

Fig. 1. Loss functions for combat type: HA-HD

We denote loss functions as follows:

$$Fl_{F(E)}(fr, toc),\ where:\ fr_{F|E} = \frac{combat\ power(F)}{combat\ power(E)}$$
$$and\ toc \in TOC = \{DD - DA,\ DD - HA,\ HD - DA,\ HD - HA\}$$

For example, in combat where participate one reference module for each side, we have the following force ratio:

$$fr_{F|E} = \frac{mMCP_F(i)}{mMCP_E(i)}$$

According to Fig. 1 we have that:

$$Fl_F(fr, HA - HD) = 0,2511 \cdot fr^{-0,762}$$
$$Fl_E(fr, HA - HD) = 0,225 \cdot fr^{0,6556}$$

At the same way, we approximated loss functions for all types of combat.

We use the function $V(fr, tot, tod) = a(tot, tod) \cdot fr^{b(tot,tod)} \cdot fh$ [3] to evaluate movement rate. Values of function $V(\cdot)$ parameters are given in the Table 3.

We assume, that movement rate is evaluated in km/day, and *fh* means hours of fight per day. The loss functions (friendly and enemy) for DD-DA and another combinations are determined in the same way and are based on [3].

Table 3. Parameters a, b

Parameters of $V(\cdot)$				tot
Hasty defese		Deliberate defense		
a	b	a	b	
0,999	0,612	0,607	0,581	OPEN, MIXED
0,521	0,846	0,455	0,545	ROUGH, URBAN
0,406	0,567	0,167	0,819	MOUNT

3.3 Mathematical Formulation of Optimization Problem of Armed Forces Structure for Defined Threat Scenario

In order to formulate problem of forces structure optimization we define the following decision variables and functions:

$\bar{X}_F = (x_F(i))_{i=1,..N^F_{RMd}}$, - structure of side F armed forces;

$\bar{X}_E = (x_E(i))_{i=1,..N^E_{RMd}}$ - structure of side E armed forces;

$x_E(i)$, $x_F(i)$ - number of reference module.

$F_1(\bar{X}_F) = K_F(\bar{X}_F) = \sum_{i=1}^{N^F_{RMd}} AMC_F(i) \cdot x_F(i)$ - annual cost of all reference modules, fixed by to $\bar{X}_F$

$$F_2(\bar{X}_F) = FlS_F(\bar{X}_F, \bar{X}_E, toc, top, tot, tod) = \\ = \sum_{i=1}^{N^F_{RMd}} staff_F(i) \cdot Fl_F(fr_{F|E}(\bar{X}_F, \bar{X}_E), toc, top, tot, tod)$$

$$CP_{F(E)}(\bar{X}_{F(E)}) = \sum_{i=1}^{N^{F(E)}_{RNd}} x_{F(E)}(i) \cdot mMCP_{F(E)}(i, top, tot, n, tod)$$

$$CP^n_{F(E)} = \sum_{i=1}^{N^{F(E)}_{RMD}} x_{F(E)}(i) \cdot mMCP^n_{F(E)}(i),\ n \in WC$$

3.3.1 Optimization of the Forces Needed for Defined Threat Scenario

The problem of armed forces structure assignment needed for success in defined type of operation, given that structure of Enemy forces are fixed before (it is known a long before such operation), could be formulated as the following multi-objective optimization problem:

$$\min_{\bar{X}_F}(F_1(\bar{X}_F),\ F_2(\bar{X}_F)), \tag{1}$$

where values of $\bar{X}_E, toc, top, tot, tod$ are fixed;

subject to the constraints:

i. $g_F(\bar{X}_F, \bar{X}_E) = Fl_F(fr_{F|E}(\bar{X}_F, \bar{X}_E), toc, top, tot, tod) \leq d_F$

ii. $g_E(\bar{X}_F, \bar{X}_E) = Fl_E(fr_{F|E}(\bar{X}_F, \bar{X}_E), toc, top, tot, tod) \geq d_E$

iii. $g_n(\bar{X}_F, \bar{X}_E) = \frac{CP_F^n(\bar{X}_F)}{CP_E^n(\bar{X}_E)} \geq d_n,\ n \in WC$

iv. $g_{aA}(\bar{X}_F, \bar{X}_E) - \frac{CP_F^T(\bar{X}_F) + CP_F^{LR}(\bar{X}_F) + CP_F^{SR}(\bar{X}_F)}{CP_E^T(\bar{X}_E) + CP_E^{AV}(\bar{X}_E) + CP_E^{BMP}(\bar{X}_E)} \geq d_{aA}$

v. $g_{AD|H}(\bar{X}_F, \bar{X}_E) = \frac{CP_F^{AD}(\bar{X}_F)}{CP_E^H(\bar{X}_E)} \geq d_{AD|H},\ n \in WC$

vi. $\begin{matrix} \bar{X}_F, \bar{X}_E \geq 0 \\ \bar{X}_F, \bar{X}_E \leq M \end{matrix}$, where M is a very large value.

3.3.2 Optimization of the Forces for Fixed Threat Scenario Under Financial Constraints

The problem of armed forces structure assignment such as it is as near as possible to structure needed for success in defined type of operation, considering financial constraints and given that structure of Enemy forces are fixed before (it is known a long before such operation), could be formulated as the following optimization problem:

$$\min_{\bar{X}_F} \sum_{m \in WC \cup \{F,E,aA,AD|H\}} (g_m(\bar{X}_F, \bar{X}_E) - d_m)^2, \quad (2)$$

where values of $\bar{X}_E, toc, top, tot, tod$ are fixed;

subject to the constraints:

$F_1(\bar{X}_F) \leq B_F,\ \begin{matrix} \bar{X}_F, \bar{X}_E \geq 0 \\ \bar{X}_F, \bar{X}_E \leq M \end{matrix}$, where M is a very large value.

Problem of armed forces structure optimization could be also formulated in another manner. We could consider gaming model where both sides may optimize their structure. In such situation, we should find structure which is the best for the most dangerous or probable decision of enemy side, considering or not our financial constraints.

We could consider problem of structure optimization, taking into account possible support of our allies. It means, that vector of friendly forces could be defined as $\bar{X}_F = \bar{X}_F^{own} + \bar{X}_F^{allies}$, where $\bar{X}_F^{own}$ defined our forces, and $\bar{X}_F^{allies}$ defined forces of our allies. We should find the best $\bar{X}_F^{own}$, assuming that $\bar{X}_F^{allies}$ is defined.

4 Methods of Problem Resolution

In order to solve the problem we defined, we adopted one of the methods of artificial intelligence - the evolutionary algorithm in order to resolve the non-smooth problem defined in p. 3. It is a subset of the evolutionary calculations of a generic population-based metaheuristic optimization algorithm. It uses mechanisms inspired by biological evolution, namely reproduction, mutation, recombination and selection. The problem is coded in a number of bit strings that are modified by the algorithm in few steps. Possible (candidate) solutions to the problem play the role of individuals in the population. Decision variables and problem functions are used directly. Evolutionary algorithms differ from classic methods in several ways: random operation (versus deterministic), population solution (versus single best solution), creating new solutions with help of mutation, combining solutions with crossover operation, selecting solutions regarding a rule "survival of the fittest". The other feature (a drawback) is that the evolutionary algorithm leads to a solution which is "better" only in comparison to other, presently known solutions (not a global one) [12–14].

"Randomness" in evolutionary algorithm means that it partially relies on random sampling. Due that fact it might be treated as a nondeterministic method. In a consequence, it may conduct to different solutions on different runs, even with unchanged model and input data. However for the optimization of AF structure that is kind of equivalence of good enough solutions, which can be used in a battle. It is a kind of walking through the search space in order to find the best solution. However, we do not know if some better one may later be found (outside the vicinity of the current solution). One or a few (with equivalent objectives) of these is "best". The algorithm should start from one of "sample points". It is guarantee that the evolutionary algorithm avoid becoming "trapped" at a local optimum.

The third step is mutation – that operation leads to periodical random changes or mutations in one or more position of the current population. After a mutation a new possible candidate solution may be either better or worse than existing population members. In the proposed method we can perform a "mutation" via three different mutation strategies. However, we should remember that the result of "mutation" may be an infeasible solution. The subsequent operation is "crossover". The evolutionary algorithm makes attempts to mix some items of existing solutions (e.g. decision variable values) trying to obtain a new better solution. The result will possess some of the features of each "parent".

Finally, following a natural selection in evolution, the evolutionary algorithm realizes a selection activity going towards ever-better solutions. After that crucial step only the "most fit" items of the population will survive. We should stress that in a constrained optimization problem the concept of "fitness" is in relation with problem of "feasibility" of a solution (i.e. whether the solution satisfies all of the defined constraints). Moreover, it partly depends on its objective function value.

In order to calibrate the evolutionary algorithm and find the best solution some parameters are open to modification: number of chromosomes in population, cross-over probability, random selection probability, chromosome mutation probability, cross-over type. Concluding, having a highly nonlinear problem and with multiple local minima, the evolutionary algorithm performs better than gradient-based methods in a task of finding a global optimum.

5 Computational Experiments

5.1 Allocation Problem Without Cost Constraints

Scenario 1: DD-DA, terrain: Mixed, Enemy: 4x Armored Brigades, 4xMechanized Brigades, 2x Artillery Brigades, 1x Missile Artillery Brigade, 1xAir-Defense Brigade, 1xHellicopter Combat Brigade, 3xTactical Missile Brigade (Table 4).

Table 4. Parameters of allocation problem

Constraint type	Friendly power	Enemy power	Ratio	Losses %		Constr. value
Friendly losses	2957,77	4118,27	0,72	0,26	<=	0,3
Enemy losses			0,72	0,30	>=	0,3
Combat power BWP KTO	306,13	465,47	0,66		>=	0,5
Armored power	1559,04	1747,63	0,89		>=	0,6
Anti-armored power	129,11	246,36	0,52		>=	0,5
LR artillery power	275,59	513,51	0,54		>=	0,5
Medium artillery power	200,16	362,49	0,55		>=	0,5
Air-defense power	104,00	200,28	0,52		>=	0,5
Helicopter power	48,00	90,00	0,53		>=	0,5
Infantry power	119,74	204,53	0,59		>=	0,5
Anti-armored/ armored	1736,15	2213,09	0,78		>=	0,4
Air-defense/ helicopter	104,00	90,00	1,16		>=	0,5
Tactical missile power	216	288	0,75		>=	0,4
Anti-armored power LR		188,17				
Anti-armored power MR		58,19				

Optimal AF structure is presented in Table 5.

Table 5. Allocation – AF structure

No	Type of module	AF structure	Ref module cost mln PLN	Combat power
1	Motorized Rifle battalion 1	3	81,4	95,5
2	Motorized Rifle battalion 2	2	91,1	98,3
3	Mount. inf. bat	3	80,1	83,9
4	Airborne bat.	0	218	66,2
5	Air cavalry bat.	0	205,5	78,18
6	Mechanized bat.	0	122,6	111,2
7	Terr. def. bat.	0	15,2	32,9
8	Armored bat. PT91	0	97,9	131
9	Armored bat. LeoPL	3	128,9	259,8
10	Armored bat. Gep.	2	113,4	214,6
11	Antitank squadron	1	69,5	99,9
12	Artillery squadr (SR) 1	2	64,9	40,8
13	Artillery squadr (SR) 2	1	61,7	36
14	Artillery squadr (SR) 3	3	55,3	29,8
15	Artillery squadr (MR)	4	71,3	26,4
16	MLRS	4	72,4	20,9
17	Artillery squadr TM 1	4	60,7	108
18	Artillery squadr TM 2	0	132,3	120
19	Air defense squadr 1	1	55,8	22,7
20	Air defense squadr 2	2	53,1	28,8
21	Attack squadr hellicop 1	1	104,7	48
22	Attack squadr hellicop 2	0	318,1	96

Number of soldiers: 15 292, Number of modules: 36, Cost of Structure: 2 791,71mln PLN, Personnel losses: 3 954

Effect and cost characteristics (Figs. 2 and 3)

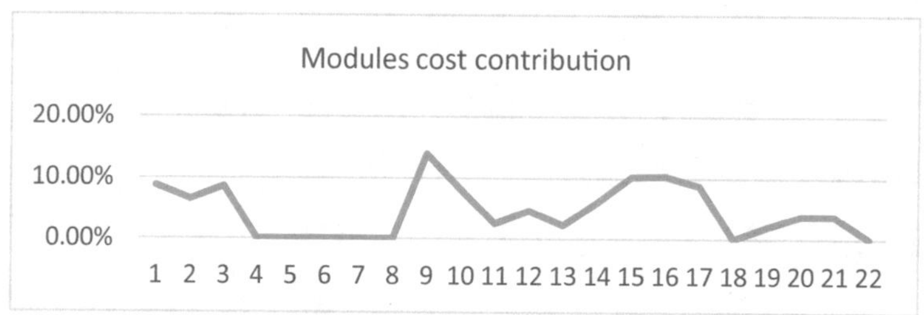

Fig. 2. Modules cost contribution in threat scenario 1 (no cost constraints)

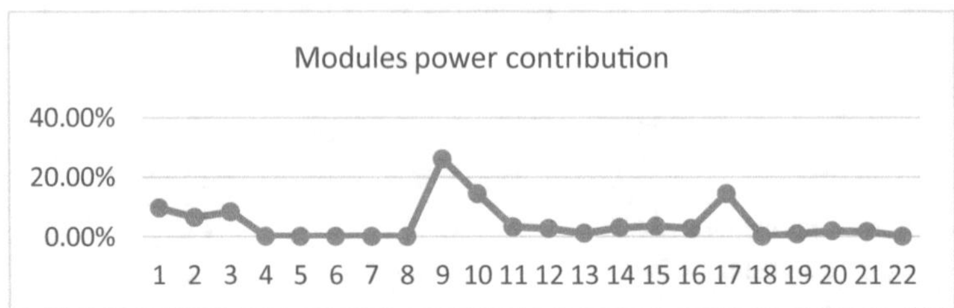

Fig. 3. Modules combat power contribution in threat scenario 1 (no cost constraints)

5.2 Allocation Problem Under Cost Constraints

Scenario 2: DD-DA, terrain: Mixed, Enemy:4x Armored Brigades, 4xMechanized Brigades, 2x Artillery Brigades, 1x Missile Artillery Brigade, 1xAir-Defense Brigade, 1xHelicopter Combat Brigade, 3xTactical Missile Brigade (Table 6).

Table 6. Parameters of allocation problem (scenario 2)

Constraint type	Friendly power	Enemy power	Ratio	Losses %	relation	Constr. value	Square of deviation
Friendly losses	2671,77	4118,27	0,65	0,33	<=	0,3	0,00043997
Enemy losses			0,65	0,25	>=	0,3	0,00039105
Combat power BWP KTO	34,70	465,47	0,07		>=	0,5	0,36
Armored power	2391,04	1747,63	1,37		>=	0,6	0,25
Anti-armored power	24,12	246,36	0,10		>=	0,5	0,25
LR artillery power	76,80	513,51	0,15		>=	0,5	0,25
Medium artillery power	0,00	362,49	0,00		>=	0,5	0,25
Air-defense power	8,96	200,28	0,04		>=	0,5	0,25
Helicopter power	0,00	90,00	0,00		>=	0,5	0,25
Infantry power	16,15	204,53	0,08		>=	0,5	0,25
Anti-armored/armored	2415,16	2213,09	1,09		>=	0,4	0,16
Air-defense/helicopter	8,96	90,00	0,10		>=	0,5	0,25
Tactical missile power	120	288	0,42		>=	0,4	0,16
Anti-armored power LR		188,17					Value of criterion:2,68
Anti-armored power MR		58,19					

Enemy paste of attack: 7,21 km/day; time of defense: 13,88 days (100 km depth of defense)

Optimal AF structure under the cost constraint F1 <= 2000 mln PLN – Table 7.

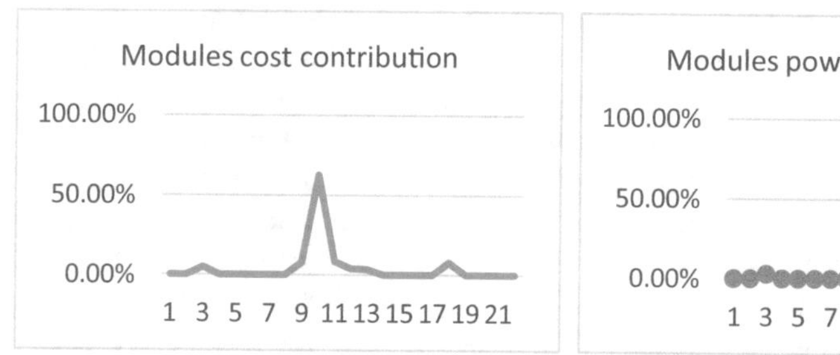

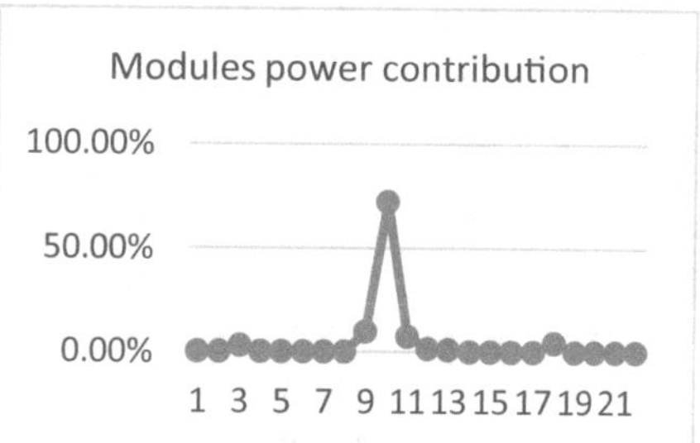

Fig. 4. Modules cost and combat contribution in threat scenario 2 (cost constraints)

Table 7. Allocation – AF structure (scenario 2)

No	Type of module	AF Structure	Ref module cost mln PLN	Combat power
1	Motorized Rifle battalion 1	0	81,4	95,5
2	Motorized Rifle battalion 2	0	91,1	98,3
3	Mount. inf. bat	1	80,1	83,9
4	Airborne bat.	0	218	66,2
5	Air cavalry bat.	0	205,5	78,18
6	Mechanized bat.	0	122,6	111,2
7	Terr. def. bat	0	15,2	32,9
8	Armored bat. PT91	0	97,9	131
9	Armored bat. LeoPL	1	128,9	259,8
10	Armored bat. Gep.	9	113,4	214,6
11	Antitank squadron	2	69,5	99,9
12	Artillery squadr (SR) 1	1	64,9	40,8
13	Artillery squadr (SR) 2	1	61,7	36
14	Artillery squadr (SR) 3	0	55,3	29,8
15	Artillery squadr (MR)	0	71,3	26,4
16	MLRS	0	72,4	20,9
17	Artillery squadr TM 1	0	60,7	108
18	Artillery squadr TM 2	1	132,3	120
19	Air defense squadr 1	0	55,8	22,7
20	Air defense squadr 2	0	53,1	28,8
21	Attack squadr hellicop 1	0	104,7	48
22	Attack squadr hellicop 2	0	318,1	96

Number of soldiers: 6 990, Number of modules: 16, Cost of Structure: 1 627,59 mln PLN, Personnel losses: 1 951

Effect and cost characteristics (Fig. 4)

6 Conclusions

The presented approach and some results of our researches were used in comprehensive analysis within Strategic Defense review of Polish Armed Forces. However data and scenarios used in the paper are something different and partially hypothetical. The analysis was focused on possibility of Armed Forces structures determination by taking into account the most probable scenarios of threat and allocation of different types of reference modules. The advantage of the approach in contrast to the others [1, 2, 8] depends on cost and combat power optimization. We proposed complete computational method to provide such analysis for defense planners. We have some extensions of the method by using specific combat simulator, which is used for much more detailed verification of the method and tools presented here. It will be described in the next publications of the authors.

References

1. Bracken, J., Kress, M., Rosenthal, R.E.: Warfare Modeling. Wiley, Hoboken (1995)
2. The taskform methodology: a technique for assessing comparative force modernization, TASC, Inc., Arlington, USA (1995)
3. Zanella, J.A.: Combat Power Analysis is Combat Power Density, School for Advanced Military Studies (SAMS), 201 Reynolds Avenue Fort Leavenworth, KS 66027-2339 (2012)
4. Allen, P.: Situational Force Scoring: Accounting for Combined Arms Effects in Aggregate Combat Models, RAND, Santa Monica (1992)
5. STUDENT TEXT 100-3, BATTLE BOOK, U.S. Army Command and General Staff College Fort Leavenworth, Kansas ACADEMIC YEAR 07 August
6. Tillman, M.E.: Optimizing force ratios to develop a course of action for the G3 (Operations Officer). Math. Comput. Model. Int. J. **23**(1–2), 55–63 (1996)
7. Brown, G.G., Dell, R.F.: Optimizing Military Capital Planning. Interfaces **34**(6), 415–425 (2004)
8. Fauske, M.F., Vestli, M., Glærum, S.: Optimization model for robust acquisition decisions in the norwegian armed forces. Interfaces **43**(4), 335–352 (2013)
9. Tagarev, T., Tsachev, T., Zhivkov, N.: Formalizing the optimization problem long-term capability planning. Inf. Secur. Int. J. **23**(1), 99–114 (2009)
10. Xiong, J., Yang, K.W., Liu, J., Zhao, Q.S., Chen, Y.W.: A two-stage preference-based evolutionary multi-objective approach for capability planning problems. Knowl.-Based Syst. **31**, 128–139 (2012)
11. Najgebauer, A., Antkiewicz, R., Chmielewski, M., Dyk, M., Kasprzyk, R., Pierzchała D., Jarosław R., Tarapata Z.: The qualitative and quantitative support method for capability based planning of armed forces development. In: Intelligent Information and Database Systems, Lecture Notes in Computer Science, vol. 9012, pp. 224–234 (2015)
12. Michalewicz, Z.: Algorytmy genetyczne + struktury danych = programy ewolucyjne. WNT, Warszawa (2003)
13. Goldberg, D.E.: Algorytmy genetyczne i ich zastosowania. WNT, Warszawa (1995)
14. Ashlock, D.: Evolutionary Computation for Modeling and Optimization. Springer, New York (2006). ISBN 0-387-22196-4

An Application for Supporting the Externalisation of Expert Knowledge

Adam Dudek[1(✉)] and Justyna Patalas-Maliszewska[2]

[1] Institute of Computer Science, University of Applied Science in Nysa, Nysa, Poland
adam.dudek@pwsz.nysa.pl
[2] Institute of Computer Science and Production Management, University of Zielona Góra, Zielona Góra, Poland
j.patalas@iizp.uz.zgora.pl

Abstract. This paper explores the problem of the externalisation of expert knowledge in order to build a new procedure for a given work place. In the first stage of our approach, the processes carried out by experts in the company are verbally described by them and recorded. Next, using Google Cloud Speech Api technology, the voice recording is converted into text. In the second stage, based on the set of the text, a Finalised Word Dictionary is created. In the third stage the steps constituting the procedure are distinguished and finally the procedure for a given work place, based on the expert's vocalised report is created. The proposed application can be a useful tool for the acquisition of new explicit knowledge within an enterprise.

Keywords: Expert knowledge · Externalisation · An application

1 Introduction

Knowledge can be divided into explicit knowledge and tacit knowledge. Explicit knowledge is identified from records within a knowledge base [1]. It is formalised and can be presented by documentation, instructions, etc. Tacit knowledge is closely related to the skills, education, and experience of employees [2]. In addition, the transformation of one type of knowledge into another type of knowledge can be the first step in the creation of new knowledge within a company [3].

The value of knowledge, within a company, can increase by consequences related to limitations in the process of acquiring expert knowledge [4]. At the centre of our Expert Knowledge Externalisation Support (EEES) approach is the expert. The selection of experts is recommended as the first step in the EEE Support model, since they should be the sources of the "best" knowledge in the company as a result of the tasks they have already completed [5]. It is known, that there are methods to identify experts within a company, such as the Generalised Expertise Measure (GEM) questionnaire [6] or assessment. In this paper, we argue that experts can be selected on the basis of the numbers of years of experience in the company [7] and we state that experts in a company are selected according to their experience in performing specific

J. Świątek et al. (Eds.): ISAT 2018, AISC 853, pp. 255–265, 2019.
https://doi.org/10.1007/978-3-319-99996-8_23

activities within a company. Experts are those employees who are engaged in tracking down the expertise, experience and opinions [8].

In this paper, we will attempt to solve the problem of the externalisation of expert knowledge, in order to build a new procedure for a given work place and thus obtain new, explicit knowledge for the company.

In the first stage of our EEES approach, the processes carried out by experts in the company are verbally described by them and recorded, using smartphones. Next, using Google Cloud Speech Api technology, the voice recording is converted into text. In order to receive the new procedure for a given work place, it is assumed that the activities performed, at a given workplace, are performed sequentially and also that it is possible to distinguish the steps constituting a procedure.

In the second stage, based on the set of the text, a Finalised Word Dictionary is created. This includes the names of the actions, along with the synonyms accorded to each of the names of the actions; the names of the sub-assemblies, along with the synonyms accorded to each sub-assembly name; the names of connecting elements, along with the synonyms accorded to each name of the connecting elements; the names of the consumables, along with the synonyms accorded to each of the names of the consumables and the names of the tools, along with the synonyms accorded to each of the names of the tools. It will then be possible to distinguish the steps and finally to determine the procedure for a given work place, based on the expert's vocalised report. The approach presented has been illustrated by the real-life example of the printing process in an offset printing company. Finally, we present the Application for Supporting the Externalisation of Expert Knowledge.

2 An Approach to Supporting the Externalisation of Expert Knowledge

In the literature, the following methods for supporting the externalisation of expert knowledge can be found: the observation of the source of the expert knowledge, in real-time; the verbal and non-verbal analysis of the problems of recording the solutions as well as the hidden training and demonstration courses [9–13]. The presented approach is new in relation to the discussed above, because it is possible to acquire new knowledge within a company: a new procedure for a given work place.

The building of models or applications supporting the externalisation of expert knowledge is particularly important for maintaining a company's unique knowledge, as well as for creating new knowledge within a company. By selecting the method for externalising expert knowledge, we should look not only for solutions that would automate this process, but for those which would also minimise the level of involvement of the expert who is the source of this knowledge [14]. The applications should allow expert knowledge to be elicited based on the recognition of patterns in images, videos and sound recordings [15].

The proposed approach provides an opportunity to build a new procedure for a given work place, based on the received voice records of the experts within a company. Our approach, therefore, consists of the following stages (see Fig. 1).

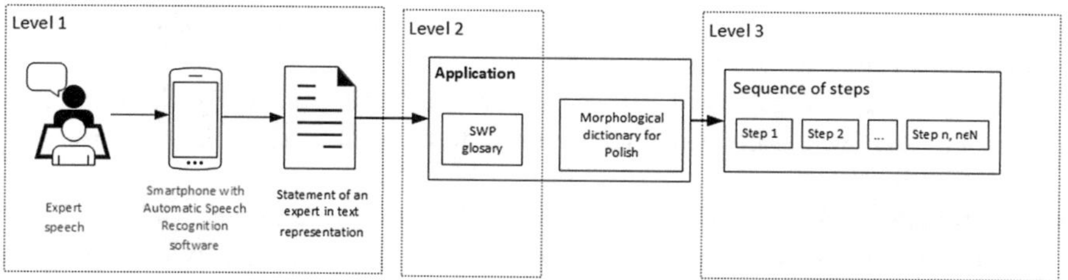

Fig. 1. An approach to supporting the externalisation of expert knowledge

Stage 1: Acquisition of recordings from an expert, concerning activities at a certain work position

The approach presented (Fig. 1) assumes that the expert knowledge could be obtained using a verbal analysis of the description of the implementation of the task. The source of knowledge is the expert's verbal statement, expressed, naturally, in his/her native Polish. It would then be proposed to use Automatic Speech Recognition to convert this expression into a form of text entry.

One of the few solutions of this class, available for the Polish language, is Google Cloud Speech Api. This solution is available in the form of an API programming interface, or as a service integrated with the Android operating system; this allows any device to be used with the system as a tool to convert the expert's statements into text. It is worth noting that although modern tools, of the ASR class, are characterised by their high accuracy, in terms of the proper recognition of words, they are not usually able to correctly interpret the punctuation marks found in vocalisations. Consequently, it is not possible to automatically divide vocalisations into individual sentences or parts thereof.

Stage 2: Building the SWP dictionary

The model assumes that in order to formulate a new procedure for a given workplace, the statement must be divided into steps. The term "step" is described by the following characteristics:

- The name of the activity performed - cz.
- The name of the components and parts supported - p.
- The names of the connecting elements used - l.
- The names of the tools used - n.
- The names of the consumables used - m.

Thus, each step in the service procedure can be presented in the form of characteristics (formula 1).

$$K = \{cz, p, l, n, m\} \tag{1}$$

The process of extracting characteristics from the textual format of vocalisations, in individual steps, consists of:

- Distinguishing the names of activities, therein, that is, distinguishing those formulations acting as predicates.
- Distinguishing the names of sub-assemblies and their connecting elements, along with the tools which the consumables used, that is, where formulations assume the role of subjects.
- Combining the designated predicates and subjects that are in their vicinity. For the sake of simplicity, it is assumed that the beginning of such a group is the predicate, while all other concepts found are assigned to it until the next predicate is found.

The object of this process is to build a Dictionary of Words and Concepts (SWP) in which basic concepts - in the process in which the solution is proposed- are applied. These concepts should be assigned to categories corresponding to elements describing the characteristics of the step. Since the model presented is to be used for texts, expressed in natural Polish, which may be characterised by its considerable degree of ambiguity, a number of synonyms have been defined for each of the basic concepts accumulated in the dictionary.

Polish - for which this particular processing model has been proposed- is characterised by its complex flection which significantly complicates its automatic processing. In order to solve this problem, its inflected forms were added for each of the synonyms stored in the SWP dictionary, in such as cases and numbers for nouns, as well as for the persons, modes and tenses of each noun. A section of one such extension, in the structure of the SWP dictionary, is presented in Fig. 2.

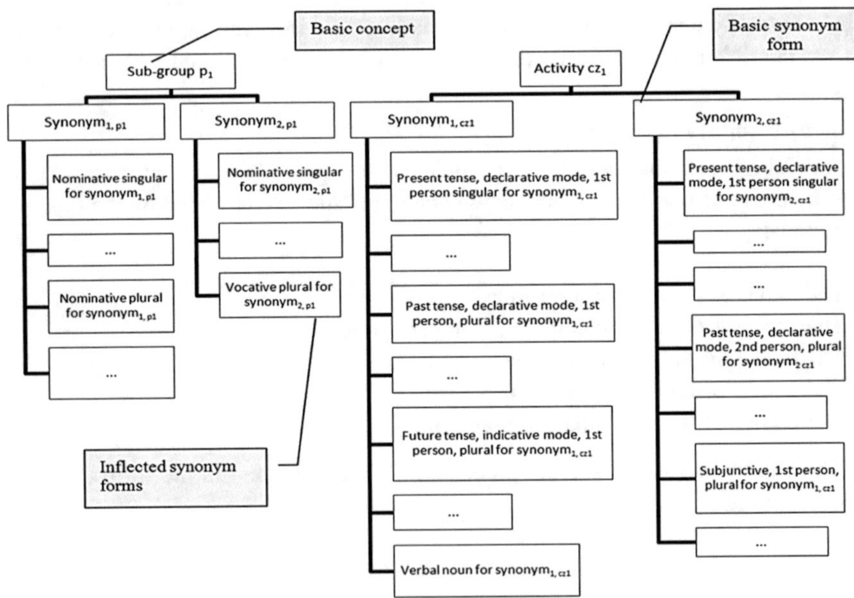

Fig. 2. Thesaurus and dictionary of inflected forms

Step 3: Extraction of steps from the expert's statement
Such an extensive SWP dictionary is the basis for the implementation of the algorithm for transforming the original speech recording. It can be expressed using the following steps:

1. Search the text analysed for the forms of inflected synonyms stored in the SWP dictionary.
2. Replace the inflected forms found in step 1 with the basic forms of the synonyms.
3. Replace instances of synonyms from step 2 with the corresponding basic concepts.
4. For each of the basic concepts designated in step 3, determine its affiliation to the appropriate category of concepts.

However, this algorithm only allows the meaning to be determined, of those concepts that have been accumulated in the SWP dictionary. The Open Language Morphological Dictionary of the Polish Language - Polimorf [16] was used for the morpho-syntactic analysis of the remaining part of the vocalisations. The above-mentioned dictionary contains some 6.6 million inflected forms. It is possible to indicate, in the document analysed, those concepts which assume the *rôle* of predicate; on this basis, therefore, the potential limits of the characteristics of the individual steps can be formed.

3 An Application for Supporting the Externalisation of Expert Knowledge

The proposed approach to the externalisation of expertise (Fig. 1) was undertaken in the form of a dedicated application for an offset machine operator's work position. Any procedure preparing a printing machine for the work position of an offset machine operator, should be prepared as part of the printing process as a whole.

Step 1: Obtaining expert recordings, of an offset machine operator at his/her work position
Two recordings of a statement by the *Head of Printing* were obtained, which after processing, using ASR (Automatic Speech Recognition) technology, are here represented textually:
Transcript 1:

"we start the offset machine then you have to check the aluminum offset plate if it is okay then you just have to place the solid board on the feeder set the lining section and adjust the machine parameters now you just need to set the dosage of auxiliaries and it is ready"

Transcript 2:

"I started the offset machine and I checked the status of the offset aluminium plate it turned out that it was the wrong offset plate and therefore I had to order a new offset plate then flashed the new offset disc and fixed the exposed image and installed the offset plate in the machine I arranged the solid board on the feed and set the lining section and adjusted parameters of the machine at the end of the adjustment I also set the dosing of auxiliaries"

In addition, supplementary information was obtained:

- The printed form is made on a thin aluminium offset plate, *viz.*, a metal sheet, of uniform, constant thickness across its surface, with a layer of light-sensitive emulsion.
- Printing is done on solid cardboard.
- There is an optically copied and chemically created drawing on a new disc.

Stage 2: Building the SWP dictionary

When implementing the assumptions of the SWP dictionary, in practice (Sect. 2), it was expanded to include the possibility of adding a drawing or a photo, symbolising the individual items stored therein. The use of such graphical symbols makes it easier for the user of the tool to interpret the sequence of steps, as presented (Fig. 3).

Fig. 3. An implementation of a SWP dictionary

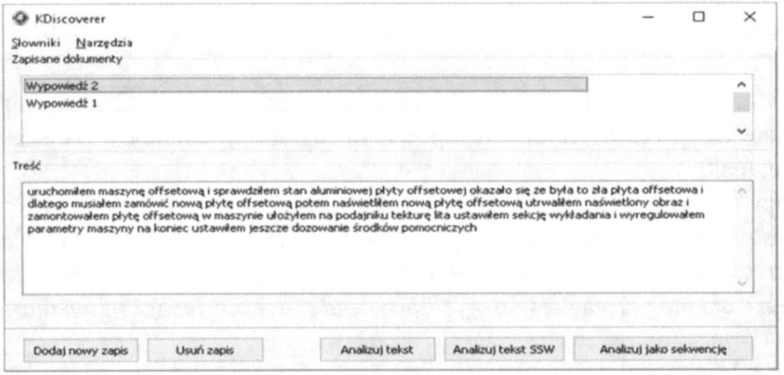

Fig. 4. Management of the collection of recorded vocalisations

The application presented above allows the text to be analysed (Fig. 4) in 3 modes:

(1) Interpretation of the content using the contents of the Polimorf dictionary only.
(2) Interpretation of the content using the concepts of both the SWP dictionary and the Polimorf dictionary.
(3) Designation and presentation of the characteristics of the sequence of steps.

Step 3: Extraction of steps from the expert's statement
Figure 5 presents the results of the analysis of the example of statement No. 1 (stage 1) carried out in mode I. It is presented in the form of a list of items that are individual words extracted from the document analysed. Each of them is characterised by indicating the main form- with its affiliation to the appropriate categories of lexemes- and determines its role in the sentence. In addition, depending on which part of the sentence it has been classified to, it is presented using different colours; such as predicates, for example, are marked in red.

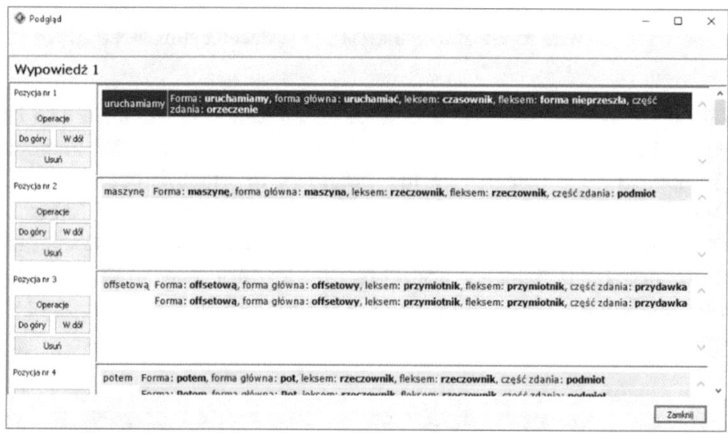

Fig. 5. The result of the analysis of a recorded vocalisation using the dictionary of inflected forms.

The application also allows the following operations, for each element on the list, to be carried out (Fig. 5):

(1) Replacement of the words indicated by any item from the SWP Dictionary.
(2) Display of synonyms for the word indicated [17].
(3) Addition of the basic form of the word indicated, as a new entry in the SWP Dictionary.
(4) Replacement of the role in the sentence of the word indicated.

Figure 6 presents the result of the analysis of the same recorded vocalisation (statement 1) in mode II, that is, using both the dictionary of inflected forms and the SWP dictionary.

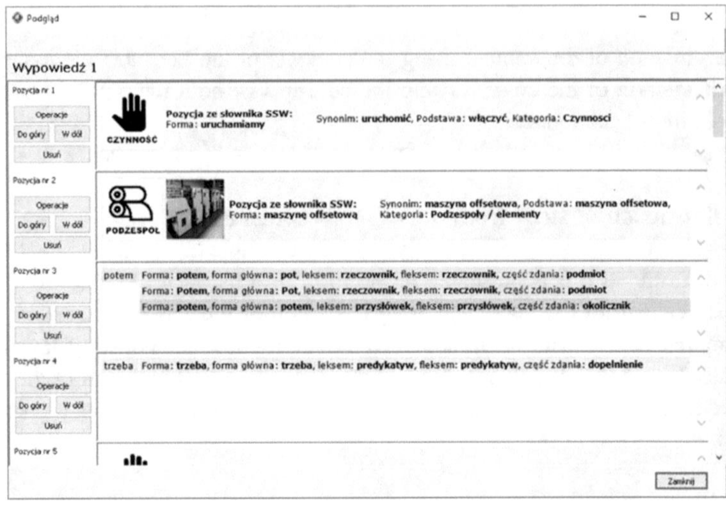

Fig. 6. The result of the analysis of the document in mode II

As with mode I, the result of the analysis is presented in the form of a list. If the word- or group of words - from the list has an equivalent in the SWP dictionary, then an icon symbolising the category in the dictionary to which it belongs is displayed, together with a graphic miniature - where such has been defined for it. In turn, analysis of the document in mode III, enables the extraction and then the presentation of the sequence of step characteristics according to the proposed model of knowledge externalisation. Figure 7 presents the result of the analysis of the recording of statement 1 in this mode.

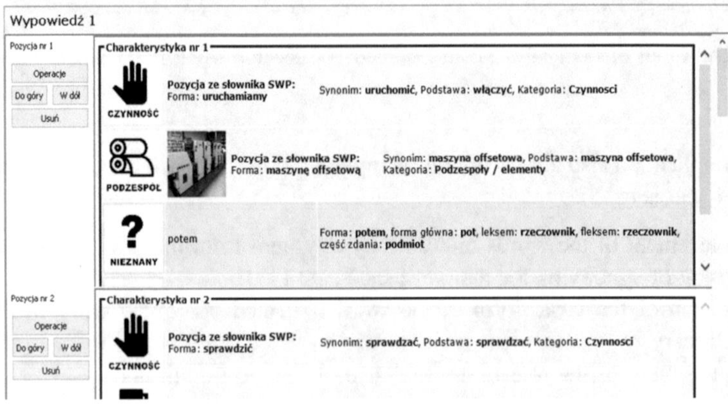

Fig. 7. Characteristics of the steps separated in the recording of the statement

The application, discussed above, allows more than one designated sequence to be analysed at a time and compared parallelly (Fig. 8).

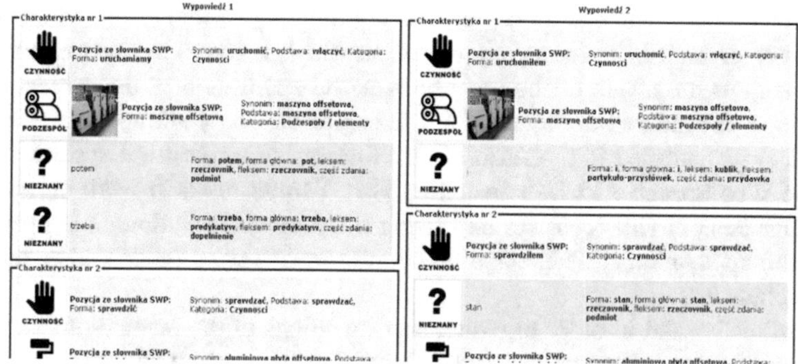

Fig. 8. Comparison of sequences obtained from two different vocalisations.

With the help of the present application, the specialist has developed - *for the work position of an offset machine operator*- a procedure for preparing the machine for printing (Fig. 9). The proposed procedure was presented using the BPMN notation, the MS Visio computer program.

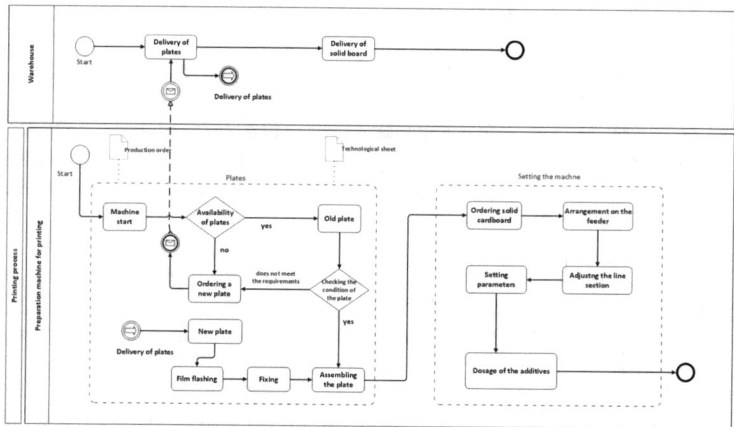

Fig. 9. The course of activities: preparation of the printing machine for the work position of an offset machine operator in the BPMN notation

4 Concluding Remarks

The application proposed, which assists the externalisation of expert knowledge, can be a useful tool for the acquisition of new knowledge within an enterprise. Applying the proposed solution, in an enterprise, enables not only expert knowledge to be obtained but also preserves it in the company in a form which is useful to other employees.

The authors are aware of the limitations in the application of the subject. Among other things, the authors are aware that the application only allows the vocalisations obtained to be analysed in the form of text files, using an Android system device or using the Web Speech API Demonstration [18]. Further research work on the part of the present authors will focus on improving the proposed solution. It is proposed to modify the application so that it can function in *multi-thread* mode and in the client-server architecture.

Additionally -and in order to reduce the limitations of the method of extraction of the characteristics of the steps presented above- the use of more advanced methods for processing natural language (NLP), that is, methods for detecting proper names, the analysis of correlations - *as well as the recognition of collocations*- have been planned [19, 20].

References

1. Patalas-Maliszewska, J.: Knowledge Worker Management: Value Assessment, Methods, and Application Tools. Springer, Heidelberg (2013)
2. Nonaka, I.: A dynamic theory of organisational knowledge creation. Organ. Sci. **5**(1), 14–37 (1994)
3. Nonaka, I., Takeuchi, H.: Creation of Knowledge in the Organisation. Poltext, Warsaw (2000)
4. Gajzler, M.: The idea of knowledge supplementation and explanation using neural networks to support decisions in construction engineering. In: 11th International Conference on Modern Building Materials, Structures and Techniques, MBMST 2013. Procedia Engineering, vol. 57, pp. 302–309 (2013)
5. Bolger, F., Wentholt, M.: Principles and practice of selecting and motivating experts. EFSA journal 2014: guidance on expert knowledge elicitation in food and feed safety risk assessment, European Food Safety Authority (EFSA), Parma, pp. 138–162 (2014)
6. Germain, M.L., Tejeda, M.J.: A preliminary exploration on the measurement of expertise: an initial development of a psychometric scale. Hum. Resour. Dev. Q. **23**(2), 203–232 (2012)
7. Alvarado-Valencia, J., Barrero, L.H., Onkal, D., Dennerlein, J.T.: Expertise, credibility of systems forecasts and integration of methods in judgmental demand forecasting. Int. J. Forecast. **33**, 298–313 (2012)
8. Gray, P.H., Meister, D.B.: Knowledge sourcing effectiveness. Manag. Sci. **50**(6), 821–834 (2004)
9. Potocoki, A, Łukasik, P.: Wybrane metody komunikacji ukierunkowane na wykorzystanie wiedzy w organizacji. In: Stabryła, A., Wawak, S. (eds.) Problemy zarządzania organizacjami w społeczeństwie informacyjnym. Wydawnictwo Mfiles, Kraków (2014)
10. Lemke, J.: Analyzing Verbal Data: Principles, Methods, and Problems, Springer International Handbooks of Education, vol. 24, pp. 1471–1484. Springer, Dordrecht (2012)

11. Salmon, K., Pipe, M.-E., Malloy, A., Mackay, K.: Do non-verbal aids increase the effectiveness of best practice verbal interview techniques? An experimental study. Appl. Cogn. Psychol. **26**(3), 370–380 (2012)
12. Zieliński, M.: Personnel conditions of creation of intelligent organization. Scientific Papers of Silesian University of Technology. Organization and Management Series, no. 1946, p. 86 (2015)
13. Pa, N.C., Taheri, L., Rusli, A.: A survey on approaches in knowledge audit in organizations. Asian Trans. Comput. **02**(05), 1–8 (2012)
14. Dudek, A., Patalas-Maliszewska, J.: A model of a tacit knowledge transformation for the service department in a manufacturing company: a case study. Found. Manag. **8**(1), 176–177 (2016)
15. Hall, P., Phan, W., Whitson, K.: The Evolution of Analytics. Opportunities and Challenges for Machine Learning in Business. O'Reilly Media, Sebastopol (2016)
16. Woliński, M., MIłkowski, M., Ogrodniczuk, M., Przepiórkowski, A., Szałkiewicz A.: PoliMorf: a (not so) new open morphological dictionary for Polish, Institute of Computer Science, pp. 860–864. Polish Academy of Sciences (2012)
17. LibreOffice: LibreOffice/Dictionaries. https://github.com/LibreOffice/dictionaries/tree/master/pl_PL. Accessed 09 May 2018
18. Google Cloud Platform: Web Speech API Demonstration. https://www.google.com/intl/pl/chrome/demos/speech.html. Accessed 09 May 2018
19. Marcińczuk, M., Kocoń, J.: Recognition of named entities boundaries in Polish texts. In: Proceedings of the 4th Biennial International Workshop on Balto-Slavic Natural Language Processing, pp. 94–99 (2013)
20. Kaczmarek, A., Marcińczuk, M.: Evaluation of coreference resolution tools for polish from the information extraction perspective. In: 5th Workshop on Balto-Slavic Natural Language Processing, Hissar, pp. 24–33 (2015)

Cognition and Decisional Experience to Support Safety Management in Workplaces

Caterine Silva de Oliveira[1(✉)], Cesar Sanin[1], and Edward Szczerbicki[2]

[1] The University of Newcastle, Newcastle, NSW, Australia
caterine.silvadeoliveira@uon.edu.au,
cesar.maldonadosanin@newcastle.edu.au
[2] Gdansk University of Technology, Gdansk, Poland
edward.szczerbicki@newcastle.edu.au

Abstract. Hazards are present in all workplaces and can result in serious injuries, short and long-term illnesses, or death. In this context, management of safety is essential to ensure the occupational health of workers. Aiming to assist the safety management process, especially in industrial environments, a Cognitive Vision Platform for Hazard Control (CVP-HC) is proposed. This platform is a Cyber Physical system, capable of identifying critical safety behaviors overcoming the limitations of current computer vision systems. In addition, the system stores experiential knowledge about safety events in an explicit and structured way. This knowledge can be easily accessed and shared and may be used to improve the user/company experience as well as to understand the company safety culture and to support a long term change process. The CVP-HC is a scalable yet adaptable system capable of working in a variety of video analysis scenarios whilst meeting specific safety requirements of companies by modifying its behavior accordingly. The proposed system is based on the Set of Experience Knowledge Structure (SOEKS) and Decisional DNA (DDNA).

Keywords: SOEKS · DDNA · Cognitive vision · Safety management

1 Introduction

Hazards are present in all workplaces and can result in serious injuries, short and long-term illnesses, or death [1]. The management of safety is essential to ensure the occupational health of workers [2]. An analysis of the effect of different types of injury prevention interventions found that the two most effective methodologies were the behavior-based and safety culture-based rubrics [3]. In this context, technologies to support the practical implementation of either approaches have emerged as a need, but the current technologies available still face considerable limitations [4–7].

Aiming to assist the safety management process, especially in industrial environments, a Cognitive Vision Platform for Hazard Control (CVP-HC) is proposed. This platform addresses the current limitations of computer vision systems and gives support for both culture change and behavior-based management approaches. The CVP-HC is a scalable yet adaptable system capable of working in a variety of video analysis

J. Świątek et al. (Eds.): ISAT 2018, AISC 853, pp. 266–275, 2019.
https://doi.org/10.1007/978-3-319-99996-8_24

scenarios whilst meeting specific safety requirements of industries by modifying its behavior accordingly. The proposed system is based on the Set of Experience Knowledge Structure (SOEKS or SOE in short) and Decisional DNA (DDNA), which were first presented by Sanin and Szczerbicki and later enhanced further for a number of dedicated domains [8, 9].

This paper is organized as follow: In Sect. 2, workplace safety management practices are presented with special focus to two main approaches: the behavior-based and culture change. In addition, the integration of both methodologies in one unique platform is suggested to achieve greater and sustaining effect in safety management. In Sect. 3 the role of Cognition to fill the gaps of current computer vision techniques that supports behavior-based strategies is explained. In Sect. 4 Decisional Experience is presented aiming to achieve a general purpose system that gives management support for the culture change process. In Sect. 5 the proposed Cognitive Vision Platform for Hazard Control (CVP-HC) based on the combination of cognition and decisional experience to support both methods is introduced. Finally, in Sect. 6 conclusions and future work are given.

2 Managing Safety in Workplaces

In industrial environments, workers are exposed to risks in a variety of unsafe situations, such as when accessing controlled zones without authorization, crossing yellow lines, not respecting safe distances from machines and areas, not wearing the required personal protective equipment (PPE), among others. The management of safety is the best way to prevent injuries in workplace [10]. An analysis of the effect of different types of approaches found that the two most effective were the behavior-based and safety culture-based [3]. These two prominent and seemingly antagonistic safety management rubrics will be explained next.

2.1 Behavior-Based and Automatic Detection of Hazards

In behavior-based safety (BBS) management, the focus of attention is directed at specific safety-related behaviors in a specific environment [11]. This approach is intended to be short-term or one shot effort, which means it might not be sufficient to provide a long term change in perception of unsafe behaviors and must be a continuous process [12]. Behavior-based safety management experts claim that the overall approach is broadly applicable to different work environments, but each application needs to be specific and tailored to the setting in question [7].

The use of sensor data and computer vision techniques can support automatic detection and tracking of workers, indicating potential dangerous behaviors or risky situations, which may be quickly corrected to avoid occurrence of accidents. Visual sensing facilities, such as video cameras, can monitor workers behavior and environment conditions without any disturbances to their movements and activities. However, besides the significant improvements in computer vision that has been done over the past few years, very little research has focused on specific requirements for industrial applications and the commercial systems available that are able to run in real time perform far from satisfactorily [5].

Furthermore, there is no such flexible and scalable system at the moment which is able to function in a broad industrial environments without the necessity of rewriting most of existing applications' code each time the circumstances or conditions change [13]. In terms of accuracy, the current vision-based systems performance designed to attend a more comprehensive diversity of scenarios and applications when operating in real life is still limited [14]. For this reason, building an automatic system capable of supporting behavior-based safety management of a variety of scenarios subject to different settings at the same time as being specific and meaningful still remains a challenge.

2.2 Culture Change and Administration of Knowledge

Culture change methodologies to safety come mostly from management and organizational behavior theory [7]. Similarly to groups of people, companies also have cultures which can be characterized, for instance, as weak or strong, rigid or flexible, functional or dysfunctional and can influence the behavior and perception of employees in regards to the significance of safety and their expectations with respect to the importance of safe work practices and incident reporting [15]. Organizations often share common practices and certain cultural values in regards to safety and may also experience similar events. However, the culture of each organization is thought to be unique and the safety behavior of their members and how it is approached must be understood in its own context [7]. It is important to point out that an organization culture can be resistant to change, leading to an exhausting and slow process.

Methods used to study and change organizational cultures tend to be largely qualitative and intuitive, which makes it difficult to generalize and share. In particular, the technology of culture change is quite diffuse, not well specified and lack explicitness. Assessments of organizational culture tend to be subjective and not easily reproduced or verified once the collection and availability of safety events occurred are most of the time scarce. Deficiency in administrating this data, also lads to limitations in measuring how programs and interventions impact particular safety problems and outcomes [7].

To date, there is no general and automatic purpose system that gives management support for the culture change process, facilitating the safe and storage of explicit knowledge that can be easily accessed to understand the culture of an organization as well as to delineate the change process in terms of stages and used to measure progress and impact of programs.

2.3 Integrative Approach to Managing Safety

Both behavior based and culture change approaches can be considered complementary to the extent that their respective strengths can be combined and their respective disadvantages tempered or eliminated [7]. By coupling these two approaches into one unique platform, the objective and experiential analysis of critical safety-related behaviors can be integrated with an organizational changing process, which can result in a more comprehensive and potentially effective methodology to manage safety.

A framework for integrating these two approaches in order to attain a greater and sustaining effect in safety management has been proposed – yet little guidance has been offered as to how these two approaches can be integrated in practice and no system has yet been developed to give support for this integration [16].

3 Cognition

To date, the creation of a general-purpose vision system with the robustness and resilience of the human visual system still remains a challenge [17]. One of the most recent tendencies in computer vision research aimed at expanding the computer vision systems to human-like capabilities is the combination of cognition and vision into cognitive computer vision. For instance, an automatic semantic and flexible annotation service able to work in a variety of video analysis with little modification to the code was proposed in work by Zambrano et al. [6]. This system is a pathway towards cognitive vision and it is composed, basically, by the combinations of detection algorithms and an experience based approximation. The results has shown scalabilities enhancements, but the tool was designed for off-line video analysis and does not have the capability of working in a real time implementation.

Cognitive vision involves functionalities for knowledge formalism, learning, recognition and categorization, reasoning about events for decision making, and goal specification, all of which are concerned with the semantics of the relationship between the visual agents and their environments i.e. context [17]. These functionalities direct cognitive vision systems towards purposeful behavior, adaptability, anticipation, and extendability – one of the main characteristics of a cognitive vision system, and stands for the capacity of accomplish certain goals, even in circumstances that were not expected or programmed for during the design process [17].

Finally, knowledge and learning are central to cognitive vision. To be readily articulated, codified, accessed and shared, knowledge must be represented in an explicit way. The "explicitness" of a knowledge depends on its nature and as well as the capacity of the system of proceduralize them [18]. In addition, cognitive systems use knowledge stored as experience to improve their performance at some specific situation. Therefore, a knowledge representation capable of storing the knowledge as experiences, facilitating sharing at the same time as allows their understanding inside different contexts is essential to cognitive vision system.

4 Decisional Experience

Reasoning, which is a computational process, to be feasible, requires systemic techniques and data structures, and in consequence, several techniques have been developed trying to represent and acquire knowledge. These kinds of technologies try to automatically collect and manage knowledge in some manner. Although these technologies involve decision-making in some way, they lack of keeping structured knowledge of the formal decision events they participate on [19]; they commonly do not store this experiential knowledge in an easy and accessible way and rarely reuse this experience. Therefore, such decisional experience is commonly disregarded, not shared, and put behind [20–23]. Little of this collected experience survives, and in some cases, over time, it becomes inaccessible due to poor knowledge engineering practices or due to technology changes in software, hardware or storage media [24].

In addition, among the features associated with knowledge engineering systems are human intelligence capabilities such as learning, reasoning and forecasting from current knowledge or experience. From an application point of view, different research projects have been presented by the scientific community involving knowledge representation and decision making technologies to extend the user's understanding; however, to our acquaintance, most of these approaches miss the potential of using knowledge based theories that might enhance the user's/organizations' experience and at the same time creating their decisional fingerprints [24].

To manage knowledge and decision events the platform utilizes Decisional DNA, a unique and single structure for capturing, storing, improving and reusing decisional experience. Besides, we make use of the Set of Experience (SOE) as part of the Decisional DNA which allows the acquisition and storage of formal decision events in a knowledge explicit form.

4.1 Set of Experience Knowledge Structure (SOEKS) and Decisional DNA

The Set of Experience Knowledge Structure (SOEKS) is a knowledge representation structure designed to obtain and store formal decision events in an explicit way. It is based on four key basic elements of decision-making actions: variables, functions, constraints, and rules. Variables are generally used to represent knowledge in an attribute-value form, following the traditional approach for knowledge representation. Given that, the set of Functions, Constraints, and Rules of SOEKS are different ways of relating knowledge variables, it is safe to say that the latter are the central component of the entire knowledge structure. Functions define relations between a dependent variable and a set of input variables; therefore, SOEKS uses functions as a way to create links among variables and to build multi-objective goals. Likewise, constraints are functions that act as a way to limit possibilities, limit the set of possible solutions and control the performance of the system in relation to its goals. Finally, rules are relationships that operate in the universe of variables and express the condition-consequence connection as "if-then-else" and are used to represent inferences and associate actions with the conditions under which they should be implemented [8].

The Decisional DNA consists is a structure capable of capturing decisional fingerprints of an individual or organization and has the SOEKS as its foundation. Multiple Sets of Experience can be collected, classified, organized and then grouped into decisional chromosomes, which accumulate decisional strategies for a specific area of an organization. The set of chromosomes comprise, finally, what is called the Decisional DNA (DDNA) of the organization [9].

5 Cognitive Vision Platform for Hazard Control (CVP-HC)

The overall architecture of the proposed Cognitive Vision Platform for Hazard Control (CVP-HC) is represented in Fig. 1. Its functioning is composed, basically by four layers: System Configuration, Central Reasoning, Experience Validation, and System Monitoring, which will be described as follows:

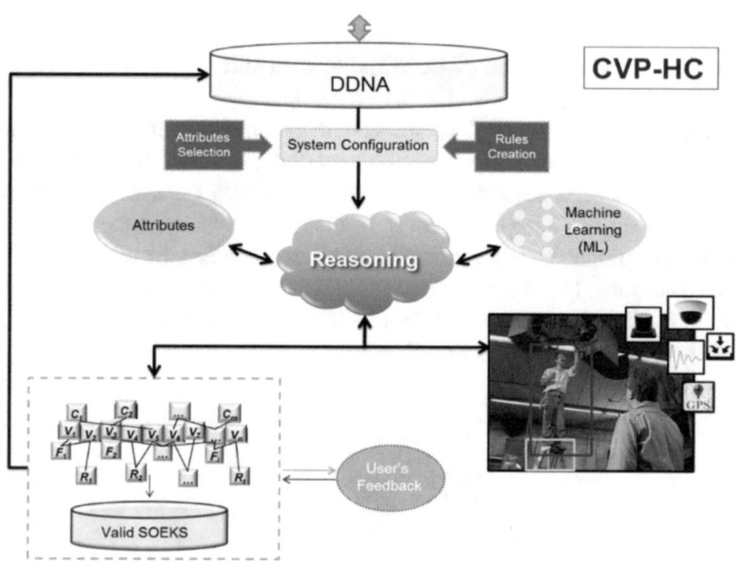

Fig. 1. Overall architecture of the Cognitive Vision Platform for Hazard Control.

System Configuration: This layer comprises the selection of the attributes according to industry requirements; labelling of machinery, areas and plants of the workplace being captured by the camera and sensors; creation of the rules and configuration of other functionalities such as frame and learning rate.

Experience Validation: In this layer, the collected experiences are compared among each other and the most redundant pruned. The remaining experiences are used for querying the user and check if given solution is reasonable. Once accepted by the user as a correct solution, the SOE is stored in the decisional DNA repository and used to retrain the system (Active Learning) to increase its specificity and accuracy.

Central Reasoning: This layer is the intelligence of the entire system. It is composed by a Convolutional Neural Network (CNN) arranged in a hierarchical structure to support detection of attributes, location and relationship among them as well as interpretation of the context and level of risk present in the recognized entity.

Monitoring: This layer represents the monitoring of workers' activities. When a hazard is identified by the system, an alert message is shown on the application with details/recommendations about the existing risk.

5.1 Data Fusion

Data fusion is the process of integrating multiple data sources to produce a more consistent, accurate, and useful information than that provided by individual data sources [25]. Data integration has two general objectives: increasing the completeness and increasing the conciseness of data that is available to provide better reasoning about a given situation [26].

The sources of data that will be integrated into the platform can vary from one company to another. The more sources integrated the more complex rules can be created and more diverse scenarios can be processed with accurate and precise output (depending also on the computer power available). However, to guarantee minimal operation of the proposed capabilities of the system, when in operation it must be composed by, at least, two main components: RGB Camera and a computer system. To guarantee flexibility the data fusion in our systems occurs within a three-step data integration process, as shown in Fig. 2.

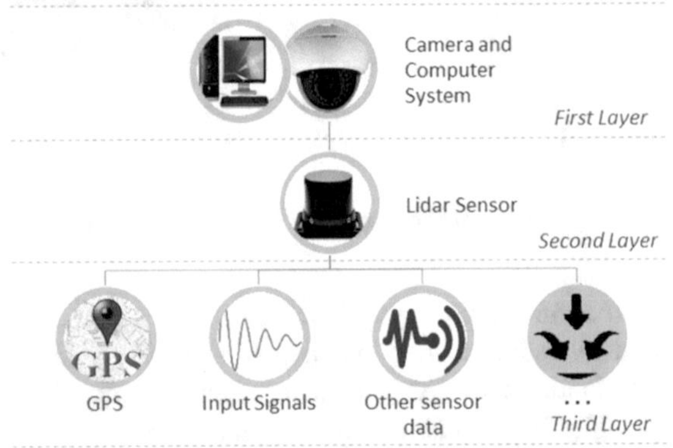

Fig. 2. Three-step data integration process.

The RGB camera gathers the visual information for recognition (using a Convolutional Neural Network) of the variables of the system. Once identified, the extracted region of interested (ROI) containing the variables are saved as part of the SOEKS. The computer system is responsible to produce the minimum context necessary to improve the interpretation of detected events (e.g. worker profile, day of the week, shift, season, industry location, industry type, location and size, etc.). These two components compose the first layer of integration. If the system is operating on this layer, only states of variables can be identified and the rules to compose the system will be limited basically to the presence/absence of variables.

On the second layer a Lidar sensor is added. The Lidar sensor feeds the system with depth information, which enables the tracking of workers, the creation of relationship among different variables and identifications of activities which are represented as functions in the SOEKS. Finally, the third layer of integration is composed for any extra sensor data available (illumination, temperature, level of oxygen, etc.), input signals (to indicate if machines are in operation or not), GPS, etc. The third layer enables the creation of unique context, and set of specific and detailed rules, as well as identification of complex events.

5.2 Knowledge Formalization

Formalizing knowledge has been a central issue of research in artificial intelligence and related areas. Even before acquiring knowledge, designers and programmers face the representation issue [18]. Computers, unfortunately, are still not capable of forming representations of the world as a human being, and even simpler, representations of just formal decision events. Adequate, efficient or effective representation of information and knowledge inside a computer is not a trivial process. For this reason, the consideration of using SOEKS as carrier for decision making is founded in the fact that experience has to be taken into consideration for developing a better cognitive vision system.

In this platform the critical safety-related behaviors are represented as rules and detected events are characterized by five main attributes: Workers, Personal Protective Equipment (PPE), Potentially Dangerous Object, Machinery or Area (PDOMA), Action/State, and Context. Workers, PPE and PDOMA are detected through cameras classified through a Convolutional Neural Network (CNN) [27] and finally represented as variables in our system. Action/State are represented as functions that model the relationship among those variables. This functions are formed by the association of RGB pixel's coordinates and the respective 3D point cloud coordinate gathered from the M8 Lidar sensor, tracked along the time, as explained previously. Unsafe events occur when the Action/States violates one of the given rules and are represented as a probabilistic value. To support formal reasoning Context is evoked when an event occurs and can increase the certainty of detected events, modify their state or influence its interpretation.

Manageable Knowledge Structure. Similar to behavior change efforts, culture change approaches are setting specific. Companies may share certain safety-related experiences, but safety events happening inside of an organization must also be understood in its own context. There appears to be considerable evidence suggesting that organizational and contextual factors are important in terms of a variety of workplace safety related outcomes [7]. In addition, organizational safety cultures are usually self-perpetuating, i.e. it is passed along, and new members are coached into the prevailing culture and beliefs which influences their behavior in regards to safety practices. This self-sustaining and perpetuating feature, also means that poor safety cultures can be resistant to change and that modifying the culture can be a very slow process [7]. Collecting information about the availability of safety equipment or about the status of hazard control activities, may provide a basis for drawing inferences about the safety culture. Following the assessment phase, a culture change program must seek to facilitate some type of analysis and planning process to crystalize the organizations safety-related values and vision, and to identify action priorities and implementation strategies for improving safety performance within the organization [3]. However, if there is a deficiency in collecting this data or if its availability is not guaranteed there is no way to ensure that culture of the organization is being impacted in any significant way. Therefore, to create unique contexts, to understand the safety culture of a company and to support an explicit long term safety culture change process that can be systematically assessed and shared, validated SOEKS from each particular organization are grouped and stored as a Decisional DNA (DDNA), a decisional safety fingerprint of a company.

6 Conclusions

In this paper we presented a Cognitive Vision Platform for Hazard Control (CVP-HC). The Cyber Physical System aims to perceives critical safety behaviors in real time giving support for the behavior-based safety strategy whilst stores information in an structured way to improve the user experience and to support an explicit long term culture change process that can be systematically assessed and shared. The CVP-HC is designed to be a scalable yet adaptable system capable of working in a variety of video analysis scenarios attending specific safety requirements of different industries by modifying its behavior accordingly without the need of recoding the system to better suit every specific application. The proposed system is based on the Set of Experience Knowledge Structure (SOEKS or SOE in short) and Decisional DNA (DDNA). It has been presented in this research the role of cognition and decisional experience to fill the gaps of current technologies that support behavior-based or culture change methodologies and make the integration of both safety approaches achievable in practice. In addition, it is explained how cognition and the decisional experience is implemented within the CVP-HC. At this point the first and second layers of data fusion is being developed. For next steps, representation of activities and functions of variables will be explored and the influence of context on interpretation of detected events will be examined.

References

1. Safe Work Australia: Australian Work Health and Safety Strategy 2012–2022. Creative Commons (2012)
2. Safetycare Australia: Recognition, evaluation and control of hazards, Victoria (2015)
3. Lund, J., Aarø, L.E.: Accident prevention. Presentation of a model placing emphasis on human, structural and cultural factors. Saf. Sci. **42**(4), 271–324 (2004)
4. Han, S., Lee, S.: A vision-based motion capture and recognition framework for behavior-based safety management. Autom. Constr. **35**, 131–141 (2013)
5. Mosberger, R., Andreasson, H., Lilienthal, A.J.: Multi-human tracking using high-visibility clothing for industrial safety. In: IEEE/RSJ International Conference on Intelligent Robots and Systems, pp. 638–644 (2013)

6. Zambrano, A., Toro, C., Nieto, M., Sotaquirá, R., Sanín, C., Szczerbicki, E.: Video semantic analysis framework based on run-time production rules – towards cognitive vision. J. Univ. Comput. Sci. **21**(6), 856–870 (2015)
7. DeJoy, D.M.: Behavior change versus culture change: divergent approaches to managing workplace safety. Saf. Sci. **43**(2), 105–129 (2005)
8. Sanin, C., Szczerbicki, E.: Set of experience: a knowledge structure for formal decision events. Found. Control Manag. Sci. **3**, 95–113 (2005)
9. Sanin, C., Szczerbicki, E.: Decisional DNA and the smart knowledge management system: a process of transforming information into knowledge. In: Gunasekaran, A. (ed.) Techniques and Tool for the Design and Implementation of Enterprise Information Systems, pp. 149–175. IGI Global, New York (2008)
10. Managing Safety (2015). http://www.safework.nsw.gov.au/health-and-safety/manage-workplace-safety/your-responsibilities
11. Krause, T.R., Hidley, J.H., Lareau, W.: Behavioral science applied to accident prevention. Prof. Saf. **29**, 21–27 (1984)
12. Skinner, B.F.: The Behavior of Organisms. Appleton-Century-Crofts, New York (1938)
13. Little, S., Jargalsaikhan, I., Clawson, K., Li, H., Nieto, M., Direkoglu, C., et al.: An information retrieval approach to identifying infrequent events in surveillance video, Dallas (2013)
14. Chen, L., Hoey, J., Nugent, C.D., Cook, D.J., Yu, Z.: Sensor-based activity recognition. IEEE Trans. Syst. Man Cybern. Part C (Appl. Rev.) **42**(6), 790–808 (2012)
15. Kotter, J., Heskett, T.: Corporate Culture and Performance. Free Press, New York (1992)
16. Wirth, O., Sigurdsson, S.O.: When workplace safety depends on behavior change: topics for behavioral safety research. J. Saf. Res. **39**(6), 589–598 (2008)
17. Vernon, D.: The space of cognitive vision. In: Cognitive Vision Systems, pp. 7–24. Springer, Heidelberg (2006)
18. Brézillon, P., Pomerol, J.C.: Contextual knowledge and proceduralized context. In: Proceedings of the AAAI-99 Workshop on Modeling Context in AI Applications, Orlando. AAAI Technical report (1999)
19. Sanin, C., Szczerbicki, E.: Experience-based knowledge representation SOEKS. Cybern. Syst. **40**(2), 99–122 (2009)
20. Blakeslee, S.: Lost on Earth: Wealth of Data Found in Space. New York Times, C1 (1990)
21. Corti, L., Backhouse, G.: Acquiring qualitative data for secondary analysis. Forum Qual. Soc. Res. **6**(2) (2005)
22. Humphrey, C.: Preserving research data: a time for action. In: Proceedings of Preservation of Electronic Records: New Knowledge and Decision-Making: Postprints of a Conference – Symposium 2003, pp. 83–89. Canadian Conservation Institute, Ottawa (2004)
23. Johnson, P.: Who you gonna call? Technicalities **10**(4), 6–8 (1990)
24. Sanin, C., Toro, C., Haoxi, Z., Sanchez, E., Szczerbicki, E., Carrasco, E., Mancilla-Amaya, L.: Decisional DNA: a multi-technology shareable knowledge structure for decisional experience. Neurocomputing **88**, 42–53 (2012)
25. Haghighat, M., Abdel-Mottaleb, M., Alhalabi, W.: Discriminant correlation analysis: real-time feature level fusion for multimodal biometric recognition. IEEE Trans. Inf. Forensics Secur. **11**(9), 1984–1996 (2016)
26. Bleiholder, J., Naumann, F.: Data fusion. ACM Comput. Surv. (CSUR) **41**(1), 1 (2009)
27. Ciresan, D.C., Meier, U., Masci, J., Gambardella, L.M., Schmidhuber, J.: Flexible, high performance convolutional neural networks for image classification. In: International Joint Conference on Artificial Intelligence (2011)

Big Data Approach to Analyzing Job Portals for the ICT Market

Celina M. Olszak(✉) and Paweł Lorek

University of Economics in Katowice, 1 Maja 50, 40-287 Katowice, Poland
celina.olszak@ue.katowice.pl,
pawel.lorek@Ue.katowice.pl

Abstract. The ability to manage Big Data (BD) is regarded as the main driving force behind organization's development, as well as maintaining its place in the market and of its innovative success. However, when we look on the number of organizations that actually use BD, it is relatively limited for the moment. It is mainly a lack of appropriate approaches, methods and tools targeted at the use of BD, but above all, there is a low level of knowledge about the importance of BD for business. This paper: (i) provides an overview of the context of BD in contemporary business; (ii) discusses the key concepts which underlay the concept of BD, (iii) examines selected methods and tools which are available for BD analysis. Based on the above, the paper then uses the job portals for the ICT market as a case to propose and to demonstrate a specific approach for BD analysis. The results from this study may be valuable for academics and ICT professionals, who search for appropriate methods and tools to manage job portals. This study provides also a value for employees who want to better know about the requirements and needs of the ICT job market as well as for employers who want to recruit more well-oriented candidates.

Keywords: Big Data · Job market · Graph exploration

1 Introduction

Big Data has nowadays a high business value. It contains various information about customers, competition, labour market, development trends of industries, products and services, as well as the public and political mood. However, a large part of this data remains hidden and is not used by the business. Many organisations make a limited use of the potential resulting from large scale data. This is mainly attributable to several reasons. Firstly, the BD paradigm has not been established well yet. Secondly, many BD phenomena and issues are characterised by a high degree of turbulence. Thirdly, in the field of science and practice, there is too little awareness of the true significance of BD in decision-making, and of benefits and potential threats arising from this fact. Fourthly, the practitioners' environment does not have methodology, recommendations and guidelines that would be a beacon for the organisation how to use BD for decision support. The proposed research is to meet the above challenges and fill the cognitive gap.

The main purpose of this study is to explore the concept of BD and to propose an approach to analyzing job portals for the ICT market. To address the objective of this

J. Świątek et al. (Eds.): ISAT 2018, AISC 853, pp. 276–285, 2019.
https://doi.org/10.1007/978-3-319-99996-8_25

study, the logic of the discourse is as follows. At the start, a critical review of the relevant literature is conducted to identify the concept of BD. Next, different methods and techniques for BD processing are presented. Finally, the idea of a BD approach to analyzing job portals for ICT market is proposed. The subject of this study are the British and Polish job portals. A comparative analysis, based on the obtained results, is presented. Natural language processing, and graph mining techniques are applied to explore job offers from the ICT sector posted at the portals. The summary of this paper includes the most important conclusions of the research. While writing the paper the following research methods are used: critical analysis of the literature, creative thinking and simulation methods with the usage of NetworkX 1.8.1 library and Gephi package (Bastian et al. 2009).

2 Big Data Concept

BD represents a wide spectrum of applications as well as a huge potential to improve and create new business opportunities (Chen et al. 2012). BD is often associated with the increase of the number of real-time data collected from social media (Facebook, Twitter, Instagram), web portals (job portals, price comparison sites), and Internet of Things applications (Himmi et al. 2017). According to Ferguson (2012) and Parise et al. (2012) BD is a capability that allows organizations to extract value from large volumes of data. Through BD, organizations could better follow the acceptance of services in the market place. In others words, BD is about analytical workloads that are associated with some combination of data volume, data velocity and data variety that may include complex analytics and complex data types.

Many authors agree (Manyika et al. 2011; Erl et al. 2015) that five main specific attributes (5V) are assigned to BD. They include: (1) Volume – the number of data is measured in peta- and zettabytes, (2) Velocity – the meteoric speed of data emergence, (3) Variety - the heterogenic nature of data, data can have different form and come from various devices and applications, (4) Veracity – data can be inconsistent, incomplete, and inaccurate, (5) Value - significant value hidden in data.

According to many scholars (Halaweh and Massry 2015; Manyika et al. 2011; O’Driscoll 2014; Das and Kumar 2013) thanks to BD, organizations can create unique value for business. This value may be manifested in improving of decision making and business processes as well as developing new business models. Some researchers (Davenport and Harris 2007; Schmarzo 2013) stress that BD means nothing like business transformation - transforming the organizations from a retrospective environment into a predictive and real-time business environment. BD analyzed in combination with traditional enterprise data (most structured and semi-structured), enables organizations not only to better understand their business, but first of all to change it and to have new sources of revenues, more stronger competitive position and greater innovation.

3 Approach to Big Data Management

Big data management is a complex and demanding task. It is even more difficult due to the influence of such factors as the aim of the processing and the type of processed data (Blazquez and Domenech 2017). Despite these considerations, some common elements can be distinguished in each BD management and processing workflow. These elements can be following (Sivarajah et al. 2017): (1) data acquisition and warehousing; (2) data mining and cleaning; (3) data aggregation and integration; (4) analysis and modeling; and (5) data interpretation.

Data acquisition and warehousing is the first stage, which provides the basis for all following steps. Due to the broad variety of data sources, the acquired data can be classified as several types. Among the common data types it can be distinguished: sensor data (Guo et al. 2014), text data (Mars and Gouider 2017), social media data (Miah et al. 2017), geospatial data (Lee and Kang 2015) and audiovisual data (Kune et al. 2015). NoSQL databases are frequently used in the first phase to acquire and store BD. Such systems just extract all data and do not categorize it or parse it by designing a schema. There exists a big challenge to generate right metadata to make a description of all data that is recorded, and the ways in which it is recorded and measured.

The characteristic feature of acquired data is high noise level. Reduction of these undesirable feature is the aim of the second step of BD management (data mining and cleaning). It is necessary to change the format of the distributed data and prepare it for further analysis. The information that can be extracted from the data depends on its quality. It means that poor quality data will almost always lead to poor results ("garbage in, garbage out"). Therefore, data cleaning (or scrubbing) is highlighted as one of the most important steps that has to be taken before data analysis is conducted. This often involves significant costs as the whole process can take from 50% to 80% of a data analyst's time together with the actual data collection costs (Reimsbach-Kounatze 2015).

The third step (data aggregation and integration) involves the preparation of a consistent data set with the form suitable for the selected analytical algorithm. Activities at this stage focus on transforming unstructured data into structured or semi-structured (Domingue et al. 2016). The purpose of this step is to build logically (not necessary physically) consistent repository that contains all the data needed for further steps of processing.

At the fourth stage (analysis and modelling), an algorithmic analysis of the data set, which has been prepared in preceding steps, is performed. The range of applicable methods is very broad. A very popular category are statistical methods (Sivarajah et al. 2017) as well as artificial intelligence methods (Fonseca and Cabral 2017). These methods are well known from the use for classic database repositories analysis. However, for the purpose of BD analysis, these methods are being modified. Modifications are usually related to introduction of parallel processing. This approach allows for efficient analysis of large data sets (Gupta et al. 2016). A wide area of applications is also text analysis with text mining methods (Kune et al. 2015).

The final stage (data interpretation) of BD management is related to the proper evaluation and interpretation of the obtained results. Visualization tools enable not only better perception and understanding of discovered data, but also evaluation of newly discovered knowledge. Despite the described conditions, it is clear that variety of BD volumes forces the need for individual approach in most cases.

4 Context of ICT Job Market and Job Portals

The ICT market has become one of the most dynamically developing labor markets in Europe and the world (Gareis et al. 2014; Husing et al. 2015; Pažur Aničić and Arbanas 2015). It has a huge impact on the innovation and competitiveness of individual businesses and the economy as a whole (Kowal and Roztocki 2013; Olszak 2014, 2016; Olszak et al. 2018). ICT sector is becoming a key driver of the knowledge-based economy. More and more jobs are created on the ICT market.

It is worth pointing out that more and more job offers related to the ICT market are available on the Internet and on various portals. Job portals offer several advantages over traditional methods, such as print advertisements or public employment agencies. For jobseekers, it reduces search and application costs, allows access to up-to date vacancy information, and improve communication with potential employers (Kuhn and Skuterud 2000). In turn, employers benefit from lower search costs of candidates, fasten recruiting, larger and more diverse pool applications, and more detailed information about applicants (Kuhn and Skuterud 2000). This justifies the reason and the need to use different BD tools to explore job portals that will continuously analyze and monitor competence profile of employees in the ICT market. Such analyzes may help employees to better and dynamically adapt to the needs of their employers. They can also provide tips on how to modify and update professional skills on a regular basis. In turn, employers may be provided with important information about the current skills of candidates and the most talented professionals.

5 Research Method

Job offers for the ICT market posted on web portals are large amount of unstructured data, constantly changing in short period of time due to a new job creation and rotation of employees. They may be also treated as a good example of BD. In this study an approach to analyze job portals for the ICT market is proposed. The workflow of analysis is based on the general methodology for analyzing Big Data as described earlier. However, this workflow has been modified and extended to better suit this specific case. The proposed approach uses different methods and tools, mainly: language processing, and graph mining techniques. Graph clustering is performed with the usage of NetworkX 1.8.1 library (networkx.github.io 2015) and the ForceAtlas2 algorithm that is available in the Gephi package (Jacomy et al. 2014; Bastian et al. 2009; gephi.github.io/toolkit 2017). All operations involving data processing are performed in the Python 2.7 environment. The subject of the analysis are 150 job offers from the British advertising agency Purely IT (www.purelyit.co.uk) and 150 from the Polish job offers portal Pracuj.pl (www.pracuj.pl).

The proposed approach is based on six stages. The first stage of the proposed approach is a selection of job offers for the software developers. In the second stage, the name of job and the associated requirements are identified and extracted from the individual offers. The collected data are next saved in tabular format. The third stage includes the unification of naming and terminology to eliminate ambiguity (e.g. the phrase "programming in C#", "knowledge of C#" has been replaced by a simple phrase "C#"). In the fourth stage of the proposed approach, processed data is organized as an undirected graph. The NetworkX 1.8.1 library is used to implement this phase. The vertices of the graph are the names of the offered jobs and the keywords describing the requirements assigned to job offers. In the fifth stage, the data structure obtained in the previous step is segmented using ForceAtlas 2 algorithm from the Gephi 0.9.1 library (Bastian et al. 2009). The last, sixth stage of this proposed approach, is the finding of the presence of clusters and their relations reflected by the position on the graph. The community analysis approach is used in this phase. It involves extracting a set of vertices characterized by a greater number of interconnections as compared with the other vertices in the graph (Bhat and Abulaish 2013). In performed visualizations, the number of connections between vertices is directly proportional to the distance between the individual graph nodes.

The proposed approach helps in the acquiring, grouping and discovering key competences and skills required on the ICT job market as well as the mapping of links of needed knowledge, and visualizing the path of acquiring to needed competences. It facilitates the employees finding linkages between their knowledge according to the required competences as well as the employers capturing the most talented candidates.

6 Results and Discussion

The conducted exploration of 300 job advertisements (from Purely IT web page and from advertising agency Pracuj.pl.) enriches our horizons and knowledge about the European ICT job market (presented in different official reports) as well as provides the detailed information on required competences and skills on this market (Fig. 1). In the illustrated visualization can be seen the presence of the major clusters. One of them is a group representing web technologies. The key skills grouped in this category are: knowledge of HTML, CSS and JavaScript technology and MySQL database. This group represents the competence of companies operating in the area of web applications. The size of the group clearly indicates that the web applications subsector is an important part of the UK IT industry.

Another cluster is a group centered around .NET technology. Among the most common appearances in this cluster of skills are the following: programming in C#, knowledge of the SQL Server database system, and knowledge of object oriented design principles (OOP node). The binder of described groups are technologies developed by Microsoft company. The most likely, part of the ICT representing this group is the group of developers of database applications in client-server architecture.

The next group consists of nodes representing technologies such as C++, Python, Java, SQL programming languages and Linux operating system. In the case of job offers related to this category, the requirement of having a formal computer science degree is more common. The relationships visible within this group are not as strict as in the other two clusters. Probably it is related to the partial representation of this

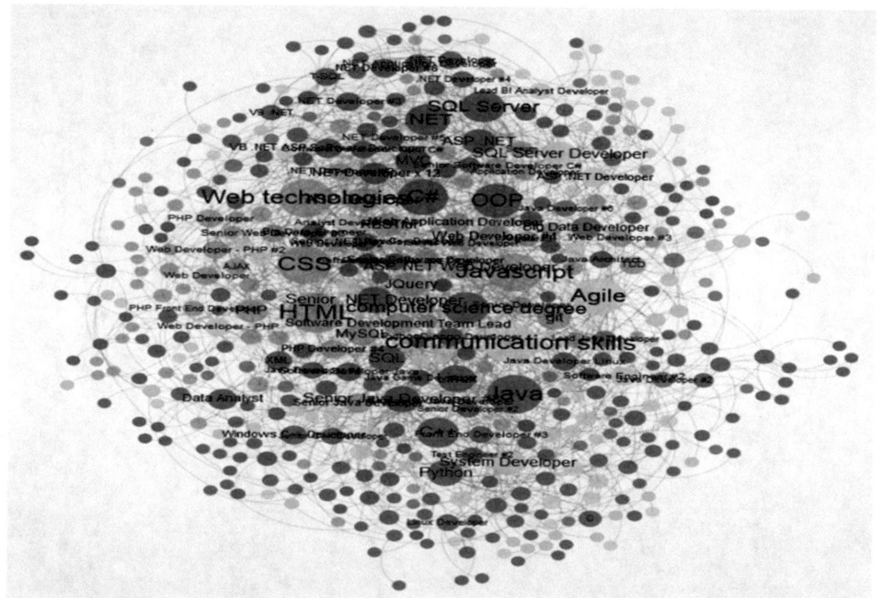

Fig. 1. Graph of competencies for the British ICT job market

cluster with subsector of applications in client-server architecture, as evidenced by the presence of Java and SQL nodes, and partial representation with the subsector of classic applications, manifested by the presence of technologies such as C++ and Python programming languages.

In addition to competences that are clearly related to specific market subsectors, there are also general market requirements. These requirements include, first of all, knowledge of Agile methodology and good communication skills for effective team working.

As in the case of the British market, the analysis for the Polish ICT sector was performed in the basis of 150 job offers from advertising agency Pracuj.pl. The result graph for the analysis is shown on Fig. 2. The nodes of the graph are labeled in Polish due to specifics of the data source.

The characteristic feature of the obtained graph is a much smaller degree of structuring and ordering than in the case of the UK market. However, also in this case, it is easy to notice the presence of cluster associated with the web technologies. The key competences in this area are: knowledge of HTML, CSS and JavaScript technologies as well as MySQL database system.

The next visible cluster is created by the nodes representing the skills associated with .NET technology. There are mainly C# language and ASP .NET technology. It is worth of noting that it is cluster of considerably smaller sizes than described above. The Java technology has a special place in the Polish ICT sector. It is undoubtedly the most popular programming language for commercial applications. In the presented illustration, it is the vicinity of soft competences such as teamwork ability (node "praca w

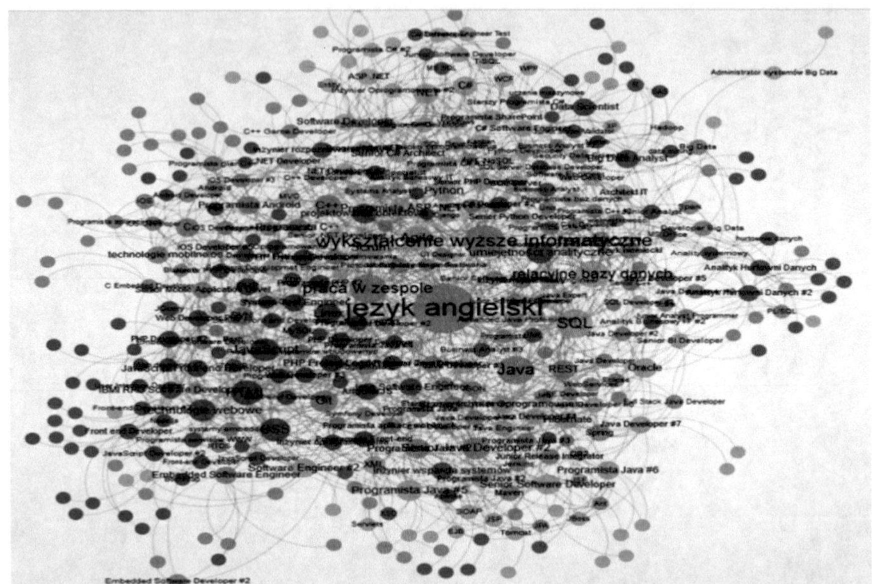

Fig. 2. Graph of competencies for the Polish ICT job market

zespole") and knowledge English language (node "język angielski"), which are the most common requirements of Polish employers. In addition to the above general trends, the periphery of the graph includes competencies in areas such as data analysis and mobile technologies.

The conducted analysis of the UK and Polish ICT labor market based on job portals, allows us to formulate following insights:

1. The UK ICT job market has a strong dichotomy for web application and client-server application subsectors, with a modest share of other subsectors. The Polish market has more clearly defined subsectors of mobile technology and data analytics.
2. The obvious difference is the greater coherence of competences within the UK market. This may prove a more rational and thoughtful formulation of the requirements by the British employers, as well as a greater degree of specialization of British firms.
3. On the Polish labor market in the ICT sector, the basic requirements (but not in all cases) are defined as soft skills (knowledge of English language and teamwork skills) as well as the requirements for formal education. On the British market these issues are also important, but the main focus is on technical skills.
4. The presence of the Java technology is definitely stronger in the Polish ICT market. In British ICT market, the .NET technology is competitive compared to Java.
5. Both the UK and Polish markets are heavily involved in the web technology subsector.

Comparing the results of the obtained competency maps, the UK ICT labor market is much more mature and ordered. This is evidenced by a greater degree of graph centralization and presence of more distinctive clusters. Probable direction of development of the Polish ICT sector will be the market with characteristic of the British market. The pace of these changes and their scope are difficult to determine. The proposed analysis should be treated as important, however as only one of the many analysis that should be taken into consideration by ICT job market.

7 Conclusions

This paper provides: (1) an overview of the context of BD; (2) discusses an approach to BD management; (3) proposes and evaluates the BD approach in analyzing job portals on the example of the British and Polish ICT labor. The importance of the findings presented are twofold: (1) an approach for BD analysis on the example of job portals for the ICT labor market is proposed, and thus offers a basis for further introspection and debate; (2) the results of the chosen case provides new insight into trends in the ICT job market. The conducted analysis of job portals with the using of BD approach may also be useful for the following groups:

1. Business executives. The results of the analyzes may be helpful in acquiring an orientation on the industry's trends and directions of expansion. Acquiring such knowledge can help to identify potential opportunities for business development, lack of education and skills by employees, identifying the range of requirements for job candidates, identifying existing development barriers (e.g. by identifying outdated technologies or occupying shrinking market segments).
2. Employees. A big challenge for ICT industry employees is the need to constantly develop and acquire new skills. This obligation is related to the need to update their knowledge in order to maintain their current place of work or to advance to higher positions. In this case, it is important to identify which competencies are most relevant for a particular career path.
3. Job candidates. With the same problem as active employees have to face candidates for job applicants. In this case, the incomplete set of the most often required competences and skills are often source of failures. With the knowledge from the results of the analyzed data, it will be possible for the candidate to become aware of his shortcomings and, consequently, to enlarge chances on the labor market by supplementing them.
4. Teachers and lecturers. The presented study may also be useful for employees in educational sector (schools and universities). Knowledge of the profiles of the most demanded employees will certainly allow for better adaptation of educational content. This will make possible to educate people with skills and knowledge adequate to current trends.
5. Researchers. The presented approach may be useful for researchers and analysts examining job portals and labor markets. The obtained results differ significantly from the traditional, quantitative indicators used for describing labor markets. This may be helpful in gaining a new insight on the labor market.

Our study makes both theoretical and practical contributions to the field. First, BD is generally an unexplored field of research. Therefore, the current study contributes to the emerging literature on BD management by investigating the issue of the analysis of job portals. Second, the current study is one of the rare studies that proposes an approach for the analysis of job portals to provide a fresh view on the analysis of job skills and competences on the ICT market. Third, this study demonstrates how graph models may be applied to analyze BD.

There are limitations associated with this research which may narrow the scope of the findings and point to potential directions for future studies. The proposed BD approach is a kind of initial analysis of job portals for the ICT market. It needs further validation and testing. Therefore, future research studies may include the use of new BD methods and techniques (to make more complex analysis) and more information, e.g., employers, employees (depending on the educational level, income, experience, individual cultural values) from different information resources.

References

Bastian, M., Heymann, S., Jacomy, M.: Gephi: an open source software for exploring and manipulating networks. In: Proceedings of the Third International Conference on Weblogs and Social Media, ICWSM 2009, pp. 361–362 (2009)

Bhat, S., Abulaish, M.: Analysis and mining of online social networks: emerging trends and challenges. WIREs Data Min. Knowl. Discov. **3**, 408–444 (2013)

Blazquez, D., Domenech, J.: Big Data sources and methods for social and economic analyses. Technol. Forecast. Soc. Chang. (2017). https://doi.org/10.1016/j.techfore.2017.07.027

Chen, H., Chiang, R.H.L., Storey, V.C.: Business intelligence and analytics: from Big Data to big impact. MIS Q. **36**(4), 1–24 (2012)

Das, T., Kumar, P.M.: Big Data analytics: a framework for unstructured data analysis. Int. J. Eng. Sci. Technol. **5**(1), 153 (2013)

Davenport, T.H., Harris, J.G.: Competing on Analytics. The New Science on Winning. Harvard Business School Press, Boston (2007)

Domingue, J., Lasierra, N., Fensel, A., van Kasteren, T., Strohbach, M., Thalhammer, A.: Big Data analysis. In: Cavanillas, J.M., Curry, E., Wahlster, W. (eds.) New Horizons for a Data-Driven Economy, pp. 63–86. Springer (2016)

Erl, T., Khattak, W., Buhler, P.: Big Data Fundamentals: Concepts, Drivers & Techniques. Prentice Hall, Boston (2015)

Ferguson, M.: Architecting a Big Data Platform for Analysis. Intelligent Business Strategies, IBM, White Paper (2012)

Fonseca, A., Cabral, B.: Prototyping a GPGPU neural network for deep-learning big data analysis. Big Data Res. **8**, 50–56 (2017)

Gareis, K., Hussing, T., Birov, S., Bludova, I., Schulz, C., Korte, W.B.: E-skills for jobs in Europe: measuring progress and moving ahead. Final Report prepared for the European Commission, Bonn (2014)

gephi.github.io/toolkit. Accessed 15 May 2017

Guo, T., Papaioannou, T.G., Aberer, K.: Efficient indexing and query processing of model-view sensor data in the cloud. Big Data Res. **1**, 52–65 (2014)

Gupta, P., Sharma, A., Jindal, R.: Scalable machine-learning algorithms for Big Data analytics: a comprehensive review. Wiley Interdiscip. Rev. Data Min. Knowl. Discov. **6**, 194–214 (2016)

Halaweh, M., Massry, A.E.: Conceptual model for successful implementation of Big Data in organizations. J. Int. Technol. Inf. Manag. **24**(2) (2015). http://scholarworks.lib.csusb.edu/jitim/vol24/iss2/

Himmi, K., Arcondara, J., Guan, J.P., Zhou, W.: Value oriented Big Data strategy: analysis & case study. In: Proceedings of 50th Hawaii International Conference on System Sciences (2017)

Husing, T., Korte, W.B., Dasnja, E.: Trends and Forecasts for the European ICR Professional and Digital Leadership Labour Markets (2015–2020), Bonn (2015)

Jacomy, M., Venturini, T., Heymann, S., Bastian, M.: ForceAtlas2, a continuous graph layout algorithm for handy network visualisation for the Gephi software. PLoS ONE **9**(6) (2014). http://journals.plos.org/plosone/article?id=10.1371/journal.pone.0098679

Kowal, J., Roztocki, N.: Information and communication technology management for global competitiveness and economic growth in emerging economics. Electron. J. Inf. Syst. Dev. Ctries. **57**, 1–12 (2013)

Kune, P., Konaghurti, P., Agarwal, A., Chillarige, R., Buyya, R.: The anatomy of Big Data computing. Softw. Pract. Exp. **46**, 79–105 (2015)

Kuhn, P., Skuterud, M.: Internet job search and unemployment durations. Am. Econ. Rev. **94**(1), 218–232 (2000)

Lee, J.G., Kang, M.: Geospatial Big Data: challenges and opportunities. Big Data Res. **2**(2), 74–81 (2015)

Manyika, J., Chui, J.M., Brown, B., Bughin, J., Dobbs, R., Roxburgh, C., Byers, A.H.: Big Data: the next frontier for innovation, competition, and productivity. McKinsey Global Institute (2011)

Mars, A., Gouider, M.S.: Big Data analysis to features opinions extraction of customer. Procedia Comput. Sci. **112**, 906–916 (2017)

Miah, S., Vu, H., Gammack, J., McGrath, M.: A Big Data analytics method for tourist behaviour analysis. Inf. Manag. **54**, 771–785 (2017)

networkx.github.io

O'Driscoll, T.: Can Big Data deliver added value. Training **51**(2), 51 (2014)

Olszak, C.M.: An overview of information tools and technologies for competitive intelligence building: theoretical approach. Issues Inf. Sci. Inf. Technol. **11**, 139–153 (2014)

Olszak, C.M.: Toward better understanding and use of business intelligence in organizations. Inf. Syst. Manag. **33**(2), 105–123 (2016)

Olszak, C.M., Bartuś, T., Lorek, P.: A comprehensive framework of information system design to provide organizational creativity support. Inf. Manag. **55**(1), 94–108 (2018)

Parise, S., Iyer, B., Vesset, D.: Four strategies to capture and create value from Big Data (2012). http://iveybusinessjournal.com/publication/four-strategies-to-capture-and-createvalue-from-big-data

Pažur Aničić, K., Arbanas, K.: Right competencies for the right ICT jobs – case study of the Croatian labor market. TEM J. **4**(3), 236–243 (2015)

Reimsbach-Kounatze, C.: The Proliferation of Big Data and Implications for Official Statistics and Statistical Agencies. OECD Digital Economy Papers 245. OECD Publishing, Paris (2015)

Schmarzo, B.: Big Data: Understanding How Data Powers Big Business. Wiley, Indianapolis (2013)

Sivarajah, U., Kanal, M.M., Irani, Z., Weerakkody, V.: Critical analysis of Big Data challenges and analytical methods. J. Bus. Res. **70**, 263–286 (2017)

www.purelyit.co.uk

www.pracuj.pl

A Parallel Algorithm for Mining High Utility Itemsets

Trinh D. D. Nguyen[1], Loan T. T. Nguyen[2,3(✉)], and Bay Vo[1]

[1] Faculty of Information Technology,
Ho Chi Minh City University of Technology (HUTECH),
Ho Chi Minh City, Vietnam
dzutrinh@gmail.com, vd.bay@hutech.edu.vn
[2] Division of Data Science, Ton Duc Thang University,
Ho Chi Minh City, Vietnam
nguyenthithuyloan@tdtu.edu.vn
[3] Faculty of Information Technology, Ton Duc Thang University,
Ho Chi Minh City, Vietnam

Abstract. High utility itemset mining (HUIM) is a popular and important mining task in recent years. The problem is considered computational expensive in terms of execution time and memory consumption. Many algorithms have been proposed to solve this problem efficiently. In this paper, we propose a parallel approach for mining HUIs, which utilizes the modern multi-core processors by splitting the search space in to disjointed sub-spaces, assign them to the processor cores and explore them in parallel. Experimental results show that the proposed algorithm outperformed the original state-of-the-art HUIM algorithm EFIM in terms of execution times and have comparable memory usage.

Keywords: High utility itemset · Pattern mining · Data mining Parallel · Multi-threaded

1 Introduction

Since the introduction of the frequent itemset mining (FIM) problem by Agrawal and Skirant in 1994 [1], the problem of analyzing customer transactions from retail stores has become a commonly studied problem in the field of data mining. The goal of FIM is to discover a group of items, called frequent itemsets, which appear frequently in a customer transaction database. Many algorithms have been proposed to solve this problem efficiently. These algorithms use of a property called *downward closure property* (or apriori property) [1]. The property states that the supersets of an infrequent itemset are infrequent and the subsets of a frequent itemset are frequent.

Considered as an extension problem of FIM, the problem of mining high utility itemsets (HUIM) has become a popular mining task in recent years [2–5]. Many algorithms have been proposed for discovering itemsets that yield a high utility (e.g. profit) within transaction databases. With HUIM, each item in the database has an external utility value or unit profit. Also, an item in a transaction has an internal utility or purchase quantity. Unlike FIM, HUIM considers that items can appear more than

J. Świątek et al. (Eds.): ISAT 2018, AISC 853, pp. 286–295, 2019.
https://doi.org/10.1007/978-3-319-99996-8_26

once in each transaction and each item in a transaction has an attached value called profit. An itemset is called high utility itemset (HUI) if its utility is no less than a user-specified minimum utility threshold (minutil). In the recent years, the problem of HUIM has becoming an essential in a wide range of applications such as cross marketing, user behavior analysis, etc.

The HUIM problem is quite challenging since the downward closure property does not hold for utility measure [1, 6]. For instance, consider the transaction database given in Table 1 while the unit profit information is given in Table 2 [7]. The database contains five transactions, each transaction provides information about the purchased quantities of items. For example, the fourth transaction indicates that a customer has purchased 5, 2, 1 and 2 unit items a, b, d and e, respectively. The table shows that the sale of each unit of item a yield \$1 profit, item b yield \$2 profit, and so on. From here, the utility (profit) of an itemset can be determined by summing up all the item's profit. For example, itemset $\{bc\}$ and $\{de\}$ were said to have a utility of \$18 and \$22, respectively in the database and their supports are 3 and 2, respectively. This analysis clearly shows that frequent itemsets are not always the most profitable ones. While the goal of FIM is to discover all itemsets that have support no less than a minimum support threshold (minsup), the goal of HUIM is to discover all the itemsets that had utility no less than a minutil. Many algorithms have been proposed to solve the HUIM problem [2], help finding the high utility itemsets [8–10]. As many efficient HUIM mining algorithms have been proposed, but when it comes to mine HUIs in large databases or when the number of HUIs found become increasingly large, the time needed to complete the task is still very long. In the recent years, modern multi-core processor architecture allows the execution of multiple tasks in parallel to enhance the performance [11]. Thus, to help improve further more performance of the HUI mining algorithms and to utilize the modern processors, parallel techniques should be considered.

Table 1. Example of a transactional database

TID	Transaction	Quantity
T_1	{b, c, d, g}	{1, 2, 1, 1}
T_2	{a, b, c, d, e}	{4, 1, 3, 1, 1}
T_3	{a, c, d}	{4, 2, 1}
T_4	{a, b, d, e}	{5, 2, 1, 2}
T_5	{a, b, c, f}	{3, 4, 1, 2}

Table 2. Unit profits

Items	a	b	c	d	e	f	g
Profit	1	2	1	5	4	3	1

The paper is organized as follows. Section 2 reviews related works on HUIM. Section 3 presents preliminaries and important definitions, formulates the problem of HUIM in transaction databases. Section 4 presents a parallel approach to the state-of-the-art algorithm EFIM for HUIM. Section 5 covers the experimental evaluations of the proposed algorithm against EFIM in terms of execution time and memory usage. Finally, conclusions are drawn and future works are discussed in Sect. 6.

2 Related Works

Unlike the itemsets from FIM, the itemsets in HUIM have their utility measure do not satisfy the monotonic or anti-monotonic property [2]. The first algorithms for mining HUIs [1, 9] were based on the idea of calculating upper bounds of the utility of itemsets. These algorithms were not complete and they may fail to discover the complete set of HUIs. To get around the fact the utility measure is not anti-monotonic and to find all HUIs, many HUIM algorithms come with a measure called Transaction Weighted Utilization (TWU) measure [5, 8–10, 12, 13] which is anti-monotonic and forms an upper bound on the utility of itemsets. The proposed algorithms use TWU to safely prune the search space since TWU also satisfies the downward closure property. Specialized data structures were incorporated into the algorithms to efficiently identify candidate HUIs and discarding non-promising items [9, 10]. These algorithms consist of two phases. First, candidates for HUIs are generated by overestimating the utility of itemsets. Then, HUIs are identified from the candidates by scanning the original transaction database once. New algorithms were then proposed to discover HUIs in a single phase [3, 4, 14]. All these algorithms are aimed at optimizing the HUIM performance via candidate pruning. A recently proposed algorithm, the EFIM algorithm [15] introduced by Tseng et al., proposed three strategies to furthermore pruning the search space in one phase. The first is called HDP aims at reducing the size of the transaction databases when long itemsets are considered. The second, HTM, reduces the search space by dynamically merging identical transactions in projected databases. The third strategy applies two upper bounds called local utility and sub-tree utility to reduce the search space. The proposed techniques used in many algorithms focus on optimizing the candidate generation process and pruning the search space. However, memory usage and running time limitations still cause scalability issue, especially when the database is becoming very large.

Parallel computing has been widely applied to improve the execution speed for many problems. One of the techniques in parallel computing is the multi-core architecture, which has been used to speed up many algorithms. But in the field of data mining, these algorithms have been rarely applied. Some parallel methods were proposed, which are based on the distributed memory system [16, 17]. For HUI mining, the PHUI-Miner algorithm [18] has been proposed, which is a parallelized version of the HUI-Miner algorithm. PHUI-Miner divided the search space of the HUIM problem and assigned to node in a cluster, which splits the work load and provided better performance.

3 Problem Definition

Given a transaction database D, $D = \{T_1, T_2, \ldots, T_n\}$. Let I be the finite set of all items within D and an itemset $X, X \subseteq I$. For each transaction T_c, $T_c \subseteq I, T_c$ has a unique identifier c called the T_{ID} (Transaction ID). Let there be an item i appearing in a transaction T. The purchase quantity of i, denoted as $q(i, T)$, and represents the number of units of item i purchased in transaction T. The unit profit of item i, denoted as $p(i)$, represents the amount of profit generated by the sale of each unit of i. The utility of an item i in T is the total profit generated by item i in that transaction, denoted as $u(i, T)$ and is defined as the product of $q(i, T)$ and $p(i)$: $u(i, T) = q(i, T) \times p(i)$ [15]. Let X be an itemset appearing in a transaction T. The utility of X in T, denoted as $u(X, T)$, is the sum of all the utilities of items in X from that transaction: $u(X, T) = \sum_{i \in X \wedge X \subseteq T} u(i, T)$. The utility of itemset X in transaction database D, denoted as $u(X)$ and is defined as the total utility of X in all transactions that contains X: $u(X) = \sum_{T \in D \wedge X \subseteq T} u(T)$ [15]. X is a high utility itemset if its utility $u(X)$ is no less than a user-specified minimum utility threshold minutil, i.e. $u(X) \geq minutil$. Given a user-defined minimum utility threshold minutil, the problem of mining HUIM is to discover all the high utility itemsets.

To pruning the candidate itemsets from the search space while mining for HUIs, the Transaction Weighted Utility (TWU) upper bound is proposed [5, 8, 9]. TWU is defined as follows: the TWU of an itemset X, denoted as $TWU(X)$, is the sum of the utility of all transactions containing X in database D. $TWU(X) = \sum_{T \in D \wedge X \subseteq T} u(T)$ [15]. For large databases, TWU become inefficient in pruning the search space [5, 7]. Tighter upper bounds are needed to further reduce the search space and increase the efficiency of HUIM. Two new upper bounds on the utility of itemsets are referred to as the local utility and sub-tree utility [15]. The new upper bounds calculation relies on a measure call remaining utility [3]. The remaining utility is defined as follows. Let $\succ$ be a total order on the set of items I, defined as the ascending order of TWU. Let there be an itemset X, appearing in transaction T. The remaining utility of X in T, denoted $ru(X, T)$, is the sum of the utilities of all items in T that succeed all items in X according to $\succ$. Thus, $ru(X, T) = \sum_{i \in T \wedge i \succ x \forall x \in X} u(i, T)$.

From the TWU definitions presented, assume that all itemsets are sorted according to $\succ$ order. The set of all extensions of an itemset X is $E(X) = \{z | z \in I \wedge z \succ x, x \in X\}$ [15]. Given an itemset X and item $z \in E(X)$, the local utility, denoted $lu(X, z)$, $lu(X, z) = \sum_{T \supset (X \cup \{z\})} [u(X, T) + ru(X, T)]$. The local utility has the following property: Given an itemset X and an item $z \in E(X)$. If $lu(X, z) < minutil$ then all itemsets containing $X \cup \{z\}$ are not HUIs [15].

Given an itemset X and an item $z \in E(X)$, utility of an itemset Y obtained by combining $X \cup \{z\}$ with any items in $E(X)$ has a utility that is not greater than $su(X, z) = \sum_{T \supset (X \cup \{z\})} \left[u(X, T) + u(z, T) + \sum_{i \in T \wedge i \in E(X \cup \{z\})} u(i, T)\right]$ [15]. This is the sub-tree utility of (X, z) and it has the following property: For an itemset X and an item $z \in E(X)$, if $su(X, z) < minutil$ then all itemsets obtained by appending items to X and z are not HUIs [15]. In other words, any itemset Y obtained by combining $X \cup \{z\}$ with any items in $E(X)$ having $su(X, z) < minutil$ is not a HUI and can be pruned. The relationship $TWU(Y) \geq lu(X, z) \geq su(X, z)$ holds and has been proved in [15]. From the proof, local utility upper bound is tighter than TWU and thus can be used to efficiently pruning the search space. For the subtree utility, if $su(X,z) < minutil$, we can prune the whole subtree of z including Y rather than only pruning the descendant nodes of Y [15].

Based on the definitions the new upper bounds, the concepts of primary and secondary items were defined, which make the subtree upper bound even tighter as follows: Given an itemset X, the primary items and the secondary items of X are defined as follow: $Primary(X) = \{z | z \in E(X) \wedge su(X,z) \geq minutil\}$, $Secondary(X) = \{z | z \in E(X) \wedge lu(X,z) \geq minutil\}$ [15]. The primary and secondary items of an itemset X contain the items that should be used to create larger itemsets constructed from itemset X according to the properties of the local utility and subtree utility, respectively. Thus, the redefined and tighter subtree upper bound is defined as follows: Let there be an itemset X and an item z. The redefined subtree utility of z w.r.t X is defined as:

$$su(X,z) = \sum_{T \supset (X \cup \{z\})} \left[u(X,T) + u(z,T) + \sum_{i \in T \wedge i \in E(X \cup \{z\}) \wedge i \in Secondary(X)} u(i,T) \right].$$

That means, items not in $Secondary(X)$ will not be included in the calculation of the subtree upper bound, and thus it is always less than or equal to the original subtree upper bound. By using these properties, fewer items need to be considered when exploring the search space and increase the efficiency of the algorithm.

Additional techniques for increasing the efficiency of the HUIs mining process were also proposed in [15], they are HDP and HTM. HDP aims at reducing the size of database with respect to a given itemset. HTM aims at merging transactions that are having identical itemsets in the projected database, reducing memory usage and increasing the efficiency of HUIM. HDP is defined as follows: Let T be a transaction and X be an itemset. The projection of X on T is defined as $T_X = \{i | i \in T \wedge i \in E(X)\}$. This is based on the observation that when an itemset X is considering during the depth-first search process, all items $i \notin E(X)$ can be ignored when scanning the database D to calculate the utility of itemsets and the upper bounds. This database is called a projected database on X, denoted D_X, and is defined as follows: $D_X = \{T_X | T_X \in D \wedge T_X \neq \emptyset\}$. To further reduce the cost of database scans during the HUIs mining process, an efficient technique for merging transaction in database D called HTM is also proposed [15]. The technique identifies identical transactions appear in D and replace them with a single new transaction, while combining their utility values. Two transactions $T_a = \{i_1, i_2, \ldots, i_m\}$ and $T_b = \{j_1, j_2, \ldots, j_n\}$ are equal, denoted as $T_a = T_b$, if $m = n$ and $\forall k = 1..n, i_k = j_k$. Consider a set of equal transactions $T_{r1} = T_{r2} = \ldots = T_{rm}$ in a (projected) database D_X. Those transactions can be merged and replaced by a new transaction $T_M = T_{r1} = T_{r2} = \ldots = T_{rm}$, the utility of each item $i \in T_M$ is defined as $u(i, T_M) = \sum_{k=1}^{m} u(i, T_{rk})$. Using the proposed definitions and properties, we can efficiently mine all the HUIs.

4 Proposed Algorithm

Parallelism is expected to relieve current HUI mining methods from the sequential bottleneck, allow them to be possibly applied to large database and improve the response time. Modern processors are now having multiple cores that can handle simultaneously many tasks. The main challenges when applying parallelism onto current HUI mining algorithms are synchronization, communication minimization, work-load balancing, data decomposition and disk I/O minimization. Zaki has showed that parallel design space spans across three main components including hardware platform, the type of parallelism exploited, and the load balancing strategy used [19], which are Distributed Memory Machine (DMM) vs Shared Memory Systems (SMP), Task vs Data Parallelism, and Static vs Dynamic Load Balancing.

Algorithm 1: the pEFIM algorithm

Input : transactional database D, $minutil$ threshold
Output : sets of all found HUIs

$X = \emptyset$; $huis = \emptyset$;
Compute $lu(X, i)$ for all items $i \in I$ by scanning D;
$Secondary(X) = \{i \mid i \in I \wedge lu(X, i) \geq minutil\}$;
Sort $Secondary(X)$ in non-decreasing order of $lu(X, i)$ values;
Scan D to remove item $i \notin Secondary(X)$ from transactions and delete empty transactions;
Sort all the transactions in D in non-decreasing order of $lu\ (X, i)$ values when reading backward;
Compute $su(X, i)$ of each item $i \in Secondary(X)$ by scanning D;
$Primary(X) = \{i \mid i \in Secondary(X) \wedge su(X, i) \geq minutil\}$;
for each item i in $Primary(X)$ **do parallel**
    $Primary_i(X) = \{i\}$;
    $huis = huis \cup Search\ (i, D, Primary_i(X), Secondary(X), minutil)$;
end;
return $huis$;

Our proposed algorithm extends the state-of-the-art HUI mining algorithm EFIM and aims to utilize the multi-core processors which are widely available today to gain higher performance. The algorithm is named pEFIM, short for parallelized EFIM. The proposed algorithm applies the Task Parallelism on SMP, using Static Load Balancing strategy. Since the Task Parallelism algorithms assign portion of the search space to separated processors using Divide and Conquer strategy, for this approach seems a natural reflection of the recursive nature of the set enumeration tree like the one shown in Fig. 1a. It takes a transaction database D, a minutil threshold, just as its original and sequential version, and outputs the set of all found HUIs in D. Considering the search space as shown as a set of enumeration tree in Fig. 1a, this search space is partitioned into separated sub-spaces, starting from the single item subsets. This space partitioning is done right as a part of the main algorithm. Since the EFIM algorithm is DFS based, each sub-space is completely disjointed, non-overlapped, and can be statically assigned to a separated thread as shown in Fig. 1b to be explored in parallel. Each thread discovers all HUIs in its own search space top-down in parallel. The recursive Search function remain almost the same as its sequential version, except now it maintains its own list of all found HUIs and return this list after the sub-space has been fully explored. After all the threads have been terminated, the main algorithm will then collect all the returned lists into a global list of all found HUIs in the transaction database.

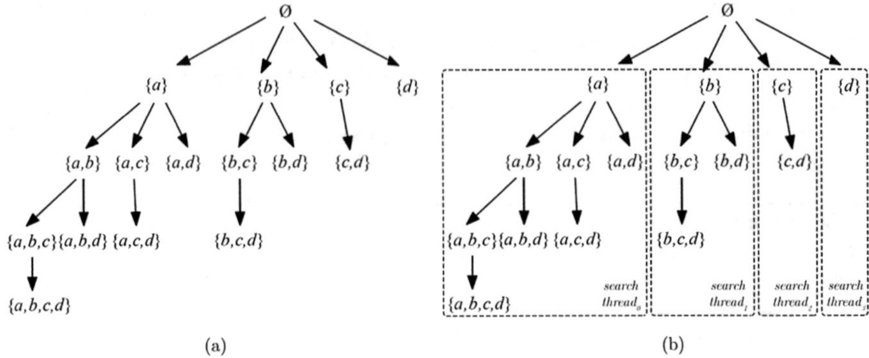

Fig. 1. (a) Search space of the algorithm (b) Search space partitioned

```
Algorithm 2 : the recursive Search algorithm
Input  : itemset X, projected database D_X, Primary(X), Secondary(X) , minutil
Output : the set of HUIs that are extensions of X.
for each item i ∈ Primary(X) do
      β = X ∪ {i}
      Scan D_X to calculate u(β) and construct D_β
      if u(β) ≥ minutil then return β
      Scan D_β to compute sup(β, z), lu(β, z) for all z ∈ Secondary(X)
      Primary(β) = {z ∈ Secondary(X) | su(β, z) ≥ minutil};
      Secondary(β) = {z ∈ Secondary(X) | lu(β, z) ≥ minutil};
      Search(β, D_β, Primary(β), Secondary(β), minutil);
end
```

The pseudo-code of pEFIM is given in Algorithm 1. The biggest change come from line 9 to 12 when compared to its sequential version. From here, the search space is partitioned into sub-space using Divide and Conquer strategy, starting from each single item i from the set of all primary items $Primary(X)$, and is explored down in parallel by calling the Search function as given in Algorithm 2. Each call of the Search function for a sub-space w.r.t an item i is attached to a separated thread. At this level of the search space, the $Secondary(X)$ is the same for all items $i \in Primary(X)$. Since the transaction database D, set of secondary items $Secondary(X)$ remain intact in each sub-space of the Search call, they are thread-safe.

Since the algorithm is designed to run on SMP architecture, load balancing is simple. These threads are placed into a fixed size thread pool, then executed in parallel, in a *first come-first served* order, until all the threads are completed.

5 Experimental Studies

A set of experiments were conducted on several transactional databases to evaluate the performance of the proposed algorithm to its original version. The experiments were conducted on a computer with a dual-core (4 threads) Intel® Core i5-5257U processor clocked at 2.7 Ghz, 8.0 GB RAM, running macOS 10.13.4. The EFIM algorithm and the databases used in the experiments were obtained from the open source SPMF package [20]. The algorithms were all implemented in Java. The experiments were performed focusing on the large databases to demonstrate the effectiveness of the parallel architecture, such as Chainstore and Accidents. The quantity of each item in each transaction was randomly generated in the [1, 10] interval. The utility of each item was randomly generated in the [2, 50] interval. The database Chainstore contains real utility values. Their characteristics are shown in Table 3.

Table 3. Database characteristics

Database	#Trans	#Items	Avg. length
Accidents	340,183	468	33.8
Chainstore	1,112,949	46,086	7.2

5.1 Runtime

In the runtime tests, for each test database, the minutil threshold was varied and the execution time of each algorithm was recorded. For each database, the pEFIM was tested against the EFIM algorithm using 2 and 4 threads, denoted as pEFIM (2c) and pEFIM (4c), respectively, each was assigned to a separated processor core. As shown in Fig. 2, the runtime of pEFIM out performs EFIM on all the test databases. Since parallelism techniques were applied to the pEFIM algorithm, it is observed that the performance of the parallel algorithm exhibits a big improvement. For real-world database Chainstore, the increased speed is up to 7 times when using 2 threads and up to 8 times using 4 threads. For Accidents, the improved performance of pEFIM (2c) and pEFIM (4c) were up to 5 times when using 2 threads and up to 6 times faster than the sequential algorithm EFIM, when using 4 threads, across the minutil thresholds tested.

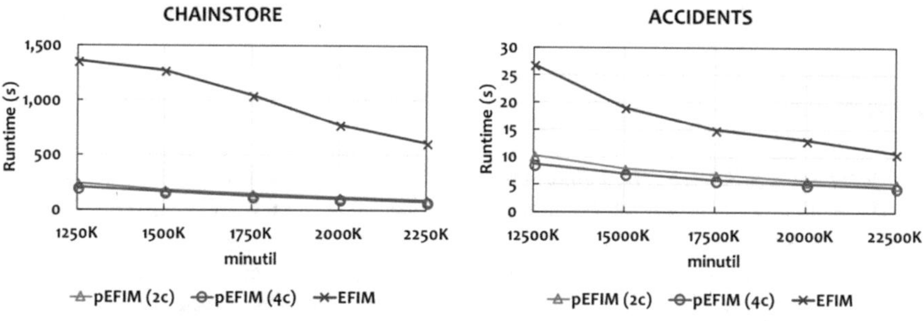

Fig. 2. Runtime performance of pEFIM using 2 and 4 threads against EFIM

5.2 Memory Usage

Average memory usages of the tested algorithm were recorded were shown in Table 4. For the Accidents database, the memory space needed for the algorithms are comparable, the pEFIM requires only a small margin more memory than the EFIM algorithm since all the working threads are sharing the same dataset. Thanks to the effectively pruning strategies introduced, only a small amount of memory is required to store the private data for each separated thread while executing. For the Chainstore database, which is a very large and sparse database, the number of subsets were formed by combining items from large *Primary* and *Secondary* sets, quickly increased when the minutil reduced, increasing number of threads to be created to explore all the subsets and thus lead to high memory consumption when executing under low minutil values.

Table 4. Average memory usage in megabytes

Database	pEFIM (2c)	pEFIM (4c)	EFIM
Accidents	798.76	795.05	781.01
Chainstore	1,098.96	1,106.23	655.28

6 Conclusions and Future Work

In this paper, we proposed an extent to the state-of-the-art HUI mining algorithm EFIM, using parallelism. This is for utilizing widely available modern multi-core processors to further increase the effectiveness and performance of the HUI mining process, especially on large databases. From the results of the experiments, the execution speed of the parallel version, pEFIM, is increased up to 6 times compared to the non-parallel algorithm EFIM on the test databases when running with 2 and 4 working threads. For the memory consumption, the parallel version requires more memory to run than the original sequential algorithm, due to the load balancing and private data space required for each thread. Since the EFIM and the pEFIM share the same techniques for pruning the search space, the overall algorithm complexity is unchanged for both versions.

As studied, it could be worth converting the parallelized algorithm into a distributed version using the Distributed Memory Machine model [19] to take advantage of the grid-computing environment. A new load balancing strategy is needed when applying Distributed Memory Machine model to help improving the algorithm's performance. Furthermore, adapting the Apache Spark into the algorithm is also considered.

References

1. Agrawal, R., Srikant, R.: Fast algorithms for mining association rules in large databases. In: VLDB 1994 Proceedings of the 20th International Conference on Very Large Data Bases, pp. 487–499 (1994)
2. Yao, H., Hamilton, H.J., Butz, C.J.: A foundational approach to mining itemset utilities from databases. In: 3rd SIAM International Conference on Data Mining, pp. 482–486 (2004)

3. Fournier-Viger, P., Wu, C.-W., Zida, S., Tseng, V.S.: FHM: faster high-utility itemset mining using estimated utility co-occurrence pruning. In: 21st International Symposium on Methodologies of Intelligent Systems, pp. 83–92 (2014)
4. Liu, M., Qu, J.: Mining high utility itemsets without candidate generation. In: 21st ACM International Conference on Information and Knowledge Management, pp. 55–64 (2012)
5. Tseng, V.S., Shie, B.-E., Cheng-Wei, W., Yu, P.S.: Efficient algorithms for mining high utility itemsets from transactional databases. IEEE Trans. Knowl. Data Eng. **25**(8), 1772–1786 (2013)
6. Han, J., Pei, J., Yin, Y.: Mining frequent patterns without candidate generation. In: SIGMOD 2000 Proceedings of the 2000 ACM SIGMOD International Conference on Management of Data, Dallas, Texas, pp. 1–12 (2000)
7. Yao, H., Hamilton, H.J.: Mining itemset utilities from transaction databases. Data Knowl. Eng. **59**(3), 603–626 (2006)
8. Liu, Y., Liao, W., Choudhary, A.: A two-phase algorithm for fast discovery of high utility itemsets. In: 9th Pacific-Asia Conference on Knowledge Discovery and Data Mining, pp. 689–695 (2005)
9. Ahmed, C.F., Tanbeer, S.K., Jeong, B.-S., Lee, Y.-K.: Efficient tree structures for high utility pattern mining in incremental databases. IEEE Trans. Knowl. Data Eng. **21**(12), 1708–1721 (2009)
10. Tseng, V.S., Wu, C.-W., Shie, B.-E., Yu, P.S.: UP-Growth: an efficient algorithm for high utility itemset mining. In: 16th ACM SIGKDD International Conference on Knowledge Discovery and Data Mining, pp. 253–262 (2010)
11. Solihin, Y.: Fundamentals of Parallel Computer Architecture. CRC Press, Boca Raton (2009)
12. Le, B., Nguyen, H., Cao, T.A., Vo, B.: A novel algorithm for mining high utility itemsets. In: 1st Intelligent Information and Database Systems, pp. 13–17 (2009)
13. Le, B., Nguyen, H., Vo, B.: An efficient strategy for mining high utility itemsets. Int. J. Intell. Inf. Database Syst. **5**(2), 164–176 (2011)
14. Krishnamoorthy, S.: Pruning strategies for mining high utility itemsets. Expert Syst. Appl. Int. J. **42**(5), 2371–2381 (2015)
15. Zida, S., Fournier-Viger, P., Lin, J.C.-W., Wu, C.-W., Tseng, V.S.: EFIM: a fast and memory efficient algorithm for high-utility itemset mining. Knowl. Inf. Syst. **51**(2), 595–625 (2017)
16. Zaki, M.J.: SPADE: an efficient algorithm for mining frequent sequences. Mach. Learn. **42**, 31–60 (2010)
17. Cong, S., Han, J., Padua, D.: Parallel mining of closed sequential pattern. In: Proceedings of ACM SIGKDD, vol. 5, pp. 562–567 (2005)
18. Chen, Y., An, A.: Approximate parallel high utility itemset mining. Big Data Res. **6**, 26–42 (2016)
19. Zaki, M.J.: Parallel and distributed association mining: a survey. IEEE Concurr. **7**(4), 14–25 (1999)
20. Fournier-Viger, P., et al.: SPMF: a Java open-source pattern mining library. J. Mach. Learn. Res. **15**(1), 3389–3393 (2014)

Use of the EPSILON Decomposition and the SVD Based LSI Techniques for Reduction of the Large Indexing Structures

Damian Raczyński(✉) and Włodzimierz Stanisławski

University of Applied Sciences in Nysa,
ul. Chodowieckiego 4, 48-300 Nysa, Poland
{damian.raczynski,
wlodzimierz.stanislawski}@pwsz.nysa.pl

Abstract. Storage of indexing structures in the Vector Space Model (VSM) form has a number of advantages. In the case when text documents are considered, the indexing structure states the Term-By-Document (TBD) matrix. Its size is proportional to the product of the indexed documents number and the keywords number. In the case of large text documents databases, the size of the indexing structure is a serious limitation. Too large TBD matrix may not be able to be stored in memory or the process of searching for documents may take too much time. The article presents a methodology that allows to reduce the size of the large TBD matrix. The operation performed on the TBD matrix is the Singular Value Decomposition (SVD). It allows to transform the original indexing structure vectors into a space with fewer dimensions. As a result of the operation, keywords used in the indexing process are generalized. This is a desirable effect, methods for generalizing the keywords are called the Latent Sematic Indexing (LSI) methods. Despite the undeniable advantages of the SVD decomposition, it has a big disadvantage. Its computational complexity is $O(n^3)$. In practice, this prevents the application of the method to a large indexing structure. The methodology presented in the article assumes the use of the Epsilon decomposition in order to divide the original TBD matrix into parts before the reduction process. The proposed modification allows the use of the SVD decomposition for the indexing structure of any size.

Keywords: Epsilon decomposition · Dimensional reduction
Application of the SVD decomposition · Latent Semantic Indexing
LSI · Large indexing structures reduction · Retrieval systems algorithms
Retrieval systems design

1 Introduction

The process of designing the modern retrieving systems is not an easy task. Such systems must meet many requirements such as compliance of results with expectations, search time, creation of found items ranking lists, etc. The basic problem, that should be solved at the very beginning of the design process, is the selection of the appropriate type of data model. The most popular models are: Boolean, fuzzy logic, VSM and

J. Świątek et al. (Eds.): ISAT 2018, AISC 853, pp. 296–307, 2019.
https://doi.org/10.1007/978-3-319-99996-8_27

probabilistic. In the case of the Information Retrieval (IR) systems, the simplest and the most effective is the VSM model [1]. A single item of the VSM model is a vector that is associated with a one indexed document. The particular element of the vector constitutes the weight associated with the occurrence of a given keyword in the document. The set of all vectors related to all indexed documents forms the TBD matrix [2]. In the case when the weights stored in the TBD matrix determine the number of keywords occurrences in the document (TF - term frequency), the matrix can look like in Fig. 1.

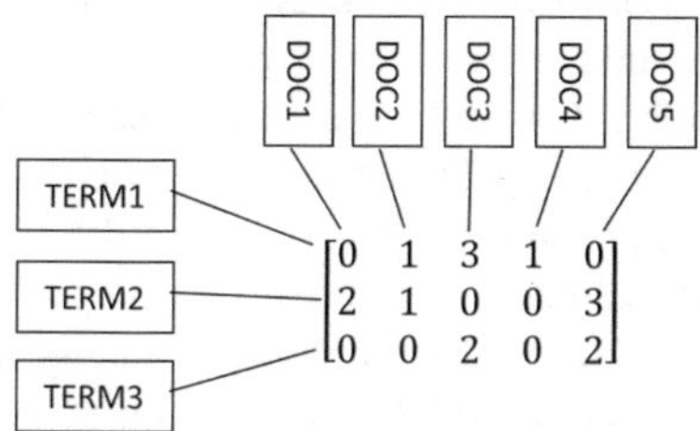

Fig. 1. A sample structure of the TF TBD matrix

The weight placed on the second position of the first vector (second row, first column) specifies that the keyword TERM2 occurs twice in the first document. From the mathematical point of view, the TBD matrix can be treated as a set of document vectors placed in the space of concepts (Fig. 2).

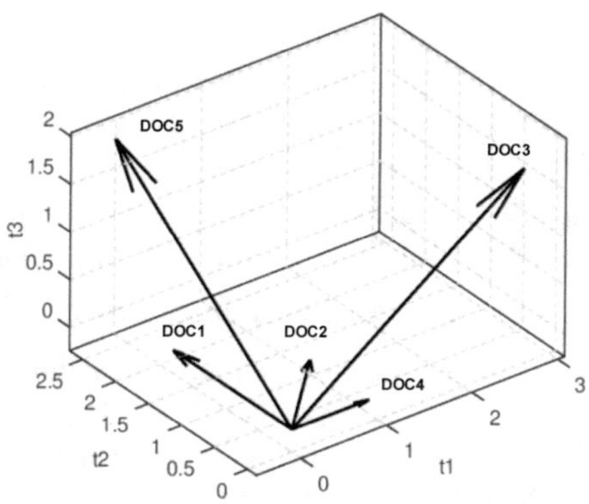

Fig. 2. A sample documents vectors in the space of concepts

The VSM model offers a big advantage for IR systems, which is a possibility to determine the similarity between the user's query and the indexed documents. The ability to compute the distance measure between documents also allows to generate a ranking list of results. The cosine distance (1) is the most frequently used measure of the distance between documents.

$$sim_{cos}(DOC_1, DOC_2) = \frac{DOC_1^T DOC_2}{|DOC_1| \cdot |DOC_2|} \tag{1}$$

The greatest similarity between two documents will take place for an angle equals to zero degrees - the cosine distance will be equal to one. In case the document vectors are perpendicular to each other, the cosine distance will be equal to zero.

Despite the many advantages of the VSM model in the field of the IR, when the indexing involves a large number of documents with multiple keywords, the size of the TBD matrix is large. Restrictions related to the speed of information retrieval or storage capacity may prevent the VSM model from being used with large databases of text documents. The reduction of the TBD matrix size can be a solution to the problem.

The use of SVD decomposition allows to reduce the size of the TBD matrix. In addition this operation reduces the dimensionality of the terms space, which is usually beneficial from the viewpoint of the IR. In practice, this means a generalization of the keywords meaning. The IR system, rather than rely on dictionary words, detects their meaning. For example, a user looking for offers related to the word "food" can get links to documents containing the words: " breakfast", " dessert", " dinner", " lunch", "meal", " snack". The IR system will generate a list of results for the user's intentions, not for the used keyword. Indexing methods using this type of strategy are called the Latent Semantic Indexing (LSI) methods. The use of the LSI method in relation to small databases of text documents with homogenous content has been well documented. Application of this method to large databases of text documents is cumbersome. The main problem is the computational complexity of the SVD algorithm. The computational complexity $O(n^3)$ prevents the use of decompositions to the large TBD matrices.

The solution to this problem may be division of the TBD matrix into two parts before performing the SVD decomposition. In the next step, the separated parts should be subjected to independent reductions. The computational complexity of the reduction process can be significantly reduced. For simplicity, assuming the division of the TBD matrix into parts of the same size, initial computational complexity $O(n^3)$ will be reduced to $O\left(k\left(\frac{n}{k}\right)^3\right)$, where k denotes the number of separated parts. Assuming that the separated parts can be reduced using parallel calculations, for example using a cluster system, the benefit of the decomposition is even greater. For example, if the TBD matrix is divided into 10 equal parts, assuming parallel calculations, the time needed for reduction can be reduced 1000 times.

The advantages of decomposing the TBD matrix before the reduction are presented in commonly available articles. In [3] there is a presentation of possible IR system work strategies: NC+SVD (Nonclustered retrieval), FC+SVD (Full clustered retrieval), PC+SVD (Partial clustered retrieval). The authors used the K-means algorithm to divide the TBD matrix into parts. In the next step, the separated parts are reduced independently of each other. The IR system will use reduced matrices as follows. Using the FC+SVD strategy, the user's query (after projection to the appropriate term space) is compared to all document vectors from reduced matrices. Using the PC+SVD strategy, the user's query is compared to the centroids associated with each reduced part of the TBD matrix. Based on this operation, the most-matched matrix is selected. The selection of the appropriate document/documents is made only with the selected matrix. In [4] the authors also used the K-means algorithm to divide the TBD matrix into parts. The authors used calculations on the graphics card to reduce the size of the matrix. Considering the best case, the use of decomposition allowed to reduce the reduction time by half. The result was obtained for dividing the TBD matrix into 4 parts. The authors in the conclusions clearly state that the main disadvantage of the methodology presented by them is the unpredictability of the K-means algorithm.

This article contains a modification of the methodology presented in [4]. We replaced the unpredictable K-means decomposition with the Epsilon decomposition algorithm which is used in the field of Control Systems. It should be noted that the authors do not know any other research on applying the Epsilon algorithm in relation to the indexing structures. The research will be carried out on the same indexing structure that was used in [2, 4] - the TBD matrix consists of 41 832 120 weights (5896 keywords and 7095 documents). Figure 3 presents the methodology for reducing the TBD matrix used in the article.

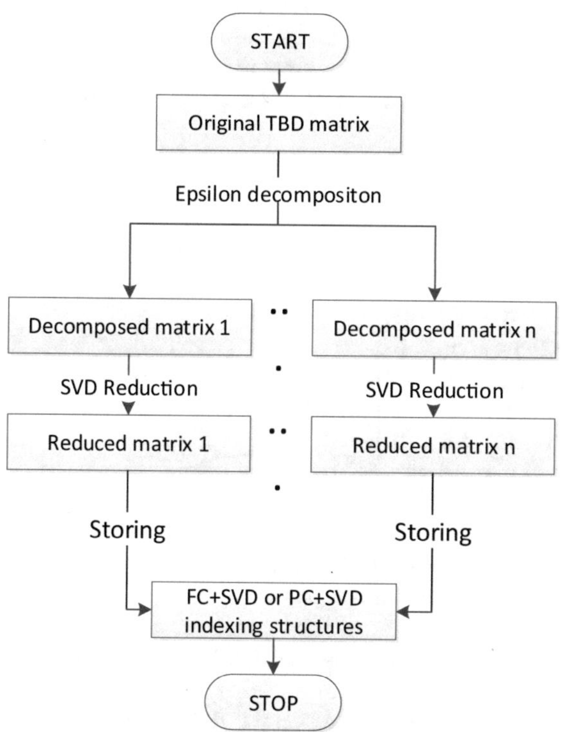

Fig. 3. The methodology for reducing the large TBD matrices using Epsilon decomposition

The authors would like to emphasize that despite many methods of features reduction, the LSI method has many advantages in case of the IR systems. The method perfectly copes with the phenomenon of the synonymy and the polysemy. Its computational complexity makes the application of the method difficult to large databases. Therefore, the authors used the Epsilon decomposition method. According to the authors' knowledge, the Epsilon decomposition has not been used for the TBD matrix so far. This is a novelty contained in this article.

2 Epsilon Decomposition

The Epsilon decomposition is an algorithm used in the field of the control systems [5–7]. The method was presented for the first time in the works of Dragoslav Šiljak. In [8] it has been demonstrated that the Epsilon decomposition can be used to decompose the LTI model into a smaller parts due to the difference in the eigenvalues of individual subsystems. In that case the decomposition was used to determine the groups strongly interrelated state variables.

The modification we introduce in current research is to use the Epsilon decomposition to find groups of closely related documents. Documents can be related to each other in many ways. The most obvious connection between two documents is when the same keywords are present in them. In the case of a group of text documents, connections between documents have the structure of a graph. For example, the TBD matrix presented in Fig. 4(left) is related to the graph of connections shown in Fig. 4 (right).

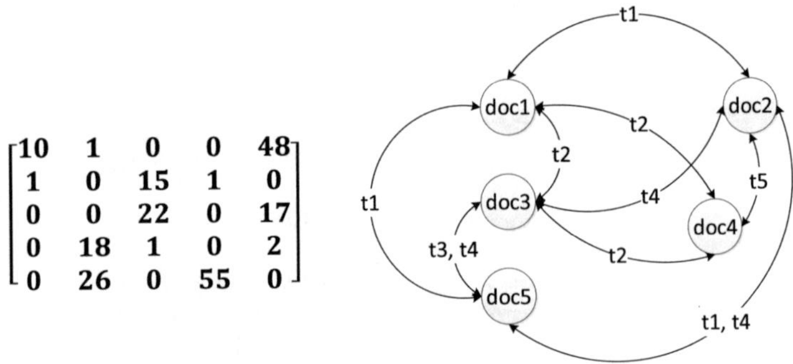

Fig. 4. The example TBD matrix (columns specify documents, lines, keywords) and the associated connection between documents graph

The graph structure contains many transitive relations. From the presented structure, it is difficult to separate groups of documents.

The Epsilon decomposition is based on the following operations. First, the matrix elements are divided by the largest element (Fig. 5a). In the next step, on the basis of an arbitrarily chosen value - epsilon (ε), all values of the matrix smaller than ε are zeroed. The remaining elements are set to one. They will define a new graph of connections between documents. For example, if we assumed $\varepsilon = 0.1$, the connection matrix is shown in Fig. 5b. In the next step, the Connected Components algorithm is performed on the connection matrix. The obtained result defines groups of related documents. For the matrix shown in Fig. 5b, the structure of the documents relations graph is shown in Fig. 5c. The structure of the graph after the algorithm has been significantly simplified. Less-important edges have been removed. In the considered case, the Connected Components algorithm will return two groups of documents (two components of graph). The first group will contain documents doc1, doc3 and doc5. There is a transitive relationship between documents doc1 and doc3. The second group is related to documents doc2 and doc4.

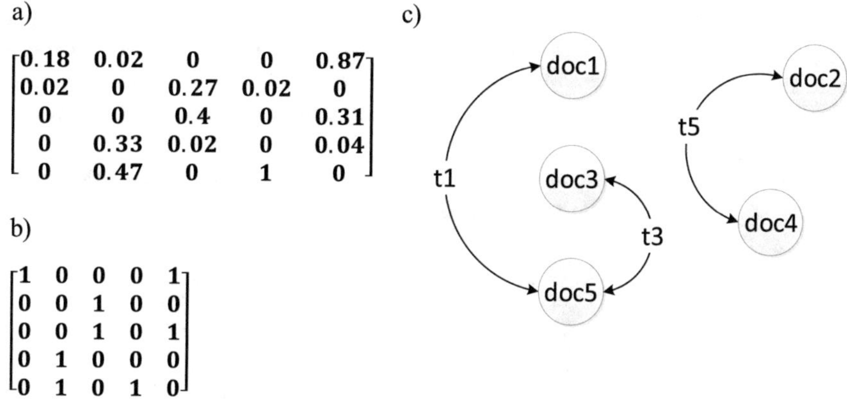

Fig. 5. The TBD matrix after normalization by the largest element (a); the connections matrix (b); the connection graph (c)

The Epsilon decomposition algorithm is a method that generates unambiguous results. In contrast to the K-means algorithm, invoking the Epsilon algorithm with the same input guarantees the same result. However, it should be noted that the number of groups returned as a result of decomposition directly depends on the assumed value of ε coefficient.

In the case of a database of text documents that are strongly interconnected, the separation of two or more groups may require a relatively high ε value. The problematic may also be the fact that the separation of two or more groups of documents does not guarantee the preservation of their similar cardinality.

For the database of closely related documents, the authors propose the use of the following modification of the Epsilon decomposition. The value of the ε coefficient should be chosen in such a way that the decomposition gives one group of strongly related documents and one group of documents that have zero vectors in the connection matrix (both groups of similar cardinality).

3 One Level Reduction

The considered collection of text documents is quite difficult for numerical methods. Numerous relations between documents (mostly occurring via transitive connections) are the main reason for this state. The data set includes 7095 text documents containing 5896 key words.

The results of the Epsilon algorithm are interpreted as follows. The first group of documents is related to vectors, which as a result of Epsilon decomposition became zero vectors (the values contained in them were zeroed in the process of creating the connection matrix). The cardinality of the group will increase as the value of ε increases. The second group of documents will be a group of closely related documents (for a given ε value). The cardinality of the group will decrease as the ε value increases. The third group (in the charts marked with the "free" label) will be a single document elements (or groups of small cardinality) that have non-zero values in the connections matrix.

Figure 6 shows the cardinality of the document groups depending on the value of the ε coefficient. The algorithm was applied to the TBD-TF matrix.

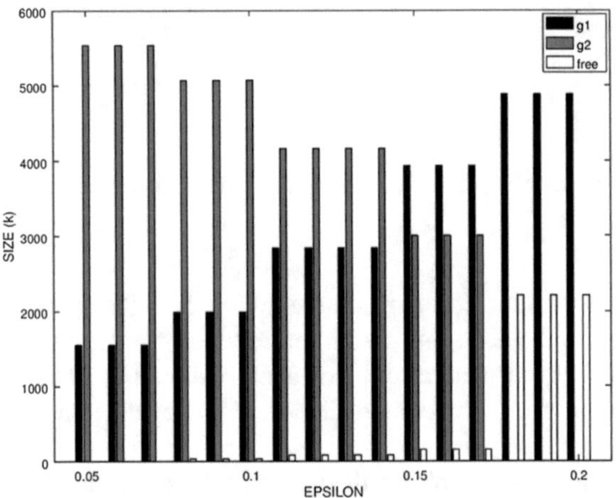

Fig. 6. The cardinality of separate groups of documents depending on the value of the ε coefficient (TBD-TF matrix)

Figure 7 shows the cardinality of the document groups depending on the value of ε coefficient when the decomposition was applied to the TBD-TFIDF matrix.

To determine the precision of an IR system containing reduced indexing data, the FC + SVD strategy was used. Similarly to the methodology included in [4], the (2) coefficient was used to measure of the precision of the IR system.

$$P = \sum\nolimits_{i=1}^{100} \begin{cases} 1 & if\ DOC_i^{org} \in LIST_i^{red} \\ 0 & if\ DOC_i^{org} \notin LIST_i^{red} \end{cases} \quad (2)$$

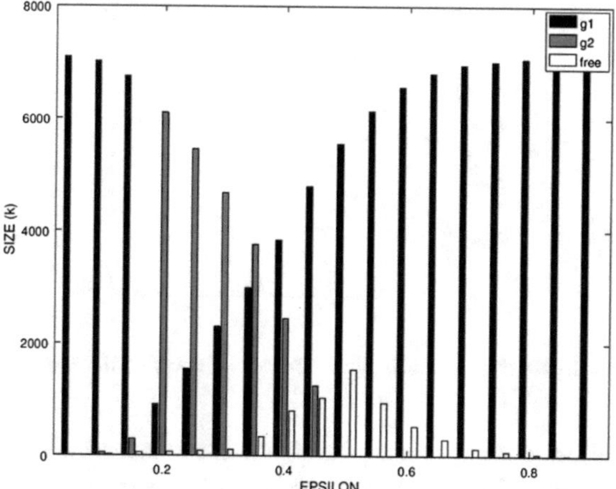

Fig. 7. The cardinality of separate groups of documents depending on the value of the ε coefficient (TBD-TFIDF matrix)

For a one hundred randomly generated user's query, the most-matched documents (DOC_1, DOC_2, …, DOC_{100}), from the original TBD matrix are determined. Then, the ranking lists, containing one hundred most matched documents ($LIST_1$, $LIST_2$, …, $LIST_{100}$), for the IR system with reduced indexing data are determined. If $DOC_i \in LIST_i$, the precision increases by 1.

The groups separated from the Epsilon decomposition of the TBD-TF matrix, for ε = 0.15, were selected for the reduction. The cardinality of the designated groups is equal: $|g_1| = 3938$, $|g_2| = 3002$, $|free| = 155$. The structure of the TBD matrix and the connections matrix after decomposition is presented in Fig. 8.

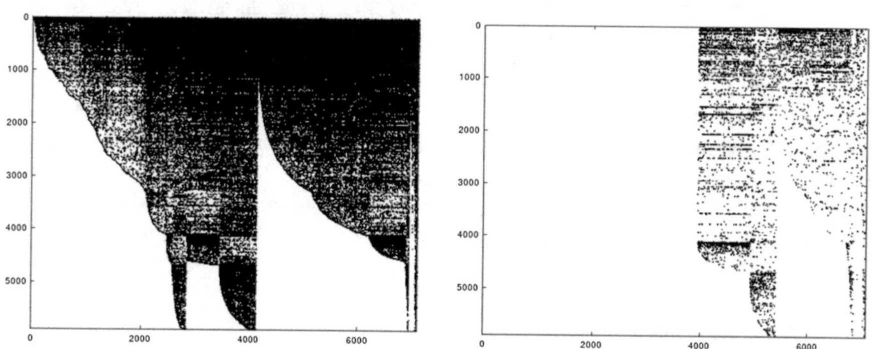

Fig. 8. The TBD (left) and connection matrix (right) after the Epsilon decomposition

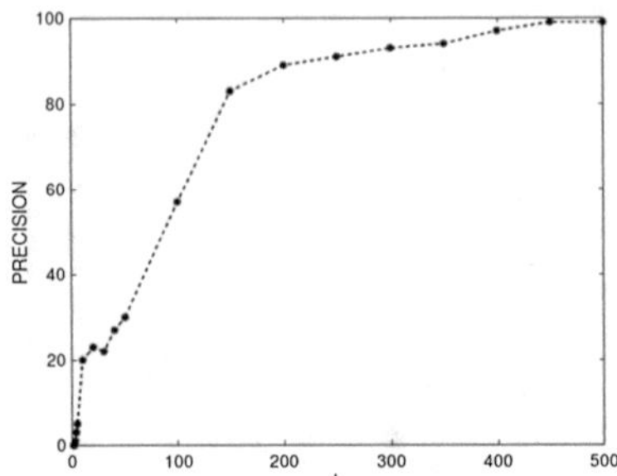

Fig. 9. The precision plot depending on the number of rows in the reduced TBD matrix (one level reduction)

The TBD matrices of the g_1 and g_2 groups were reduced independently. The TBD matrix of the *free* group, due to the small number of documents, has not been reduced. Figure 9 shows the precision of the reduced IR system depending on the number of rows in the reduced TBD matrices.

As expected, precision increases as the number of the reduced TBD matrix rows increases.

4 Two Level Reduction

The Epsilon decomposition allowed to divide the TBD matrix into two relatively equal parts. If the method is applied to large indexing structures, the division into two parts is not sufficient. In the created methodology, we propose to further divide the separated parts into smaller matrices.

On the separate groups g_1 and g_2 from the previous section the Epsilon decomposition is performed once again. When using Epsilon decomposition to the first group, for the ε coefficient = 0.3, the division into following subgroups was obtained: $|g_{1,1}| = 1554$, $|g_{1,2}| = 2374$, $|free_1| = 10$. The use of the Epsilon decomposition to the second group, with the ε coefficient = 0.24, gives the following subgroups: $|g_{2,1}| = 1558$, $|g_{2,2}| = 1231$, $|free_2| = 213$. In the next step, the groups $g_{1,1}$, $g_{1,2}$, $g_{2,1}$, $g_{2,2}$ are independently reduced. The precision of the IR system using reduced indexed structure is presented in Fig. 10.

As in the case of one level reduction, the relationship between the size of the reduced structures and the precision of the IR system is clearly visible.

5 Time Analysis

The MATLAB R2011a computing environment was used to determine the reduction time. The time needed to realize the Epsilon decomposition in the considered cases oscillated around 3 s in the case of full TBD matrix, and 1 s in the case of decomposing the groups. Figure 11 shows the time of indexing structures reduction depending on the number of decomposed parts.

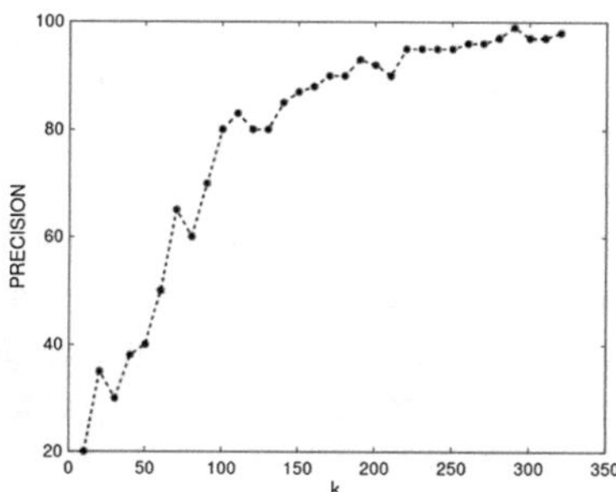

Fig. 10. The precision plot depending on the number of rows in the reduced TBD matrix (two level reduction)

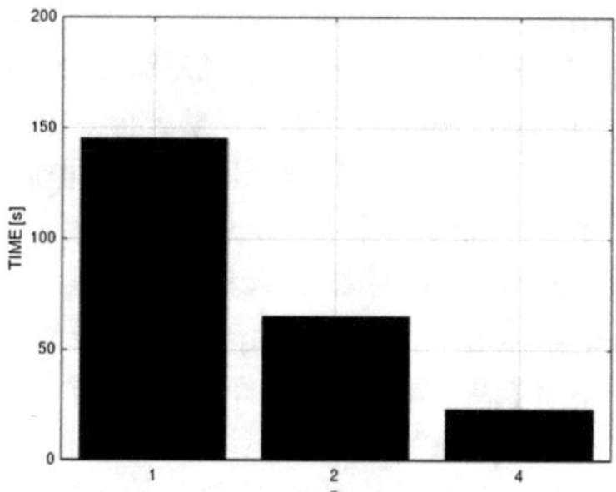

Fig. 11. The time required to reduce the TBD matrix, depending on the number of decomposed parts

The time required for the reduction decreases with the number of separated parts. Due to the fact that Epsilon decomposition was able to separate parts of similar size, the time reduction is quite significant. Comparing the reduction time of the entire TBD matrix with the time of the decomposed matrix reduction (into 4 parts), we obtained an almost seven-fold increase in speed.

6 Conclusion

The automatic methods of indexing text documents are an important element of today's world. Thanks to them, for example, search engines can work, without which browsing the web would be virtually impossible. The researches on the evolution of indexing methods are an important problem of modern science.

In the field of the IR systems one of the most important data model is the VSM model. It is difficult to use when applying to large databases of text documents. In the literature it was proposed the usage of the K-means algorithm to divide the original TBD matrix into parts. Then the separated parts can be reduced. The reduction process can additionally improve the quality of the search system by generalizing keywords.

[3, 4] The LSI technique is still used and developed, what is confirmed by the constantly appearing articles. For example, the LSI method is used for: web service retrieval [9], auto-tagging articles [10], cross-lingual articles compare [11], Malay Hadith translated document retrieval [12].

In the article, we adopted an algorithm known from the field of Control Systems - Epsilon decomposition, for grouping of text documents. From the information we have, this is the first attempt to use the Epsilon algorithm in relation to index structures. The advantage of using Epsilon decomposition is the unambiguous result. With the same input data (the TBD matrix and the ε factor), the same groups of documents will always be generated. In addition, the Epsilon algorithm does not have high computational complexity. The main operation performed by the decomposition is the graph algorithm Connected Components, which has computational complexity **O(V+E)** (V - number of vertices, E- number of edges).

We proposed a comprehensive methodology for decomposition and reduction of large-scale indexing structures. We have also proven its effectiveness in application to the TBD matrix containing 41 832 120 weights (5896 keywords and 7095 documents). The matrix was previously used in [2, 4], so we are able to compare the results. Ultimately, we can state that the use of Epsilon decomposition gives equally good results (in our case even better) as the K-means algorithm.

The considerations presented in the article open a new, possible direction of research related to the use of Epsilon decomposition. First of all, it is important to determine how Epsilon decomposition behaves at higher levels of decomposition (the article presents considerations for two levels). It is also important to determine how the decomposition behaves for other index structures, however it should be emphasized that the database used in the article is quite problematic (many transitive connections between documents). It would also be interesting to use the Epsilon decomposition together with the K-means algorithm - for example at other decomposition levels.

The main goal of the article is to determine the usefulness of Epsilon decomposition in application to the LSI method (which is a novelty). The authors showed that the proposed methodology is effective by comparing the results to an analogous decomposition carried out using the K-means algorithm [4]. However, it seems valuable to compare the proposed methodology in combination with other techniques of the features reduction.

References

1. Zhao, Y., Shi, X.: The application of vector space model in the information retrieval system. In: Zhang, W. (eds.) Software Engineering and Knowledge Engineering: Theory and Practice, Advances in Intelligent and Soft Computing, vol. 162, pp. 43–49. Springer, Heidelberg (2012)
2. Raczyński, D., Stanisławski, W.: SVD based Latent Semantic Indexing with use of the GPU computations. Int. J. Soft Comput. Math. Control (IJSCMC) **6**(2/3), 1–14 (2017)
3. Gao, J., Zhang, J.: Clustered SVD strategies in Latent Semantic Indexing. Inf. Process. Manag. **41**(5), 1051–1063 (2005)

4. Raczyński, D., Stanisławski, W.: Decomposition and reduction of indexing structures with use of the GPU computations. In: Grzech, A., Świątek, J., Wilimowska, Z., Borzemski, L. (eds.) Information Systems Architecture and Technology: Proceedings of 37th International Conference on Information Systems Architecture and Technology – ISAT 2016 – Part II, Advances in Intelligent Systems and Computing, vol. 522, pp. 225–237. Springer (2017)
5. Zečević, A., Šiljak, D.: Control of Complex Systems. Structural Constraints and Uncertainty. Springer, London (2010)
6. Šiljak, D.: Decentralized Control of Complex Systems. Academic Press, New York (1991)
7. Sezer, M., Šiljak, D.: Nested epsilon decompositions of linear systems: weakly coupled and overlapping blocks. SIAM. J. Matrix Anal. Appl. **12**(3), 521–533 (1991)
8. Raczyński, D., Stanisławski, W.: Use of the modified EPSILON decomposition for the LTI models reduction. In: Świątek, J., Borzemski, L., Wilimowska, Z. (eds.) Information Systems Architecture and Technology: Proceedings of 38th International Conference on Information Systems Architecture and Technology – ISAT 2017, ISAT 2017, Advances in Intelligent Systems and Computing, vol. 656, pp. 3–16. Springer, Cham (2018)
9. Czyszczoń, A., Zgrzywa, A.: Latent Semantic Indexing for web service retrieval. In: Hwang, D., Jung, J.J., Nguyen, N.T. (eds.) Computational Collective Intelligence, Technologies and Applications, ICCCI 2014, Lecture Notes in Computer Science, vol. 8733, pp. 694–702. Springer, Cham (2014)
10. Rattanapanich, R., Sriharee, G.: Auto-tagging articles using Latent Semantic Indexing and ontology. In: Nguyen, N.T., Attachoo, B., Trawiński, B., Somboonviwat, K. (eds.) Intelligent Information and Database Systems, ACIIDS 2014, Lecture Notes in Computer Science, vol. 8397, pp. 153–162. Springer, Cham (2014)
11. Saad, M., Langlois, D., Smaïli, K.: Cross-lingual semantic similarity measure for comparable articles. In: Przepiórkowski, A., Ogrodniczuk, M. (eds.) Advances in Natural Language Processing, NLP 2014, Lecture Notes in Computer Science, vol. 8686, pp. 105–11. Springer, Cham (2014)
12. Rahman, N.A., Mabni, Z., Omar, N., Hanum, H.F.M., Rahim, N.N.A.T.M.: A parallel Latent Semantic Indexing (LSI) algorithm for malay hadith translated document retrieval. In: Berry, M., Mohamed, A., Yap, B. (eds.) Soft Computing in Data Science, SCDS 2015, Communications in Computer and Information Science, vol. 545, pp. 154–163. Springer, Singapore (2015)

Minimax Decision Rules for Identifying an Unknown Distribution of a Random Variable

Ireneusz Jóźwiak[1(✉)] and Jerzy Legut[2]

[1] Faculty of Computer Science and Management,
Wrocław University of Science and Technology, 50-370 Wrocław, Poland
ireneusz.jozwiak@pwr.wroc.pl
[2] Faculty of Pure and Applied Mathematics,
Wrocław University of Science and Technology, 50-370 Wrocław, Poland
Jerzy.Legut@pwr.wroc.pl

Abstract. We consider a problem of identifying an unknown distribution of a random variable based on its single observation. We present known results of constructing a minimax decision rules for one-dimensional case equivalent to a problem of equitable optimal partitioning of a measurable space. An example of finding a minimax decision rule for two-dimensional case is given.

Keywords: Decision rule · Optimal partition of a measurable space

1 Introduction

Suppose we are given a continuous random variable X having one of the known distribution described by density functions $f_i : [0,1] \to \mathbb{R}_+$ $i \in I := \{1, ..., n\}$. We don't know which is the true distribution of X. We consider a classification problem (cf. [5]) in which after one observation of $X(\omega)$ (realisation of random variable X) we are to decide which is the true distribution of X. Denote by $\mathcal{B}$ the σ-algebra of measurable subsets of the unite interval $[0,1]$. By a partition $P = \{A_i\}_{i=1}^n$ of $[0,1]$ we mean a collection of $\mathcal{B}$-measurable disjoint subsets $A_1, \ldots, A_n$ whose union is equal to $[0,1]$. Let $\mathscr{P}$ stand for the set of all measurable partitions $P = \{A_i\}_{i=1}^n$ of $[0,1]$.

Definition 1. *A partition $P = \{A_i\}_{i=1}^n \in \mathscr{P}$ is called a decision rule if in case of $X(\omega) \in A_i$, we guess that X has density function f_i, $i \in I$.*

Our objective is to minimize the largest probability of misclassification

$$\max_{i \in I} Pr(X \notin A_i |\, \mathrm{dist} X = f_i),$$

over all measurable partitions $P = \{A_i\}_{i=1}^n \in \mathscr{P}$. Denote by

$$R = \inf \left\{ \max_{i \in I} Pr(X \notin A_i |\, \mathrm{dist} X = f_i) : \{A_i\}_{i=1}^n \in \mathscr{P} \right\}$$

J. Świątek et al. (Eds.): ISAT 2018, AISC 853, pp. 308–317, 2019.
https://doi.org/10.1007/978-3-319-99996-8_28

a minimal possible risk of misclassification. We obtain (cf. [6])

$$R = \inf\left\{\max_{i\in I}(1-\mu_i(A_i)) : \{A_i\}_{i=1}^n \in \mathscr{P}\right\} = 1 - \sup\left\{\min_{i\in I}\mu_i(A_i) : \{A_i\}_{i=1}^n \in \mathscr{P}\right\},$$

where μ_i denote a non-atomic probability measure defined by

$$\mu_i(A) = \int_A f_i\,dx, \quad A \in \mathcal{B},\, i \in I. \tag{1}$$

Definition 2. *A partition $P^* = \{A_i^*\}_{i=1}^n \in \mathscr{P}$ is said to be a minimax decision rule if*

$$R = 1 - \min_{i\in I}\mu_i(A_i^*).$$

The existence of minimax decision rules follows from a result of Dvoretzky et al. [3]. Unfortunately finding minimax decision rules in practice for arbitrary density functions f_i is not easy in general case. Jóźwiak and Legut [6] presented an algorithm for construction of the minimax rules for exponential distributions for $n = 2$. The problem of finding minimax decision rules is equivalent to finding equitable optimal partitions of the measurable space $([0,1],\mathcal{B})$ among n players (cf. [10]). Hence all known methods of obtaining optimal partitions can be immediately applied to finding minimax decision rules. Dall'Aglio et al. [1] used a linear programming algorithm of obtaining equitable optimal partitions for densities being simple functions. Other results will be discussed in next sections. An interesting problem is to estimate the probability of committing an error using the decision rule. The estimating inequalities were first presented by Elton et al. [4] and then by Legut [7].

A general form of the minimax decision rules could be helpful in some cases for finding constructive methods of equitable optimal partitioning of a measurable space. Let $S = \{\vec{s} = (s_1,\ldots,s_n) \in \mathbb{R}^n, s_i \geq 0,\, i \in I,\, \sum_{i=1}^n s_i = 1\}$ be the $(n-1)$-dimensional simplex. Let nonatomic measures $\mu_1,\ldots,\mu_n$ defined on a measurable space $(\mathcal{Z},\mathcal{B}_\mathcal{Z})$ are absolutely continuous with respect to the same measure ν (e.g. $\nu = \sum_{i=1}^n \mu_i$). Denote by $f_i = d\mu_i/d\nu$ the Radon-Nikodym derivatives, i.e.

$$\mu_i(A) = \int_A f_i\,d\nu, \text{ for } A \in \mathcal{B}_\mathcal{Z} \text{ and } i \in I.$$

For $\vec{p} = (p_1,...,p_n) \in S$ and $i \in I$, define the following measurable sets

$$B_i(\vec{p}) = \bigcap_{k=1,k\neq i}^{n} \{z \in \mathcal{Z} : p_i f_i(z) > p_k f_k(z)\},$$

$$C_i(\vec{p}) = \bigcap_{k=1}^{n} \{z \in \mathcal{Z} : p_i f_i(z) \geq p_k f_k(z)\}.$$

Legut and Wilczyński [10] using a minmax theorem of Sion (cf. [2]) proved the following theorem presented here in less general form

Theorem 1. *There exists a point $\overrightarrow{p^*} \in S$ and a corresponding equitable optimal partition $P^* = \{A_i^*\}_{i=1}^n$ satisfying*

(i) $B_i(\overrightarrow{p^*}) \subset A_i^* \subset C_i(\overrightarrow{p^*})$,
(ii) $\mu_1(A_1^*) = \mu_2(A_2^*) = \ldots = \mu_n(A_n^*)$.

Moreover, any partition $P^* = \{A_i^*\}_{i=1}^n$ which satisfies (i) and (ii) is equitable optimal.

2 One-Dimensional Case

2.1 Piecewise Linear Density Functions

In this section we present a method based on a result of Legut [8] of obtaining the minimax decision rules for piecewise linear density functions.

Suppose we are given n piecewise linear functions defined on the unit interval $[0, 1)$

$$f_i(x) = \sum_{j=1}^{m} (c_{ij}x + d_{ij}) I_{[a_j, a_{j+1})}(x), \quad \int_0^1 f_i(x)\, dx = 1, \quad i \in I,$$

where $\{[a_j, a_{j+1})\}_{j=1}^m$ is a partition of the interval $[0, 1)$ such that

$$[0,1) = \bigcup_{j=1}^{m} [a_j, a_{j+1}), \;\; a_1 = 0, \; a_{m+1} = 1, \; a_{j+1} > a_j \; j = 1, ..., m.$$

By $I_A(x)$ we denote here the indicator of the set $A \in \mathcal{B}$.

We assume that

$$c_{ij}x + d_{ij} \geq 0 \quad \text{for all} \quad x \in [a_j, a_{j+1}), \, i \in I, \, j = 1, ..., m.$$

Consider nonatomic probability measures $\mu_1, ..., \mu_n$ defined by (1). Throughout this paper and without loss of generality we consider only left side closed and right side open intervals unless they are otherwise defined. Consider partitions of each interval $[a_j, a_{j+1})$, $j = 1, ..., m$ into n subintervals by cuts in points $b_k^{(j)}, k = 1, ..., n-1, j = 1, ..., m$ such that

$$[a_j, a_{j+1}) = \bigcup_{k=1}^{n} [b_{k-1}^{(j)}, b_k^{(j)}),$$

where $b_0^{(j)} = a_j, \; b_n^{(j)} = a_{j+1}, \; b_{k+1}^{(j)} \geq b_k^{(j)}, \; k = 1, ..., n-1, \; j = 1, ..., m.$

If $b_{k-1}^{(j)} = b_k^{(j)}$ for some $k = 1, ..., n$ we put $[b_{k-1}^{(j)}, b_k^{(j)}) = \emptyset$.

For simplicity we will also denote $B_{kj} := [b_{k-1}^{(j)}, b_k^{(j)})$, $k = 1, ..., n$, $j = 1, ..., m$.

Now we construct an assignment of each subintervals B_{kj} to each $i \in I$. Let p_j, q_j, $j = 1, ..., m$ be integers satisfying $0 \le p_j \le q_j \le n$ and

$$\#\{i : i \in I, c_{ij} < 0\} = p_j,$$

$$\#\{i : i \in I, c_{ij} = 0\} = q_j - p_j,$$

$$\#\{i : i \in I, c_{ij} > 0\} = n - q_j,$$

where by $\#A$ we denote the number of elements of a finite set A.

For each interval $[a_j, a_{j+1})$, $j = 1, ..., m$, we define permutations $\sigma_j : I \longrightarrow I, j = 1, ..., m$ satisfying the following conditions:

1. If $p_j > 0$ we define $\sigma_j(k) \in \{i : i \in I, c_{ij} < 0\}$ for $k = 1, ..., p_j$ such that

$$\frac{d_{\sigma_j(k)j}}{c_{\sigma_j(k)j}} \ge \frac{d_{\sigma_j(k+1)j}}{c_{\sigma_j(k+1)j}}, \quad k = 1, ..., p_j - 1$$

2. If $q_j - p_j > 0$ we define $\sigma_j(k) \in \{i : i \in I, c_{ij} = 0\}$ for $k = p_j + 1, ..., q_j$ such that

$$\sigma_j(k) \le \sigma_j(k+1), \quad k = p_j + 1, ..., q_j - 1$$

3. If $n - q_j > 0$ we define $\sigma_j(k) \in \{i : i \in I, c_{ij} > 0\}$ for $k = q_j + 1, ..., n$ such that

$$\frac{d_{\sigma_j(k)j}}{c_{\sigma_j(k)j}} \ge \frac{d_{\sigma_j(k+1)j}}{c_{\sigma_j(k+1)j}}, \quad k = q_j + 1, ..., n - 1$$

Permutations σ_j, $j = 1, ..., m$ define one-to-one assignment of the subintervals $B_{ij} \subset [a_j, a_{j+1})$, $i \in I, j = 1, ..., m$ such that the subinterval $B_{\sigma_j^{-1}(i)j}$ is assigned to $i \in I$. Finally we obtain a partition $\{B_i\}_{i=1}^n$ of the unit interval defined by

$$B_i = \bigcup_{j=1}^{m} B_{\sigma_j^{-1}(i)j}, \; i \in I.$$

Legut [8] proved the following theorem presenting an algorithm for obtaining an minimax decision rule.

Theorem 2. *Let the collection of numbers* $z^*, \{c_k^{(j)}\}, k = 1, ..., n-1, \; j = 1, ..., m$ *be a solution of the following nonlinear programming (NLP) problem*

$$max\ z$$

subject to quadratic constraints

$$z = \sum_{j=1}^{m} \mu_i(B_{\sigma_j^{-1}(i)j}) = \sum_{j=1}^{m} \int_{B_{\sigma_j^{-1}(i)j}} f_i dx, \quad i = 1, ..., n,$$

with respect to variables z, $\{b_k^{(j)}\}$ $k = 1, ..., n-1, j = 1, ..., m$ *satisfying the following inequalities*

$$0 = a_1 \le b_1^{(1)} \le ... \le b_{n-1}^{(1)} \le a_2,$$

$$a_2 \le b_1^{(2)} \le ... \le b_{n-1}^{(2)} \le a_3,$$

$$...$$

$$a_m \le b_1^{(m)} \le ... \le b_{n-1}^{(m)} \le a_{m+1} = 1.$$

Then, the partition $\{C_i\}_{i=1}^n$ *of the unit interval* $[0, 1)$ *defined by*

$$C_i = \bigcup_{j=1}^{m} C_{\sigma_j^{-1}(i)j}, \; i \in I,$$

where

$$C_{\sigma_j^{-1}(i)j} = [c^{(j)}_{\sigma_j^{-1}(i)-1}, c^{(j)}_{\sigma_j^{-1}(i)}),$$

and $c_0^{(j)} = a_j$, $c_n^{(j)} = a_{j+1}$, $j = 1, ..., m$ *is a minimax decision rule and* $R := 1 - z^*$ *is the minimal possible risk.*

An example of computing a minimax decision rule for distributions described by piecewise linear density functions is presented in [8].

2.2 Densities with Piecewise MLR Property

Assume now that

$$f_i(x) > 0, \quad \text{for all} \quad x \in [0, 1),$$

and

Assumption 1. *There exists a partition* $\{[a_j, a_{j+1})\}_{j=1}^m$ *of the interval* $[0, 1)$, *where* $a_1 = 0$, $a_{m+1} = 1$, *such that the densities* f_i *satisfy strictly monotone likelihood ratio (SMLR) property on each interval* $[a_j, a_{j+1}), j \in J := \{1, ..., m\}$, *i.e. for any* $i, k \in I$, $i \neq k$, *the ratios* $\dfrac{f_i(x)}{f_k(x)}$ *are strictly monotone on each interval* $[a_j, a_{j+1})$.

Legut [9] proved the following

Proposition 1. *If the density functions* f_i, $i \in I$, *are differentiable and the set*

$$D := \{x \in (0, 1) : f_i'(x) f_k(x) = f_i(x) f_k'(x), \; i, k \in I, \; i \neq k\}$$

is finite then Assumption 1 is satisfied.

Define absolutely continuous and strictly increasing functions $F_i : [0,1] \rightarrow [0,1]$ by

$$F_i(t) = \int_{[0,t)} f_i \, dx, \quad t \in [0,1], \quad i \in I. \tag{2}$$

We need the following proposition (cf. [9]) to define similar permutations to those presented in Sect. 2.

Proposition 2. *Suppose the densities f_i satisfy Assumption 1. Then for any numbers θ_1, θ_2 satisfying $a_j \leq \theta_1 < \theta_2 < a_{j+1}$, $j \in J$, and any $i, k \in I$, $i \neq k$ the one of the two following inequalities*

$$\frac{F_i(t) - F_i(\theta_1)}{F_i(\theta_2) - F_i(\theta_1)} < \frac{F_k(t) - F_k(\theta_1)}{F_k(\theta_2) - F_k(\theta_1)} \tag{3}$$

$$\frac{F_i(t) - F_i(\theta_1)}{F_i(\theta_2) - F_i(\theta_1)} > \frac{F_k(t) - F_k(\theta_1)}{F_k(\theta_2) - F_k(\theta_1)} \tag{4}$$

holds for each $t \in (\theta_1, \theta_2)$.

The inequalities (3) and (4) mean that there is a strict relative convexity relationship between the functions F_i and F_k, $i \neq k$, defined by (2). If the inequality (3) holds, then F_i is strictly convex with respect to F_k. The relation of strict relative convexity induces on each interval (a_j, a_{j+1}) a strict partial ordering of the functions F_i. Let $F_i \prec_j F_k$ denote that F_i is strictly convex with respect to F_k on (a_j, a_{j+1}). For each $j \in J$ define permutation $\sigma_j : I \longrightarrow I$, such that

$$F_{\sigma_j(k+1)} \prec_j F_{\sigma_j(k)}.$$

Legut [9] proved the following

Theorem 3. *Let a collection of numbers z^*, $\{x_k^{*(j)}\}, k = 1, ..., n-1$, $j \in J$, be a solution of the following nonlinear programming (NLP) problem*

$$max\ z$$

subject to constraints

$$z = \sum_{j=1}^{m} \left[F_i(x^{(j)}_{\sigma_j(i)}) - F_i(x^{(j)}_{\sigma_j(i)-1}) \right] \quad i = 1, ..., n,$$

with respect to variables z, $\{x_k^{(j)}\}$, $k = 1, ..., n-1, j \in J$, satisfying the following inequalities

$$0 = a_1 \leq x_1^{(1)} \leq ... \leq x_{n-1}^{(1)} \leq a_2,$$

$$a_2 \leq x_1^{(2)} \leq ... \leq x_{n-1}^{(2)} \leq a_3,$$

$$...$$

$$a_m \leq x_1^{(m)} \leq ... \leq x_{n-1}^{(m)} \leq a_{m+1} = 1.$$

Then the partition $\{A_i^*\}_{i=1}^n \in \mathscr{P}$ *of the unit interval* $[0,1)$ *defined by*

$$A_i^* = \bigcup_{j=1}^{m} \left[x_{\sigma_j(i)-1}^{*(j)}, x_{\sigma_j(i)}^{*(j)}\right), \; i \in I,$$

where $x_0^{*(j)} = a_j$, $x_n^{*(j)} = a_{j+1}$, $j \in J$, *is a minimax decision rule for the measures* μ_i, $i \in I$ *and* $R := 1 - z^*$ *is the minimal possible risk.*

2.3 Example 1

Suppose a random variable X has one of the following distributions described by three density functions

$$f_1 = 12\left(x - \frac{1}{2}\right)^2, \; f_2 = 2x, \; f_3 \equiv 1, \quad x \in [0,1).$$

We use the algorithm described in Theorem 3 to obtain an minimax decision rule. First we need to divide the interval $[0,1)$ into some subintervals on which the densities f_i, $i = 1,2,3$, separably satisfy SMLR property. For this reason we consider the following ratios

$$\frac{f_1(x)}{f_2(x)} = \frac{12\left(x - \frac{1}{2}\right)^2}{2x}, \quad \frac{f_1(x)}{f_3(x)} = 12\left(x - \frac{1}{2}\right)^2, \quad \frac{f_2(x)}{f_3(x)} = 2x, \quad x \in (0,1)$$

It is easy to verify that the densities f_i, $i = 1,2,3$, satisfy the SMLR property on intervals $[0, \frac{1}{2})$ and $[\frac{1}{2}, 1)$. Denote $F_i(t) = \int_0^t f_i(x)\,dx$, $i = 1,2,3$. It follows from Proposition 2 that functions

$$\frac{F_i(t) - F_i(0)}{F_i(\frac{1}{2}) - F_i(0)}, \; i = 1,2,3, \quad t \in \left[0, \frac{1}{2}\right)$$

and

$$\frac{F_i(t) - F_i(\frac{1}{2})}{F_i(1) - F_i(\frac{1}{2})}, \; i = 1,2,3, \quad t \in \left[\frac{1}{2}, 1\right)$$

are strictly convex or strictly concave with respect to each other on the intervals $[0, \frac{1}{2})$ and $[\frac{1}{2}, 1)$ respectively. Now we establish the proper order of assigments of the subintervals of $[0, \frac{1}{2})$ and $[\frac{1}{2}, 1)$ to each $i = 1,2,3$. Easy calculations give the following ineaqualities

$$\frac{F_1(t) - F_1(0)}{F_1(\frac{1}{2}) - F_1(0)} > \frac{F_3(t) - F_3(0)}{F_3(\frac{1}{2}) - F_3(0)} > \frac{F_2(t) - F_1(0)}{F_2(\frac{1}{2}) - F_2(0)}, \quad \text{for all} \quad t \in \left[0, \frac{1}{2}\right)$$

and

$$\frac{F_3(t) - F_3(0)}{F_3(1) - F_3(\frac{1}{2})} > \frac{F_2(t) - F_1(0)}{F_2(1) - F_2(\frac{1}{2})} > \frac{F_1(t) - F_1(0)}{F_1(1) - F_1(\frac{1}{2})}, \quad \text{for all} \quad t \in \left[\frac{1}{2}, 1\right).$$

Hence, we obtain permutations

$$\sigma_1 = \begin{pmatrix} 1\,2\,3 \\ 1\,3\,2 \end{pmatrix} \quad \text{and} \quad \sigma_2 = \begin{pmatrix} 1\,2\,3 \\ 3\,2\,1 \end{pmatrix}.$$

Now we are ready to formulate an NLP problem as in Theorem 3

$$max\ z$$

subject to constraints

$$z = F_1(x_1^{(1)}) - F_1(0) + F_1(1) - F_1(x_2^{(2)}),$$
$$z = F_2(\frac{1}{2}) - F_2(x_1^{(2)}) + F_2(x_2^{(2)}) - F_2(x_2^{(1)}),$$
$$z = F_3(x_1^{(2)}) - F_3(x_1^{(1)}) + F_3(x_2^{(1)}) - F_3(\frac{1}{2}),$$

with respect to the variables $z, \{x_k^{(j)}\}$ $k = 1, 2, j = 1, 2$, satisfying the following inequalities

$$0 \leq x_1^{(1)} \leq x_1^{(2)} \leq \frac{1}{2} \leq x_2^{(1)} \leq x_2^{(2)} \leq 1.$$

Solving the above NLP problem we obtain

$$z^* = 0.4843,\ x_1^{*(1)} = 0.1426,\ x_1^{*(2)} = a_2 = 0.5,\ x_2^{*(1)} = 0.6269,\ x_2^{*(2)} = 0.9367.$$

Hence, we get the minimax decision rule $\{A_i^*\}_{i=1}^3 \in \mathscr{P}$ where

$$A_1^* = [0, x_1^{*(1)}) \cup [x_2^{*(2)}, 1), \quad A_2^* = [x_2^{*(2)}, x_2^{*(1)}) \quad \text{and} \quad A_3^* = [x_2^{*(1)}, x_1^{*(1)}).$$

3 Two-Dimensional Case

Consider a continuous random variable X having one of two known distributions defined on the unite square $[0, 1]^2$ by linear density functions:

$$f_i(x, y) = a_i x + b_i y + c_i, \text{ for } (x, y) \in [0, 1]^2. \tag{5}$$

The following proposition follows from Theorem 1, where we set $(\mathcal{Z}, \mathcal{B}_{\mathcal{Z}}) = ([0, 1]^2, \mathcal{B}_{[0,1]^2})$ and $n = 2$:

Proposition 3. *Let $\mu_i, i = 1, 2$ be two measures defined on measurable space $([0, 1]^2, \mathcal{B}_{[0,1]^2})$ defined by density functions $f_i : [0, 1]^2 \to \mathbb{R}_+, i = 1, 2$. Then a partition $\{A_i^*\}_{i=1}^2$ is a minimax decision rule if and only if there exist a number $p \in (0, 1)$ such that*

- $A_1^* = \{(x, y) \in [0, 1]^2 : pf_1(x, y) > (1 - p)f_2(x, y)\}$,
- $A_2^* = [0, 1]^2 \setminus A_1^*$,

$\mu_1(A_1^*) = \mu_2(A_2^*)$.

Hence, to find a minimax decision rule for density functions (5) we need to solve the equation

$$\iint_{A_1(p)} (a_1x + b_1y + c_1)dxdy = \iint_{[0,1]^2 \setminus A_1(p)} (a_2x + b_2y + c_2)dxdy$$

with respect to variable $p \in (0,1)$, where

$$A_1(p) = \{(x,y) \in [0,1]^2 : p(a_1x + b_1y + c_1) > (1-p)(a_2x + b_2y + c_2)\}.$$

3.1 Example 2

Suppose we observe a random variable X which has one of the two following densities:

$$f_1(x,y) = x + y \quad \text{and} \quad f_2(x,y) = -x/2 - y/2 + 3/2. \tag{6}$$

Let

$$\begin{aligned} A_1(p) &= \{(x,y) \in [0,1]^2 : p(x+y) > (1-p)(-x/2 - y/2 + 3/2)\} \\ &= \left\{(x,y) \in [0,1]^2 : y > -x + 3\frac{1-p}{1+p}\right\} = \{(x,y) \in [0,1]^2 : y > -x + r\}, \end{aligned} \tag{7}$$

where

$$r = 3\frac{1-p}{1+p}. \tag{8}$$

First for $r \le 1$ we solve the following equation with respect to r:

$$1 - \int_0^r dx \int_0^{-x+r} (x+y)dy = \int_0^r dx \int_0^{-x+r} (-x/2 - y/2 + 3/2)dy.$$

Hence we have

$$1 - \frac{r^3}{3} = \frac{1}{12}r^2(9 - 2r).$$

The only approximate positive solution of the above equation is $r \approx 1.04063$ which contradicts our assumption that $r \le 1$. Assume now that $1 < r \le 2$. In this case we need to solve the following equation:

$$\int_{1-r}^1 dx \int_{-x+r}^1 (x+y)dy = 1 - \int_{1-r}^1 dx \int_{-x+r}^1 (-x/2 - y/2 + 3/2)dy.$$

After simple calculation we obtain the following equation

$$\frac{1}{3}(r-2)^2(1+r) = 1 + \frac{1}{12}(r-2)^2(2r-7)$$

and approximate solution $r \approx 1.04235 \in (1, 2]$. From the formula (8) we get $p \approx 0.48429$. Finally, using (7) for calculated r we obtain the minimax decision rule $A_1^*, A_2^* = [0, 1]^2 \setminus A_1^*$ for the densities (6). Hence we get the risk

$$R \approx 1 - \iint_{A_1^*} (x + y) dx dy \approx 0.37565.$$

The method presented in above example can be applied also for more complex densities.

References

1. Dall'Aglio, M., Legut, J., Wilczyński, M.: On finding optimal partitions of a measurable space. Mathematica Applicanda **43**(2), 193–206 (2015)
2. Aubin, J.P.: Mathematical Methods of Game and Conomic Theory. North-Holland Publishing Company, Amsterdam (1980)
3. Dvoretzky, A., Wald, A., Wolfowitz, J.: Relations among certain ranges of vector measures. Pacific J. Math. **1**, 59–74 (1951)
4. Elton, J., Hill, T., Kertz, R.: Optimal partitioning ineaqualities for non-atomic probability measures. Trans. Amer. Math. Soc. **296**, 703–725 (1986)
5. Hill, T., Tong, Y.: Optimal-partitioning ineaqualities in classification and multi hypotheses testing. Ann. Stat. **17**, 1325–1334 (1989)
6. Jóźwiak, I., Legut, J.: Decision rule for an exponential reliability function. Microelectron. Reliab. **31**(1), 71–73 (1990)
7. Legut, J.: Inequalities for α- optimal partitioning of a measurable space. Proc. Amer. Math. Soc. **104**, 1249–1251 (1988)
8. Legut, J.: Optimal fair division for measures with piecewise linear density functions. Int. Game Theory Rev. **19**(2), 1750009 (2017)
9. Legut, J.: How to obtain an equitable optimal fair division, working paper
10. Legut, J., Wilczyński, M.: Optimal partitioning of a measurable space. Proc. Amer. Math. Soc. **104**, 262–264 (1988)

Artificial Intelligence Methods and Algorithms

Decision Making Model Based on Neural Network with Diagonalized Synaptic Connections

R. Peleshchak[1(✉)], V. Lytvyn[2], I. Peleshchak[2], R. Olyvko[2], and J. Korniak[3]

[1] Department of Physics, Ivan Franko Drohobych State Pedagogical University Drohobych, Lviv, Ukraine
rpeleshchak@ukr.net
[2] Information Systems and Networks Department, Lviv Polytechnic National University, Lviv, Ukraine
vasyl17.lytvyn@gmail.com
[3] University of Information Technology and Management in Rzeszow, Rzeszów, Poland

Abstract. In this paper, we propose a decision-making model based on the architecture of a three-layer perceptron with diagonal weighted synaptic connections between the neurons of the input, the latent and the original layers. The evolution of the model is carried out as a task of adaptation of the neural network, which consists of procedures for correction of the number of synaptic connections between the neurons of the input hidden and output layers due to the diagonalization of the matrices of synaptic connections in the basis of the input vector vectors. It is shown that the time of decision making in the diagonalized three-layer neural network is smaller in comparison with the time in the non-diagonalized.

Keywords: Decision-making · Model
Diagonalized three-layer neural network · Matrices of synaptic connections
Time of decision making

1 Introduction

To date, the methods of processing intellectual information in conditions of uncertainty have acquired considerable scientific and practical interest. In particular, among these methods are fuzzy models. They allow us to describe processes using natural languages and linguistic variables with the help of a clear mechanism of fuzzy logical conclusion. Therefore, fuzzy models are widely used to solve the problems of identification, recognition, decision support.

In order to adapt the model to the fuzzy input information [1], the theory of fuzzy sets is used quite actively, which implies the representation of the quantitative values of the model parameters in the form of linguistic variables, which are estimated by fuzzy terms [2]. Of course, the theory of fuzzy sets has its disadvantages, in particular, such as subjectivity when forming the functions of belonging to fuzzy sets. To overcome this

J. Świątek et al. (Eds.): ISAT 2018, AISC 853, pp. 321–329, 2019.
https://doi.org/10.1007/978-3-319-99996-8_29

problem adaptive neuro-fuzzy systems were created. They allow you to identify model parameters using experimental data. One of the most popular neuro-fuzzy systems is Adaptive Neuro-Fuzzy inference system (ANFIS) created by Robert Jang [3]. ANFIS is a universal approximator because of using neural network units and ANFIS provides good logic inference because of using fuzzy logic. But it has the problem of dimensionality in case of big number of input variables. Existing methods of ANFIS learning, namely gradient, hybrid gradient methods [3] are intended to identify parameters of neurons in hidden layers but they say nothing about changing the structure of ANFIS. In the paper [4] proposed an immune approach that allows not only to identify all ANFIS parameters, but also to find the optimal ANFIS structure. This approach reduces the number of neurons in hidden ANFIS layers using artificial immune systems.

The second approach, which allows changing the structure of ANFIS, is the method of diagonalizing the matrix of weight synaptic connections in the neural network [5].

One of the approaches to analyzing the quantitative and qualitative characteristics of the behavior of the object and the preparation of the necessary data for the organization of the strategy of management and decision-making in the management of incomplete information is the neuron-fuzzy technology for the formation of linguistic causation estimates, which is presented in [6]. Mathematically, neural networks (NN) can be considered as a class of methods for statistical simulation, which in turn can be divided into three classes: probability density estimation, classification and regression. It is assumed that the decision support system (DSS) can be fully realized on the NN. In contrast to the traditional use of NN to solve only the problems of recognition and the formation of images [7], the DSS agreed to solve the following tasks: recognition and formation of images; obtaining and preserving knowledge; evaluation of qualitative characteristics of images; decision-making. The neural network solution of the set tasks involves the analysis and implementation of the most productive ways of processing initial experimental data, the formation of training and test samples, the construction of neural network structures, analysis, processing and visualization of the results [8]. Consequently, modern requirements to control systems necessitate the introduction of intellectual DSS and adaptive methods of multidimensional analysis [9].

The purpose of this work is to construct a three-layer neuro-fuzzy network architecture with diagonalized weight synaptic connections to create a decision support system.

2 Diagonazation Matrix of Synaptic Connections

For diagonalization matrix of synaptic connections and memory of the prototype, the input image is a three-layer neural network of direct distribution (Fig. 1), we write the input image in the form of a deterministic vector

$$\vec{V} = (V_1, V_2, \ldots, V_n), \tag{1}$$

where $V_n = \vec{e}_n \vec{V}$ – projection $\vec{V}$ on $\vec{e}_n$ ($\vec{e}_n$ – n-th basis vector of the coordinate system).

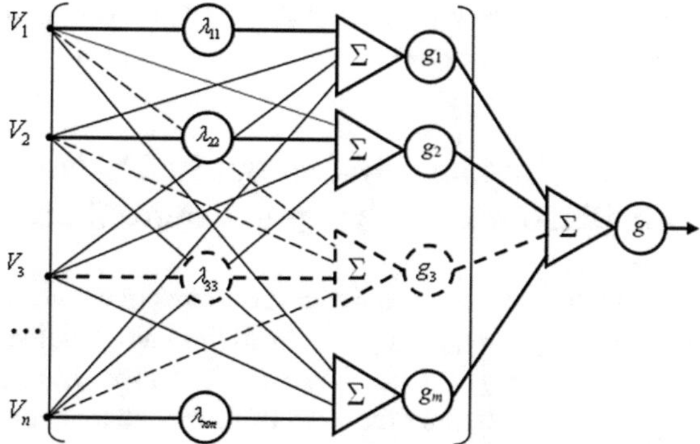

Fig. 1. As an example, schematic representation of the three-layer neural network of direct distribution with a diagonal and not diagonal weight coefficients of synaptic connections λ_{nm}.

To remember the prototype image (information signal) applied to the synaptic connection λ_{nm} (synaptic connections from sources $V_1, V_2, \ldots, V_n$ to neurons *1, 2, 3,…, N*) constraints

$$\lambda_{nm} = V_n \cdot V_m, \lambda_{nm} \neq \lambda_{mn}, n \neq m \tag{2}$$

and form the matrix $\hat{\lambda}$ with a deterministic matrix elements

$$\langle \lambda_{nm} \rangle = \langle V_n \rangle \cdot \langle V_m \rangle \tag{3}$$

and bring it to diagonal form with real eigenvalues

$$\tilde{\lambda}_{nm}(V_1, V_2, \ldots, V_n) = \beta_n(V_1, V_2, \ldots, V_n) \cdot \delta_{nm} \tag{4}$$

To bring the $\hat{\lambda}$ matrix to diagonal form we reduce it to the symmetrical shape and make a linear transformation

$$\hat{\hat{\lambda}} = \hat{U}^{-1}\hat{\lambda}\hat{U}, \tag{5}$$

where $\hat{U}-$ the matrix consists of the basis vectors $\vec{u}_m$ in matrix $\hat{\lambda}$, that is $\hat{U} = (\vec{u}_1, \vec{u}_2, \ldots, \vec{u}_n)$;

$$\hat{\lambda}\vec{u}_m = \beta_m \vec{u}_m. \tag{6}$$

In the basis of eigenvectors of the $\vec{u}_m$ matrix of a linear transformation $\hat{\hat{\lambda}}$ is a diagonal view, and on the main diagonal are located the valid eigenvalues of the matrix $\hat{\hat{\lambda}}$.

$$\tilde{\lambda}_{nm}(V_1, V_2, \ldots, V_n) = \beta_n(V_1, V_2, \ldots, V_n) \cdot \delta_{nm} \tag{7}$$

where δ_{nm}– the Kronecker symbol; $\beta_n(V_1, V_2, \ldots, V_n)$– actual eigenvalues of a diagonal matrix of synaptic connections.

3 Model of Decision Support System on the Basis of Neural Network with Diagonalized Synaptic Connections

For decision-making, the model has an analytical relationship between the values of the initial vector of states $Y^n = \{y_1, y_2, \ldots, y_n\} \subset Y$ and the known values of the vector of the input characteristics $V^* = \{V_1, V_2, \ldots, V_m\} \subset V$. The relationship between the values of the initial vector of states and the vector of input characteristics is carried out with the aid of a three-layer perceptron with diagonalized synaptic connections (Fig. 1). The number of neurons in the hidden layer is equal to the number of classes of decisions. Activation functions in the hidden layer of the neural network are selected as sigmoid:

$$g_m = f(u_m) = \frac{1}{1 + e^{-\alpha_m(\beta_m V_m + \sum_{i=N+1}^{M} \lambda_{im} V_i + \lambda_{0m})}}, \tag{8}$$

where $g_m,\ m = \overline{1, M}$ – output signal of m-th neuron from hidden layer, that consist of M neurons, that has N inputs; $V_n,\ n = \overline{1, N}$ – n-th component of the input characteristic vector; λ_{im} – weight coefficient of the n-th input characteristic V_n that received at the input of the m-th neuron of the hidden layer; λ_{0m} – offset value; α_m – coefficient determines the steepness of the activation function $f(u_m)$; β_m – the actual eigenvalues of the diagonal matrix of synaptic ties between the neurons of the input and the hidden layers. Elements β_m contain information about the vector of input signals (images) and are directly used for the training of the neural network [10] (Fig. 2).

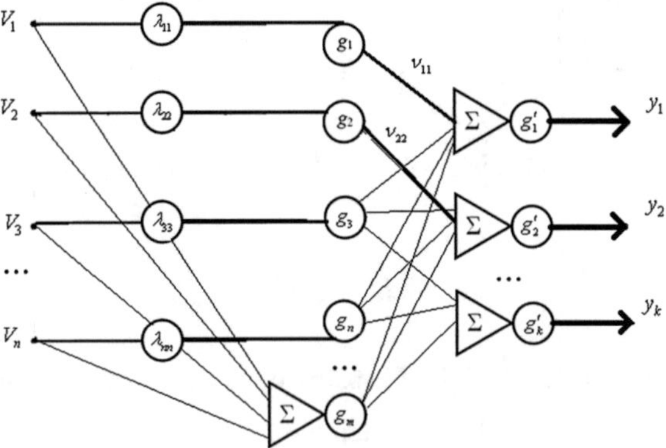

Fig. 2. Architecture of a three-layer perceptron with diagonalized synaptic connections.

Let the $\lambda_{im}, \tilde{\lambda}_{nm}$ are stochastic variables and do not depend on the connections that exist between other neurons $\lambda_{pl}(p \neq i; l \neq m)$, then their statistical properties are $\lambda_{im}, \tilde{\lambda}_{nm}$ completely determined by the distribution function $f(\lambda_{im}), f\left(\tilde{\lambda}_{nm}\right)$ on connection between the n-th and m-th neurons. Suppose that $f(\lambda_{im}), f\left(\tilde{\lambda}_{nm}\right)$ is the Gaussian distribution function $f(\lambda_{im}) = \frac{1}{\tilde{\lambda}_{im}\sqrt{2\pi}} \exp\left(-\frac{\lambda_{im}^2}{2\tilde{\lambda}_{im}^2}\right)$, which is defined by two parameters: mean value $\bar{\lambda}_{im} = \langle \lambda_{im} \rangle$ and variance $\tilde{\lambda}_{im} = \langle \lambda_{im}^2 \rangle - \langle \lambda_{im} \rangle^2$.

The function of activating neurons in the source layer of the neural network has a threshold appearance. The neurons of the output layer determine which class to assign the solution to.

$$y_k = \phi\left(\gamma_k g_k + \sum_{j=P+1}^{M} v_{jk} \cdot g_j + v_{0k}\right) = \begin{cases} 1, & \text{if } \left(\gamma_k g_k + \sum_{j=P+1}^{M} v_{jk} \cdot g_j + v_{0k}\right) \geq 0, \\ 0, & \text{if } \left(\gamma_k g_k + \sum_{j=P+1}^{M} v_{jk} \cdot g_j + v_{0k}\right) < 0, \end{cases} \quad k = \overline{1, K} \tag{9}$$

where v_{jk} – weight coefficients; v_{0k} – bias; K – number of NN outputs; γ_k – the actual eigenvalues of the diagonal matrix of synaptic connections between the neurons of the hidden and the original layers [10–12].

4 Learning Fuzzy Neural Network with Diagonal Synaptic Connections

Teaching neural networks is based on the use of appropriate training samples

$$\begin{aligned} V^{(1)} &= (V_1^{(1)}, V_2^{(1)}, \ldots, V_n^{(1)})^T, y_1^r \\ V^{(2)} &= (V_1^{(2)}, V_2^{(2)}, \ldots, V_n^{(2)})^T, y_2^r \\ &\cdots \\ V^{(k)} &= (V_1^{(k)}, V_2^{(k)}, \ldots, V_n^{(k)})^T, y_k^r \\ &\cdots \\ V^{(K)} &= (V_1^{(K)}, V_2^{(K)}, \ldots, V_n^{(K)})^T, y_K^r \end{aligned} \tag{10}$$

The goal of the training of the neural network is to adjust the weighted synaptic coefficients λ_{im} and v_{jk} by the criterion of minimizing E on the training sample (10):

$$E = \sum_{k=1}^{K} E_k \rightarrow \min(\vec{L}) \tag{11}$$

where $E_k = \frac{1}{2}\left[\phi\left(\gamma_k g_k + \sum_{j=P+1}^{M} v_{jk} g_j + v_{0k}\right) - y_k^r\right]^2$; y_k^r – required output values of the neural network; $\vec{L} = (\lambda_{im}, v_{jk}, \lambda_{0m}, v_{0k}, \alpha_m)$ – the vector of the parameters of the neural network $n = \overline{1,N}$, $m = \overline{1,M}$, $k = \overline{1,K}$ (N, M, K the number of neurons in the input, hidden and output layers, respectively).

The optimization problem is solved by the gradient method using the relation:

$$\lambda_{im}(t+1) := \lambda_{im}(t) - \eta E'(\lambda_{im}(t)) \tag{12}$$

$$v_{jk}(t+1) := v_{jk}(t) - \eta E'(v_{jk}(t)) \tag{13}$$

where $0 < \eta < 1$ – coefficient of learning speed; $E'(\lambda_{im}(t)), E'(v_{jk}(t))$ – gradients of function E to $\lambda_{im}(t)$ and $v_{jk}(t)$ respectively.

5 The Algorithm of Teaching a Three-Layer Neural Network with Diagonal Synaptic Connections

1. Network initialization: weighting factors and network shifts take small random values. The rate of learning $\eta (0 < \eta < 1)$ is given, the desired value of the mean square error of training $E_{\max}$.
2. Issued $k = 1$.
3. Sequential input of the neural network is provided by the training vectors from the training sample. Introducing the next study pair $(V^{(k)}, y_k^r)$ and calculating derivatives $E'(\lambda_{im}(t)), E'(v_{jk}(t))$.
4. Updated the synaptic weights of a neural network: $\lambda_{im}(t+1) := \lambda_{im}(t) - \eta E' (\lambda_{im}(t)), v_{jk}(t+1) := v_{jk}(t) - \eta E'(v_{jk}(t))$.
5. Calculated $E_k = \frac{1}{2}\left[\phi\left(\gamma_k g_k + \sum_{j=P+1}^{M} v_{jk} g_j + v_{0k}\right) - y_k^r\right]^2$.
6. If $k < K$, then $k := k + 1$ and the transition to 3, otherwise 7.
7. $E = \sum_{k=1}^{K} E_k$. If $E \geq E_{\max}$ then a new training cycle begins with the transition to 2. If $E < E_{\max}$, then the completion of the learning algorithm.

6 The Architecture of the Diagonal Neural Network to Choose the Optimal Operating System that Is Used for the Local Computer Network

The problem of designing and analyzing a Local Area Network (LAN) was considered as a test. This problem is an example of a hard-to-formalize task, which requires anintegrated approach. When designing a LAN, it is necessary to determine the output parameters that the network must satisfy and the initial conditions (input parameters) that are set before the design process.

LAN analysis allowed to distinguish the input characteristics, the most significant of which are following: V_1 – network cost; V_2 – number and location of users; V_3 – how easy to install and change network configuration; V_4 – network bandwidth; V_5 – network reliability; V_6 – network security; V_7 – the possibility of expanding the network.

We have the following basic output parameters based on the design requirements of LAN: (1) operation system (OS); (2) network topology; (3) network technology. OS options: (1) Open Enterprise Server (OES); (2) Microsoft Windows (7, 8, 10); (3) UNIX systems(Solaris, FreeBSD); (4) GNU/Linux systems; (5) IOS; (6) ZyNOS produced by ZyXEL. Network topology options: (1) star; (2) bus; (3) ring; (4) tree; (5) fully connected; (6) mesh; (7) hybrid. Network technology options: (1) Fast Ethernet; (2) Token Ring; (3) FDDI.

The implementation of this task using method of simple sorting is not applicable, since the combinatorial capacity of the sorting is several orders of magnitude. Moreover, most combinations of options will never be implemented. Therefore, the implementation was performed using evolutionary NN with AIS.

The task was conditionally divided into three parallel tasks. Each task solves the problem by one of the output parameters. In this case, each of the output parameters can be determined by the values of not all input, but only a few of them. For each of the three tasks, a NN was created. Each NN contains a certain number of inputs and outputs (Fig. 3). So, for selecting the OS –7 and 6, for selecting a network topology – 6 and 7, for selecting network technology – 6 and 3 respectively. NN has one hidden layer that contains 15 or more neurons.

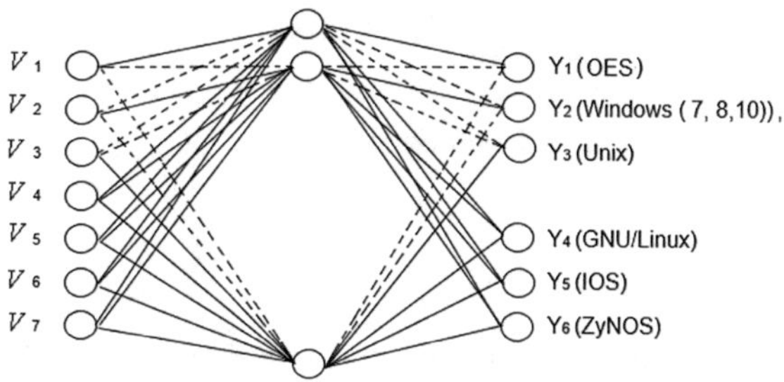

Fig. 3. The structure of the diagonalized neural network to determine the OS

The Fast Ethernet computer network with the physical topology "star" with a transmission speed of 100 Mbit has been used to simulate the obtained results. It consists of 10 quad-core Intel Core 2 Quad CPUs Q8200 @ 2.33 GHz with a GeForce GTX 460 graphics card. Parallelization is performed using OpenMP and MS MPI technologies. A number of experiments were performed to analyze the procedure of parallel and distributed NN learning on a multiprocessor system [13]. The analysis showed that MPI technology allows distributed NN learning accelerates 11 times for the “star” topology of data transmissions. However, with a significant increase in the number of processors in the system, the time for data transfer between them increases. Therefore, when you achieve a certain increase in performance, a further increase in the number of processors in the system produces the opposite effect: performance begins decrease. The OpenMP standard allows to accelerate NN parallel learning up to 7 times. Accelerated data processing allows intelligent DSS to operate in real-time.

Thus, the conducted studies showed that multilayered modular neuron networks of the perceptron type with immune training should be used to predict possible solutions for choosing LCN parameters. At the same time, the stability of the results obtained is high.

In particular, the decision-making time will be reduced by a relatively non-diagonal neural network

$$P_t = \left(1 - \frac{N_1}{N_3}\right) * \left(1 - \frac{N_2}{N_4}\right) * 100\% \tag{14}$$

N_1, N_2 - the number of non-diagonal elements of the matrix of synaptic connections between the incoming and the hidden layer and between the hidden and the output layer respectively; N_3, N_4 is the total number of elements of the matrix of synaptic connections between the input and the hidden layer and between the hidden and the output layer respectively.

Using the formula (14) determine by what percentage will decrease while the choice of OS to work on a local computer network with a diagonalized neural network relative to a non-diagonalized one.

$$\left(1 - \frac{15}{21}\right) \cdot \left(1 - \frac{12}{18}\right) \cdot 100\% = 9,52\% \tag{15}$$

The time of choice will be reduced by 9,52%.

7 Conclusions

As a result of the research, a decision support model based on the architecture of a three-layer perceptron with diagonal weighted synaptic connections between the neurons of the input, the latent and the initial layers is constructed.

Diagonalization of the matrix of weight synaptic connections allowed to simplify the structure of the neural network, increase the speed of adjustment of weight synaptic connections in the process of teaching the neural network by reducing the number of weighted synaptic relationships between the neurons, which contributes to a significant reduction in the time of decision-making. In particular, the decision-making time will be reduced by a relatively non-diagonal neural network $P_t = \left(1 - \frac{N_1}{N_3}\right) * \left(1 - \frac{N_2}{N_4}\right) * 100\%$, N_1, N_2 – the number of non-diagonal elements of the matrix of synaptic connections between the incoming and the hidden layer and between the hidden and the output layer respectively; N_3, N_4 is the total number of elements of the matrix of synaptic connections between the input and the hidden layer and between the hidden and the output layer respectively.

References

1. Zhu, B., Xu, Z.: Consistency measures for hesitant fuzzy linguistic preference relations. IEEE Trans. Fuzzy Syst. **22**(1), 35–45 (2014)
2. Stecenko, D.O.: Development of intelligent algorithms for control of the bragocracy installation. Technol. Audit Prod. Reserves **6/1**(14), 51–54 (2013)
3. Jang, J.-S.R.: ANFIS: adaptive-network-based fuzzy inference system. IEEE Trans. Syst. Cybern. **23**, 665–685 (1993)
4. Korablev, N., Sorokina, I.: Immune approach for neuro-fuzzy systems learning using multiantibody model. In: ICARIS, Springer Lecture Notes in Computer Science, vol. 6825, pp. 395–405 (2011)
5. Lytvyn, V., Peleshchak, I., Peleshchak, R.: The compression of the input images in neural network that using method diagonalization the matrices of synaptic weight connections. In: 2nd International Conference on Advanced Information and Communication Technologies (AICT), pp. 66–70 (2017)
6. Stecenko, D., Zigunov, O., Smitjuh, J.: Intelligent processing of data in the system of automated control of the technological complex of bragorectification. Technol. Audit Prod. Reserves **2**(1)(16), 49–52 (2016)
7. Jarrett, K., Kavukcuoglu, K., Ranzato, M.: What is the best multi-stage architecture for object recognition. In: IEEE 12th International Conference on Computer Vision, pp. 2146–2153 (2016)
8. Lee, H., Grosse, R., Ranganath, R.: Convolutional deep belief networks for scalable unsupervised learning of hierarchical representations. In: Proceedings of the 26th Annual International Conference on Machine Learning, pp. 609–616 (2009)
9. Gladun, V., Velichko, J.: Instrumental complex of support of decision-making on the basis of the network model of the domain. In: Sb. Papers Science-Practice Conference with International Participation Decision Support Systems, Theory and Practice, pp. 126–128 (2012)
10. Lytvyn, V., Peleshchak, I., Peleshchak, R.: Increase the speed of detection and recognition of computer attacks in combined diagonalized neural networks. In: 4th International Scientific-Practical Conference Problems of Infocommunications, Science and Technolohy, pp. 152–155 (2017)
11. Lytvyn, V., Vysotska, V., Peleshchak, I., Rishnyak, I., Peleshchak, R.: Time dependence of the output signal morphology for nonlinear oscillator neuron based on Van der Pol model. Int. J. Intell. Syst. Appl. **10**, 8–17 (2018)
12. Chaplya, Y., Chernukha, O., Bilushchak, Y.: Contact initial boundary-value problem of the diffusion of admixture particles in a two-phase stochastically inhomogeneous stra-tified strip. J. Math. Sci. **183**(1), 83–99 (2012)
13. Axak, N.: Development of multi-agent system of neural network diagnostics and remote monitoring of patient. East.-Eur. J. Enterp. Technol. **4/9**(82), 4–11 (2012)

Computational Investigation of Probabilistic Learning Task with Use of Machine Learning

Justyna Częstochowska[1], Marlena Duda[1], Karolina Cwojdzińska[1], Jarosław Drapała[1(✉)], Dorota Frydecka[2], and Jerzy Świątek[1]

[1] Faculty of Computer Science and Management, Wrocław University of Science and Technology, ul. Ignacego Łukasiewicza 5, 50-371 Wrocław, Poland
jaroslaw.drapala@pwr.edu.pl

[2] Department of Psychiatry, Wrocław Medical University, 10 Pasteur Street, 50-367 Wrocław, Poland

Abstract. Probabilistic Learning Task is a game that serve psychiatrists and psychologists to measure some cognitive abilities of people having various cognitive disorders. Mathematical models together with machine learning techniques are routinely used to summarize large amount of data produced by players during the game. Parameters of mathematical models are taken to represent behavioral data gathered during the game. However, there is no study of reliability of those parameters available in literature. We investigate how much one can trust the values of models parameters. We proposed a specific method to assess reliability of models parameters, that makes use of the game sessions of human players and their virtual counterparts.

Keywords: Reinforcement learning · Maximum likelihood method Model selection

1 Introduction

Reinforcement Learning Games rely on repetitions of many trials including decisions made by a player followed by reward or punishment delivered by the game [18]. The decision is about selecting only one among few options. Those games are widely used by cognitive scientists and psychiatrists for scientific purposes.

Researchers use Reinforcement Learning Games as tools to investigate learning abilities of human brain and to analyse decision making strategies of people having some cognitive disorders [15]. They propose hypotheses concerning: – the way brains make decisions under uncertain conditions [1,10], – mechanisms of learning from experience [4], – how behaviour is affected by disorders [7,9]. Typical disorders include: schizophrenia, ADHD, drug addiction, parkinson's disease, Tourette's syndrome [15].

J. Świątek et al. (Eds.): ISAT 2018, AISC 853, pp. 330–339, 2019.
https://doi.org/10.1007/978-3-319-99996-8_30

All game events are recorded and serve as behavioral data sets. The problem is large volume of those data. It prevents researches from direct interpretation of the game outcome [3]. Therefore, in the field of computational psychiatry, a standard method to deal with this problem is to fit a computational model to the game data (also called behavioral data) and to use resulting model as a concise representation of these data [17]. Computational models express mathematically hypotheses posed and tested by researches and include some free parameters that allow models to be fit to the behavioral data [5]. Thus, computational model is a mixture of both the general hypothesis and the behavioral data of a particular player [2,11]. In other words: the model represents the point of view of the data interpretation [12,16], whereas at the same time the model is adjusted to those data. This methodological loop makes the analysis of outcomes of Reinforcement Learning Games a non-trivial task [3].

Mathematical framework of Reinforcement Learning Games is Markov Decision Process, so the general form of computational models is mainly drawn from the Reinforcement Learning theory [19]. Researches propose modifications of basic equations according to their knowledge of processes that take part in the brain [6,8]. Model parameters are estimated in such a way, that the model is best fit to the behavioral data of a given player. Typically, the maximum likelihood method or bayesian methods are employed with extensive use of numerical optimization routines [3]. When many hypotheses - and hence models - compete, model selection methods are used [4,6]. The most popular approach is to use criteria such as AIC and BIC, [3].

As can be seen from the description above, important ingredient of Reinforcement Learning Games analysis are machine learning techniques [13]. The problem is, that machine learning results are very often used carelessly. The only critical paper known to authors is [5]. Other authors seem to use the reinforcement learning models as fully reliable interpretation tool.

This contribution aims at critical assessment of results of parameters of computational models estimation. On the basis of the models learned from behavioral data many conclusions were made in literature, but even visual inspection of the model parameters space reveals that those results may not be conclusive at all. Unfortunately, authors are not used to report distribution of the model parameters. Instead, they only rely on values of the pseudoR^2, AIC or BIC.

We focus on the most representative game, namely the Probabilistic Learning Task. The next section introduces the rules of the game. Further on, the mathematical description of computational models is given together with the parameter estimation algorithm. The idea of the so called virtual players is pointed out and its role for analysis of reliability of the parameter estimation algorithm is explained. Further on, the analysis with use of the real behavioral data and their artificial counterparts is performed. Eventually, the conclusions are drawn.

2 Probabilistic Learning Task

Probabilistic learning tasks are a kind of games where the player learns on the basis of the rewards and punishments received during successive trials of the game. At each trial the player selects one stimulus among two presented on the screen and receives feedback (reward or punishment) with probability assigned to the stimulus. Thus, each stimulus may return reward as well as punishment, but with different probabilities. The higher the probability of returning reward, the better the stimulus. Probability of receiving reward is further called contingency. The player's aim is to collect as many rewards as possible.

At the beginning of the game, the player does not know contingencies, so she must learn it from punishments and rewards gained during the course of the game. Typically, Japanese Hiragana characters are used as stimuli. Three pairs of stimuli (AB, CD, EF) appear on the screen in a random order. During the whole game, each pair is shown 30 times. The player's task is to pick up one stimuli, which she thinks is better. Reward is indicated as the blue "well done" message and punishment as the red "bad choice" message.

As told before, the rewards for choosing a given picture are awarded with different probabilities. In the case of a pair AB, the probability is (80/20), for the pair of CD (70/30), and for the pair of EF (60/40), see Fig. 1.

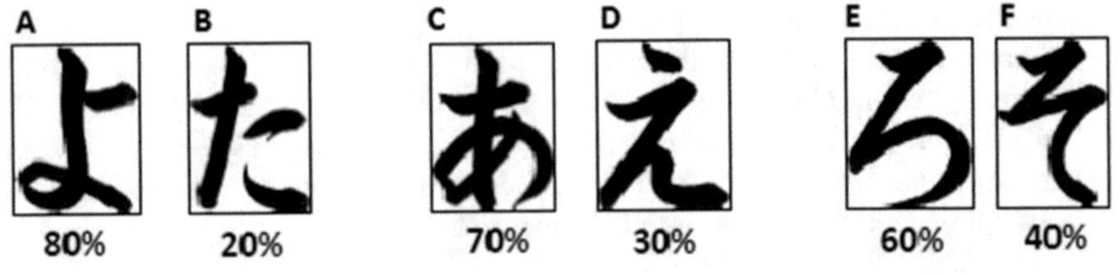

Fig. 1. Stimuli pairs: Japanese Hiragana characters with contingencies.

The player has limited time to pick up the picture. If she does not manage to make decision within 5 s, the trial is wasted. Between each appearing pair of stimuli, a control screen showing small green circle is presented for a while. Typically, the game takes about half an hour.

3 Mathematical Models

3.1 Q–*learning* Model

Current knowledge of how brains learn from experience is mainly derived from fMRI (eng. *functional magnetic resonance imaging*). Measurements strongly support the use of reinforcement learning theory to model the behavior of human playing the probabilistic learning task [1]. The most fundamental in the field is the so called Q-*learning* model described be two equations [19]. The first equation accounts for decision making:

$$p(A) = \frac{1}{1 + \exp(\frac{Q_B - Q_A}{T})}. \quad (1)$$

Symbols Q_A and Q_B stand for the *expected rewards*, which are simply estimations of probabilities that reward will be gained for picking up the stimulus represented by the Q-value. Stimuli A and B are paired, $p(A)$ is probability that A is chosen and in consequence $p(B) = 1 - p(A)$. One may recognize in Eq. (1) the softmax function. There is free parameter T that shapes the function. Low values of T make the softmax function look like the step function, representing conservative decision maker, which selects stimulus having even slightly greater Q-value than the other stimulus. The higher the T value, the more likely the decision maker selects the stimulus having lower Q-value, which characterizes how prone one is to make hazardous decisions.

The second equation embodies the learning process, which is about adjusting the value of Q_X in response to feedback received just after selection of stimuli X:

$$Q_X \leftarrow Q_X + \alpha(r - Q_X), \quad (2)$$

where $r \in \{-1, 1\}$ is the reward gained at the current trial and X stands for selected stimulus (A, B or others), α is a step size and represents the learning speed. The higher the α, the larger the adjustment of Q-value. The Eq. (2) is applied at each trial for the selected stimulus only. Expected rewards of another stimuli is left untouched at the current trial.

The model includes only two parameters: T and α.

3.2 Rescorla–Wagner Model

Important extension of the basic Q-*learning* equations is to differentiate the player's reaction to rewards and punishments. This distinction is supported by the evidence from fMRI and knowledge of dopamine influence on the learning process [4]. Commonly used Rescorla-Wagner model introduces two different learning speeds: α_{Gain} referring to player's sensitivity to rewards and α_{Loose} referring to player's sensitivity to punishments. The only modification lies in extending Eq. (2) as follows:

$$\alpha = \begin{cases} \alpha_{Gain} & \text{if } r = 1 \\ \alpha_{Loose} & \text{if } r = -1 \end{cases}, \quad (3)$$

meaning that after the reward, α_{Gain} is used in place of α in Eq. (2) and α_{Loose} after the punishment. Intuitively, it may be interpreted as how quickly one learns from rewards (α_{Gain}) or punishments (α_{Loose}).

The model contains three parameters: T, α_{Gain} and α_{Loose}.

3.3 Parameter Estimation and Model Selection

Maximum likelihood method is classical approach to fitting model to data [3]. The log-likelihood of behavioral data D conditioned on the model parameters Θ is an objective function:

$$LLE(\Theta, D) = \ln p(D|\Theta) = \sum_{i=1}^{N} \ln p\left(d_i|\Theta\right), \tag{4}$$

where N is the total number of trials. The data set includes information recorded during the whole game (see the Table 1).

Table 1. Example of the game session.

Stimulus pair	3	2	1	2	3	1
Stimulus left	5	4	1	3	6	2
Stimulus right	6	3	2	4	5	1
Action	1	0	1	1	0	1
Reward	1	1	−1	−1	1	−1
Response time	1.48	1.67	1.87	1.56	1.08	2.12

Note, that probabilities of decisions made by the player are not independent. According to Eq. (1), their values depend on expected rewards that vary from trial to trial as described by Eq. (2).

Parameter estimation reduces to unconstrained optimization task, where the likelihood function (4) is maximized with respect to model parameters Θ. We use the Nelder-Mead optimization procedure for this purpose [14].

When more than one model is proposed to describe the data, the question arises: which one is better? In machine learning, typical procedure is to compare models under consideration taking into account not only the fit to the data but also the complexity of the model [3,13]. Selection criteria include both: the fit to the data and complexity (here complexity of the model reduces to the number of its free parameters). We take the Akaike Information Criteria for model selection [13]:

$$AIC = 2q - 2LLE, \tag{5}$$

where LLE is the value of log-likelihood function and q stands for the total number of model parameters. The lower the AIC value, the better for the model.

AIC allows for comparisons between models, but the reference point is delivered by another criteria, pseudoR^2, that measures improvement of the model over the random guessing [6]:

$$\text{pseudoR}^2 = (LLE - r)/r, \tag{6}$$

where r is the value of LLE for the model representing the random guessing:

$$r = N * \log 0.5.$$

4 The "Virtual Twin" Trick

To verify reliability of the model parameters estimates, we propose the following procedure.

We take the game session of a human player and estimate parameters of Q-*learning* model and Rescorla-Wagner model. Afterwards those parameters are fed to the system generating artificial game session. The model is treated as a "virtual player" and is expected to mimic the behavior of the human player, since they both have the same values of parameters. The game session of the virtual player is generated and parameters are estimated again. As a result, we obtain two sets of parameter estimates: the first one for the human player and second one for her "virtual twin". Subsequently, we analyze the discrepancy between values of two sets of parameters: those obtained from human players with those obtained from reinforcement learning models.

The whole idea is depicted in Fig. 2.

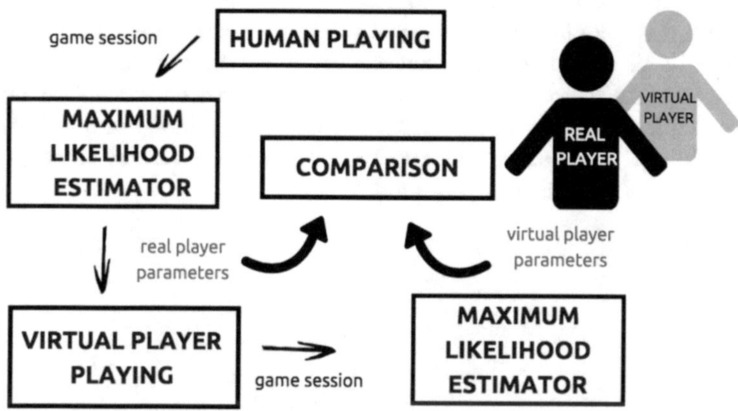

Fig. 2. The idea behind the so called "virtual twin".

5 Simulation Study and Discussion

The computational investigation of reliability of parameter estimates was performed with use of the data set containing the game sessions of human players. Behavioral data was collected from two groups. The first one consists of twenty-two players, mainly college students about twenty years old, but also middle-aged men and women, all of them mentally healthy. The second group consists of thirty-six schizophrenic patients of Medical University in Wroclaw. Medical investigations typically make use of the control group and the study group, but for the purpose of our numerical study we can put aside this division. Therefore, we mix both groups of players into a single data set, because we focus on properties of numerical method regardless of medical issues. Our purpose is to make the best use of the available data.

As stated before, estimates of both Q-*learning* and Rescolra-Wagner models were obtained with use of numerical optimization routine, where the objective function was LLE described by Eq. (4). Unfortunately, due to the shape of this function, the final result is very vulnerable for even slightest change in optimization starting points. Therefore, to make the probability of getting stuck in a local optimum as low as possible, we used multiple runs of the numerical search, each starting from different initial point. Initial values were chosen from the grid: the range for T was $[1, 10]$ with interval 0.1 and the range for all α values was $[0.1, 1]$ with interval 0.05.

It is worth to mention that starting points were the same both for human player parameters estimates and associated virtual twin parameters. Figure 3 illustrates typical example of the objective function to be maximized for the two dimensional Q-*learning* model. However, it must be stressed, that this is rather "nice" example of the log-likelihood function. For most players the hill is not so narrow and the confidence intervals for values of parameters are considerably wider. Results of models comparison are given in Table 2. The table presents model fits for data belonging from real players and virtual players, as indicated by Akaike's information criterion (AIC), pseudoR^2 and maximum log-likelihood estimate (LLE). It is clear that the Rescolra-Wagner model is better in all respects than the basic Q-*learning* model.

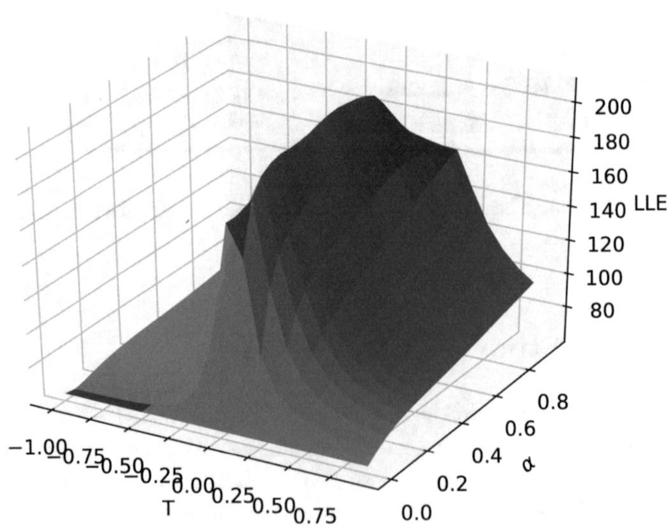

Fig. 3. Plot of the log-likelihood function in the space of α and T

Table 2. Results of models comparison.

Model	No. of parameters	*AIC*	pseudoR^2	*LLE*
Q-*learning*	2	107.14	0.17	−56.09
Rescorla-Wagner	3	98.45	0.23	−47.91

The distribution of two-dimensional Q-*learning* model parameters estimates is illustrated in Fig. 4. Similar illustration of three-dimensional Rescorla-Wagner model parameters estimates is given in Fig. 5.

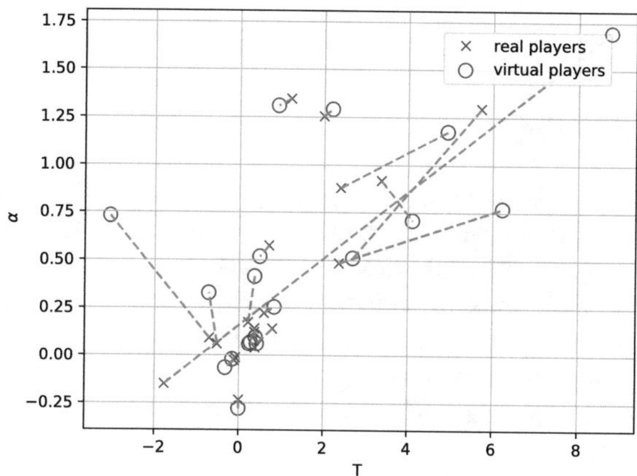

Fig. 4. Estimated parameters of Q-*learning* model for human players and associated "virtual twins"

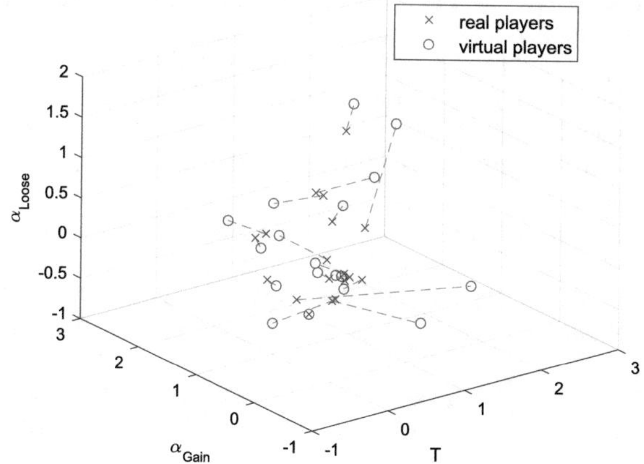

Fig. 5. Estimated parameters of Rescorla-Wagner model for human players and their "virtual twins"

Note the large discrepancy between parameters of human players and their "virtual twins". Moreover, many estimates are negative, whereas negative values of models parameters make no sense. Many of our players' parameters could not be easily used to describe a person as they exceed interpretable intervals both for T and α. Al those observations are strong evidence for poor reliability of the parameter estimates. But both those sets of parameters values were supposed to originate from decision makers following similar strategies! One could derive mathematical explanation of these effect from analysis of properties of the LLE that was performed in [3]. Large confidence intervals results from non-linearity of Eq. (1). And here is the question, whether model equations describe properly strategy followed by humans? And is it possible to state, how much of an effect should we ascribe to non-linearity of LLE and how much to inadequate model equations?

6 Final Remarks

The contribution of this paper is critical view on the use of reinforcement learning models for modeling behavior of human solving probabilistic learning task. Moreover, we proposed "virtual trick" twin as a tool to investigate reliability of model parameters estimates provided by machine learning methods. We claim that this is important result in the field, because researches use to avoid the problem by resorting to constrained optimization routines to find parameters of the models or by hiding values of parameters and revealing only model selection criteria.

References

1. Botvinick, M.M., Niv, Y., Barto, A.C.: Hierarchically organized behavior and its neural foundations: a reinforcement learning perspective. Cognition **113**(3), 262–280 (2009). https://doi.org/10.1016/j.cognition.2008.08.011
2. Collins, A.G., Frank, M.J.: How much of reinforcement learning is working memory, not reinforcement learning? a behavioral, computational, and neurogenetic analysis. Eur. J. Neurosci. **35**(7), 1024–1035 (2012). https://doi.org/10.1111/j.1460-9568.2011.07980.x
3. Daw, N.D.: Trial-by-trial data analysis using computational models. Decis. Making, Affect, Learn. Atten. Perform. XXIII **23**, 3–38 (2011)
4. Daw, N.D., Doya, K.: The computational neurobiology of learning and reward. Curr. Opin. Neurobiol. **16**(2), 199–204 (2006). https://doi.org/10.1016/j.conb.2006.03.006
5. Dayan, P., Niv, Y.: Reinforcement learning: the good, the bad and the ugly. Curr. Opin. Neurobiol. **18**(2), 185–196 (2008). https://doi.org/10.1016/j.conb.2008.08.003
6. Doll, B.B., Jacobs, W.J., Sanfey, A.G., Frank, M.J.: Instructional control of reinforcement learning: a behavioral and neurocomputational investigation. Brain Res. **1299**, 74–94 (2009). https://doi.org/10.1016/j.brainres.2009.07.007

7. Doll, B.B., Waltz, J.A., Cockburn, J., Brown, J.K., Frank, M.J., Gold, J.M.: Reduced susceptibility to confirmation bias in schizophrenia. Cognitive Affect. Behav. Neurosci. **14**, 715–728 (2014). https://doi.org/10.3758/s13415-014-0250-6
8. Doya, K.: Modulators of decision making. Nat. Neurosci. **11**(4), 410–416 (2008)
9. Frank, M.J.: Dynamic dopamine modulation in the basal ganglia: a neurocomputational account of cognitive deficits in medicated and nonmedicated parkinsonism. J. Cogn. Neurosci. **17**, 51–72 (2005). https://doi.org/10.1162/0898929052880093
10. Frank, M.J.: Hold your horses: a dynamic computational role for the subthalamic nucleus in decision making. Neural Netw. **19**(8), 1120–1136 (2006). https://doi.org/10.1016/j.neunet.2006.03.006
11. Frank, M.J., Moustafa, A.A., Haughey, H.M., Curran, T., Hutchison, K.E.: Genetic triple dissociation reveals multiple roles for dopamine in reinforcement learning. Proc. Natl. Acad. Sci. **104**(41), 16311–16316 (2007). https://doi.org/10.1073/pnas.0706111104
12. Humphries, M.D., Khamassi, M., Gurney, K.: Dopaminergic control of the exploration-exploitation trade-off via the basal ganglia. Front. Neurosci. **6**, 9 (2012). https://doi.org/10.3389/fnins.2012.00009
13. Jerome, F., Hastie, T., Tibshirani, R.: The Elements of Statistical Learning. Springer series in statistics, New York (2001)
14. Kiusalaas, J.: Numerical Methods in Engineering with Python 3. Cambridge University Press, Cambridge (2013)
15. Maia, T.V., Frank, M.J.: From reinforcement learning models to psychiatric and neurological disorders. Nat. Neurosci. **14**(2), 154–162 (2011). https://doi.org/10.1038/nn.2723
16. Montague, P.R., Hyman, S.E., Cohen, J.D.: Computational roles for dopamine in behavioural control. Nature **431**(7010), 760–767 (2004). https://doi.org/10.1038/nature03015
17. Montague, P.R., Dolan, R.J., Friston, K.J., Dayan, P.: Computational psychiatry. Trends Cogn. Sci. **16**(1), 72–80 (2012). https://doi.org/10.1016/j.tics.2011.11.018
18. Niv, Y.: Reinforcement learning in the brain. J. Math. Psychol. **53**(3), 139–154 (2009). https://doi.org/10.1016/j.jmp.2008.12.005
19. Sutton, R.S., Barto, A.G.: Reinforcement Learning: An Introduction. MIT press, Cambridge (1998)

Evaluation of the Prediction-Based Approach to Cost Reduction in Mutation Testing

Joanna Strug[1(✉)] and Barbara Strug[2]

[1] Faculty of Electrical and Computer Engineering, Cracow University of Technology, ul. Warszawska 24, 31-155 Krakow, Poland
joanna.strug@pk.edu.pl

[2] Department of Physics, Astronomy and Applied Computer Science, Jagiellonian University, Lojasiewicza 11, 30-348 Krakow, Poland
barbara.strug@uj.edu.pl

Abstract. Mutation testing is the most effective technique for assessing the quality of test suites, but it is also very expensive in terms of computational costs. The cost arises from the need to generate and execute a large number of so called mutants. The paper presents and evaluates a machine learning approach to dealing with the issue of limiting the number of executed mutants. The approach uses classification algorithm to predict mutants execution results for a subset of the generated mutants without their execution. The evaluation of the approach takes into consideration two aspects: accuracy of the predicted results and stability of prediction. In the paper the details of the evaluation experiment and its results are presented and discussed. The approach is tested on four examples having different number of mutants ranging from 90 to over 300. The obtained results indicate that the predicted value of the mutation score is consistently higher then the actual one thus allowing for using the results with high confidence.

Keywords: Mutation testing · Machine learning
Mutant classification · kNN classifier

1 Introduction

The main goal of software testing is finding faults in a tested system [12]. It is therefore important to ensure, that tests used to this end are capable of detecting the existing faults, otherwise testing results will be useless. Mutation testing [1,3] is known as the most effective technique for accurate assessing and measuring test suites quality in terms of their ability to detect faults in a system [18].

In mutation testing a test suite ability to detect faults is checked by executing the original system and a number of its mutants (copies of the system, each containing one, small modification inserted basing on a specific rule called mutation operator) against tests from the suite under assessed to see if the mutants behave

J. Świątek et al. (Eds.): ISAT 2018, AISC 853, pp. 340–350, 2019.
https://doi.org/10.1007/978-3-319-99996-8_31

differently from the original system. When different behaviour is observed, for any of the tests, the mutant is said to be killed (detected) by the test suite, otherwise it stays alive (undetected). Basing on the mutants execution results, a test suite ability to detect faults is expressed by its mutation score - a ratio of the number of mutants the suite have killed over the total number of all, non-equivalent [6] mutants generated for the system. Test suites achieving high score (ideally 1 or close to 1) are reported to be the most effective at detecting real faults in software systems [2,7].

In spite of its effectiveness, mutation testing has a serious issue that limits its widespread acceptation as a primary test quality assessment technique. Its application is very expensive, as it requires the generation and execution of a large number of mutants [19]. While the mutant generation cost can be significantly reduced [6,19], the execution of mutants still remains computationally expensive [21] despite the effort put in solving the problem. This paper presents and evaluates an approach dealing with the issue of reducing the number of executed mutants by applying machine learning methods [16]. Specifically, in the approach a classification algorithm is used to predict the mutant execution results (killed or alive) for a set of mutants basing on their similarity to a set of executed mutants. The research presented in this paper focus mainly on evaluating the approach considering the accuracy of the obtained results and stability of the prediction, but we have also refined the previously used strategies for calculating the distance between mutants.

The rest of the paper is organized as follows. Section 2 discusses some related works dedicated to the cost reduction problem. Section 3 describes the key aspects of the approach. Section 4 presents the experiment evaluating the approach and its results and Sect. 5 concludes the paper.

2 Related Work

The problem of reducing the computational cost of applying mutation testing was addressed by many researchers. The proposed approaches aim either at the mutant generation phase (e.g. selective mutation [13,20]) or the mutant execution phase (e.g. parallel processing [10], week mutation [4], mutant sampling [1], mutant clustering [5,8], machine learning based mutation testing [14–17,21]).

While application of techniques such as selective mutation can effectively reduce the number of generated mutants [19,21], the second phase, mutant execution, remains very expensive, as it required the generated mutants be executed against (possibly) all tests from the assessed suite.

Some techniques, such as parallel mutation or week mutation are able reduce the time needed for the execution of mutants, but still required all mutants be executed. So, most of the techniques focusing on the second phase tends to look for ways to limit the number of mutants that has to be executed. The random mutation sampling was one of the first approaches to limit the number of executed mutants [1]. Its authors proposed to select a set percentage of mutants and consider the results obtained from executing only the selected mutants as valid for the entire set. This approach was further refined to select a set percentage of mutants of each type [20]. Other approaches adapted clustering algorithms [5,8] and machine learning methods [14–17,21]. The methods described in [5,8] consisted in using clustering algorithms to divide the mutants into sets and then execute only some mutants that were representative for the defined sets. In the work in [5] the mutants were classified basing on a domain specific information. The authors of [8] clustered mutants that were expected to provide the same results under a test. In our previous research presented in [14–16] we have proposed an approach using classification algorithm to predict mutant execution results for some mutants basing on their similarity to a selected set of mutants for which the execution results were known. In the research graph representations of the mutants were used. The approach provided satisfactory classification results, but data preparation required some effort. Use of bytecode representation as proposed in [17] significantly reduced the data preparation costs. The concept of applying machine learning methods, specifically a random forest classification, was recently presented also in [21]. In the approach the classification model was built basing on mutation testing results obtained for some earlier, existing versions on a project. The results of both research, ours and the authors of [21], lead us to believe that application of machine learning methods can improve significantly the efficiency of applying mutation testing with only minor loss in its effectiveness.

3 Classification-Based Reduction of Mutants Execution

The approach, as proposed in [17] uses machine learning to reduce the number of executed mutants. It uses a kNN classification algorithm [11]. To apply the algorithm the entire set of mutants is divided into two sets:

1. a training set containing mutants for which the execution results (killed or alive) will be known and assigned to them in a form of labels, and
2. a test set containing mutants which execution results will be predicted basing on their similarity to the mutant from the training set.

Application of the approach requires only the mutants belonging to the training set (for brevity called here training mutants) be executed to provide the labels, the one belonging to the test set (called here test mutants) are not executed. Thus, it allows to reduce the mutants execution costs proportionally to the size of the test set and the number of tests in the suite under assessment.

When the approach is used, the mutation score of an assessed test suite is calculated as the ratio of the actually killed mutants and mutants predicted as killed over the total number of non-equivalent mutants.

The following two subsections provide some details concerning the representation of mutants, the classification process and its expected results.

3.1 Mutant Representation

At a source code level a mutant is a copy of a system that differs from the original system by one, small modification defined by a given mutation operator. In the research java programs and their bytecode representations were used. The classification was performed on the bytecode. The bytecode, if compared with a source code, has much simpler and much more regular structure, thus its analyzing is less expensive in term of computational costs. Moreover, it is straightforward attainable.

```
public int findmax()
  {
     int i = 0;
     int max = values[i];
     for (i = 1; i < size; i = i + 1) {
        if (values[i] > max) {
           max = values[i];
        }
     }
     return max;
  }
```

a) original

```
public int findmax()
{
   int i = 0;
   int max = values[i];
   for (i = 1; i < size; i = i + 1) {
      if (!(values[i] > max)) {
         max = values[i];
      }
   }
   return max;
}
```

b) COI mutant

```
public int findmax()
{
   int i = 0;
   int max = values[i];
   for (i = 1; i < size; i = i + 1) {
      if (values[i] > ++max) {
         max = values[i];
      }
   }
   return max;
}
```

c) AOIS mutant

Fig. 1. A part of a findmax method (a), a COI (b) and AOIS (c) mutants of the method

A part of a example program source code and its two mutants are show in Fig. 1. The mutant from Fig. 1(b) was generated using COI mutation operator [9], and the one in Fig. 1(c) was obtained by applying AOIS operator [9]. In Fig. 2 bytecode representations of the program and the mutants from Fig. 1 are presented. At the bytecode level the mutations transfers to change and/or addition or removal of some instructions. At the code level, the COI operator negated the condition in if statement, and the AOIS operator incremented max in the statement. At the bytecode level the first mutation changed the instruction if_icmple to if_icmpgt and the second mutation resulted in additional instruction iinc (both changes are underlined in Fig. 2(b) and (c)).

3.2 Classification of Mutants

The goal of the classification is to predict whether the test mutants can or cannot be killed by any test from the assessed suite without actually executing them. The kNN algorithm compares each of the test mutants with the training mutants and identifies k training mutants to which the classified test mutant is most similar. Then the algorithm labels the classified test mutant accordingly to the most common label among the identified k training mutants.

The training set is selected as a predefined percentage of the entire set of mutants. The comparison of mutants is done basing on a distance between them. The distance, for each pair of mutants, is calculated taking into account the number of differences in their bytecode representations and the positions of the differences.

```
public int findmax();
 Code:
   0: iconst_0
   1: istore_1
   2: aload_0
   3: getfield     #3
   6: iload_1
   7: iaload
   8: istore_2
   9: iconst_0
  10: istore_1
  11: iload_1
  12: aload_0
  13: getfield     #2
  16: if_icmpge    43
  19: aload_0
  20: getfield     #3
  23: iload_1
  24: iaload
  25: iload_2
  26: if_icmple    36
  29: aload_0
  30: getfield     #3
  33: iload_1
  34: iaload
  35: istore_2
  36: iload_1
  37: iconst_1
  38: iadd
  39: istore_1
  40: goto         11
  43: iload_2
  44: ireturn
```

a) original

```
public int findmax();
 Code:
   0: iconst_0
   1: istore_1
   2: aload_0
   3: getfield     #3
   6: iload_1
   7: iaload
   8: istore_2
   9: iconst_0
  10: istore_1
  11: iload_1
  12: aload_0
  13: getfield     #2
  16: if_icmpge    43
  19: aload_0
  20: getfield     #3
  23: iload_1
  24: iaload
  25: iload_2
  26: if_icmpgt    36
  29: aload_0
  30: getfield     #3
  33: iload_1
  34: iaload
  35: istore_2
  36: iload_1
  37: iconst_1
  38: iadd
  39: istore_1
  40: goto         11
  43: iload_2
  44: ireturn
```

b) COI mutant

```
public int findmax();
 Code:
   0: iconst_0
   1: istore_1
   2: aload_0
   3: getfield     #3
   6: iload_1
   7: iaload
   8: istore_2
   9: iconst_0
  10: istore_1
  11: iload_1
  12: aload_0
  13: getfield     #2
  16: if_icmpge    46
  19: aload_0
  20: getfield     #3
  23: iload_1
  24: iaload
  25: iinc         2, 1
  28: iload_2
  29: if_icmple    39
  32: aload_0
  33: getfield     #3
  36: iload_1
  37: iaload
  38: istore_2
  39: iload_1
  40: iconst_1
  41: iadd
  42: istore_1
  43: goto         11
  46: iload_2
  47: ireturn
```

c) AOIS mutant

Fig. 2. A part of a findmax method (a), a COI (b) and AOIS (c) mutants of the method

3.3 Classification Results

Application of mutation testing concludes with a mutation score for an assessed test suite. When applying the classification, the final mutation score is calculated using both, the actual mutants execution results (for the training mutants) and the predicted execution results (for the test mutants). The predicted mutation score for a test suite (denoted by $MS^P(T)$) is expressed as follows (Eq. 1):

$$MS^P(T) = (|M_K| + |M_K^P|)/(|M| - |M_E|), \text{where} \tag{1}$$

- $|M_K|$ - denotes the number of training mutants killed by T
- $|M_K^P|$ - denotes the number of test mutants predicted as killed by T
- $|M|$ - denotes the total number of generated mutants
- $|M_E|$ - denotes the number of equivalent mutants

4 Experimental Evaluation

Application of any cost reduction technique may cause the mutation testing be less effective. The goal of this experiment is to evaluate the approach to see how accurate the predicted results ($MS^P(T)$) are with respect to the actual mutation score (denoted by $MS(T)$) obtained by executing all mutants.

The following subsections describe briefly the experimental procedures and the measures used to evaluate the results, as well as shows and discussed the results obtained for experimental programs.

4.1 Experimental Procedures

The experiment was carried for four, small experimental programs. Table 1 gives, for each example its name, function, the size of its entire set of mutants ($|M|$) and the actual mutation score ($MS(T)$) obtained for a test suite provided for the program.

Table 1. The examples summary

Name	Description	$\|M\|$	MS(T)
Search	Searches for a given value in a set (linear search)	90	0,888
Max.	Finds maximal value in a se	103	0,806
BinSearch	Searches for a given value in a set (binary search)	263	0,806
Triangle	Determines a type of triangle	338	0,800

The experiment, for each example followed the same procedure.

Generation and Execution of Mutants. The mutants for all examples, were generated and executed using mujava [9]. The tool requires a java program and a test suite be provided and performs both tasks: generation of mutants using mutation operators from a predefined set operators designed for java, and execution of the mutants with the provided test suite. The results of running the tool are the mutants, a mutation score for the test suite and lists of killed and alive mutants.

Data Preparation. The classification algorithm requires two inputs: a list of training mutants labeled as killed or alive and a distance matrix for all mutants. The list and the distance matrix were produced basing on the output of mujava by a prototype, custom tool. For the experimental purpose the list of labeled mutants consisted of all mutants, not only the training ones, because the actual execution results for the test mutants were needed to evaluate the accuracy of the classification results. Evaluation measures. The experiment aimed in particular at evaluation the approach considering two aspects accuracy and stability of the classification results.

Accuracy. Basically the accuracy calculated for a test suite T (denoted by A(T)) shows how distant, on average, the predicted mutation score for the suite ($avgMS^P(T)$) is from its actual mutation score (MS(T)). It is expressed as follows (Eq. 2):

$$A(T) = |avgMS^P(T) - MS(T))|, \text{where} \tag{2}$$

- $MS(T)$ – denotes the actual mutation score for a test suite T
- $avgMS^P(T)$ – denotes the mean of all mutation scores predicted in all runs of the classification experiment

Stability. The stability of the results (denoted by S(T)) shows how close to the average predicted mutation score are the results obtained in each individual run of the classification experiment. It is expressed by the standard deviation of the predicted values of mutation score.

Experimental Setup and Experimental Results. The classification was performed also with a help of a prototype, custom tool implementing the kNN algorithm. The experimental parameters were set up as follows: the size of the training set was set to 70% of all mutants, k was set to 7. The setup was selected as a trade-off between accuracy and efficiency of the classification.

The classification was run 100 times for each examples to gather the data for the evaluation of the approach, in particular for measuring the accuracy and stability of results. The results for all examples are presented in Table 2 and shown in charts in Figs. 3, 4, 5 and 6. Table 2 presents the actual mutation score, the average predicted mutation score, accuracy and stability calculated for each example. Each of the charts in Figs. 3, 4, 5 and 6 shows the distribution of predicted mutation scores for one example. The horizontal lines mark the actual mutation score (the lower one) and the average predicted mutation score.

Table 2. Classification results

Name	MS(T)	$avgMS^P(T)$	A(T)	S(T)
Search	0.888	0.922	0.033	0.012
Max.	0.806	0.862	0.056	0.015
BinSearch	0.806	0.854	0.048	0.013
Triangle	0.800	0.862	0.055	0.011

Discussion of the Results. As it can be observed from the results shown in Table 2 and Figs. 3, 4, 5 and 6 the predicted mutation score is slightly higher than the actual one. Thus it seems that the predicted results would always lead to a small overestimation of the test suite quality. However as the difference is

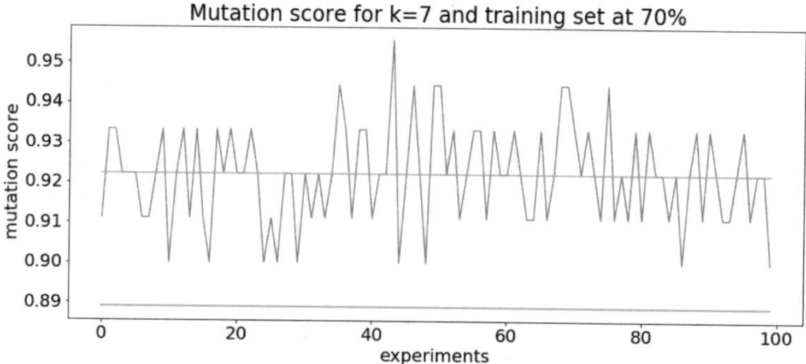

Fig. 3. Distribution of $MS^P(T)$ for Search.

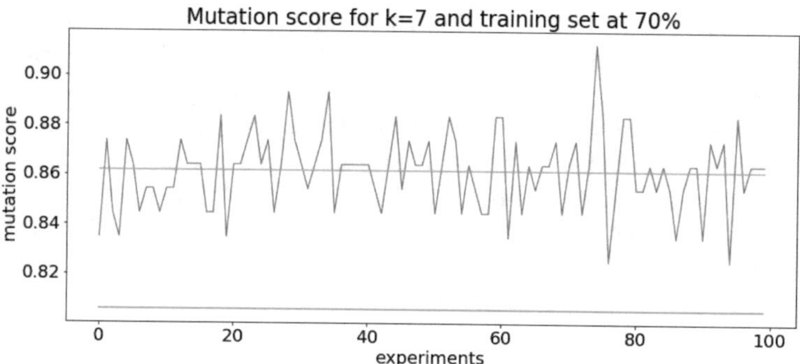

Fig. 4. Distribution of $MS^P(T)$ for Max.

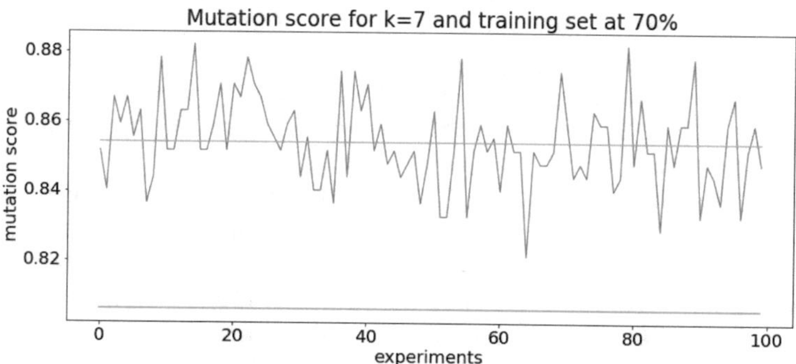

Fig. 5. Distribution of $MS^P(T)$ for BinSearch.

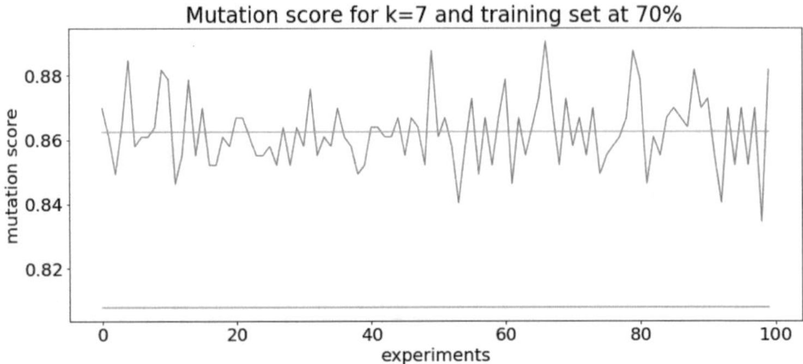

Fig. 6. Distribution of $MS^P(T)$ for Triangle.

relatively consistent (as can be seen in the accuracy measure) it could be easy to estimate the actual quality from the predicted one. In real world application of the approach repeating the prediction process 100 times would not be feasible, so we need to be able to make predictions based on just one run. To evaluate the difference between the results in subsequent runs in the experiment we have measured the stability of the predicted value of the mutation score. This value, shown in Table 2, is consistently rather good - it ranges from 0.011 to 0.015. It means that the actual difference between subsequent experiments is small and thus the results of any experiment can be safely used.

5 Conclusions

Mutation testing measures the quality of test suites with a very high accuracy, but its application, especially execution of mutants is expensive. As indicated in [21] some loss of the accuracy can be acceptable if the efficiency can be improved. This paper focuses on the evaluation of a classification-based approach to the problem of the reduction of costs of mutation testing. The approach helps to reduces the number of executed mutants depending on the program for which they are generated.

In the evaluation the focus has been placed on two aspects: accuracy, defined as the difference between the predicted value of the mutation score and the actual one, and stability, defined as the standard deviation of the predicted mutation score. These measures allow us to assess the reliability of results one the basis of just one run. The approach has been tested on four different programs. The obtained results indicate that the approach is reliable and can be used to predict the mutation score for a given test suite on the basis of executing only a sample of mutants and predicting the execution results for the remaining ones. As the best results were obtained for a sample of 70% mutants it means the number of mutants to execute can be reduced by a third.

In the future experiments we plan to extend our evaluation of the approach quality by adding different measures. One of the possibilities seems to be a more quantitative assessment of the relation between the efficiency and the effectivness of the approach. Especially we plan to include execution time in our evaluation process.

It is also planned to refine the way a training set is sampled by taking a proportional number of mutants of each type (generated by a given type of mutation operators) instead of a random sampling.

In the approach Java bytecode was used to calculate the distance between mutants. Even that not all languages compile to a bytecode many of the most often used, general purpose languages (for example as *C#*) have some form of it. Hence the approach has a potential to be applied in other contexts too.

References

1. Acree, A.T.: On Mutation. PhD Thesis, Georgia Institute of Technology, Atlanta, Georgia (1980)
2. Andrews, J.H., Briand, L.C., Labiche, Y.: Is mutation an appropriate tool for testing experiments? In: Proceedings of the ICSE05, pp. 402–411. IEEE (2005)
3. DeMillo, R.A., Lipton, R.J., Sayward, F.G.: Hints on test data selection: help for the practicing programmer. Computer **11**(4), 34–41 (1978)
4. Howden, W.E.: Weak mutation testing and completeness of test sets. IEEE Trans. Softw. Eng. **8**, 371–379 (1982)
5. Ji, C., Chen, Z., Xu, B., Zhao, Z.: A novel method of mutation clustering based on domain analysis. In: Proceedings of the 21st ICSEKE09, Boston, USA (2009)
6. Jia, Y., Harman, M.: An analysis and survey of the development of mutation testing. IEEE Trans. Softw. Eng. **37**, 649–678 (2011)
7. Just, R., Jalali, D., Inozemtseva, L., Ernst, M.D., Holmes, R., Fraser, G.: Are mutants a valid substitute for real faults in software testing? In: Proceedings of ACM SIGSOFT Symposium on the Foundations of Software Engineering, Hong Kong, China, pp. 654–665 (2014)
8. Ma, Y.-S., Kim, S.-W.: Mutation testing cost reduction by clustering overlapped mutants. J. Syst. Softw. **115**, 18–30 (2016)
9. Ma, Y., Offutt, J., Kwon, Y.R.: MuJava: a mutation system for Java. In: Proceedings of ICSE06, Shanghai, China, pp. 827–830 (2006)
10. Mathur, A.P., Krauser, E.W.: Mutant Unification for Improved Vectorization. Purdue University, West Lafayette, Indiana, Technical report SERC-TR-14-P (1988)
11. Mitchell, T.: Machine Learning. Mcgraw-Hill Education, New York City (1997)
12. Myers, G., Sandler, C., Badgett, T.: The Art of Software Testing. Wiley, Hoboken (2011)
13. Offutt, A.J., Rothermel, G., Zapf, C.: An experimental evaluation of selective mutation. In: Proceedings of ICSE, pp. 100–107 (1993)
14. Strug, J., Strug, B.: Machine learning approach in mutation testing. In: LNCS, vol. 764, pp. 200–214 (2012)
15. Strug, J., Strug, B.: Classifying mutants with decomposition Kernel. In: Proceedings of ICAISC2016, LNCS, vol. 9692, pp. 644–654 (2016)
16. Strug, J., Strug, B.: Using classification for cost reduction of applying mutation testing. In: Proceedings of FedCSIS2017, pp. 99–108 (2017)
17. Strug, J., Strug, B.: Cost reduction in mutation testing with bytecode-level mutants classification. In: Proceedings of ICAISC2018, LNCS, vol. 10841, pp. 714–723. Springer (2018)

18. Thierry, T.C., Papadakis, M., Traon, Y.L., Harman, M.: Empirical study on mutation, statement and branch coverage fault revelation that avoids the unreliable clean program assumption. In: Proceedings of IEEE/ACM ICSE17, Buenos Aires, Argentina, pp. 597–608 (2017)
19. Usaola, M.P., Mateo, P.R.: Mutation testing cost reduction techniques: a survey. IEEE Softw. **27**(3), 80–86 (2010)
20. Wong, W.E., Mathur, A.P.: Reducing the cost of mutation testing: an empirical study. JSS **31**(3), 185–196 (1995)
21. Zhang, J., Zhang, L., Harman, M., Hao, D., Jia, Y., Zhang, L.: Predictive mutation testing. In: Proceedings of ISSTA2016, Saarbrücken, Germany, pp. 342–353 (2016)

Optimization of Decision Rules Relative to Length - Comparative Study

Beata Zielosko(✉) and Krzysztof Żabiński

Institute of Computer Science, University of Silesia in Katowice,
39, Będzińska St., 41-200 Sosnowiec, Poland
{beata.zielosko,kzabinski}@us.edu.pl

Abstract. The paper presents a modification of a dynamic programming approach employed for decision rules optimization with respect to their length. There are two aspects taken into account: (i) consideration on the length of approximate decision rules and (ii) consideration on the size of a directed acyclic graph constructed by the modified algorithm.

Keywords: Decision rules · Length · Optimization
Dynamic programming approach

1 Introduction

Knowledge representation in data mining area can be expressed in many ways. Nevertheless, decision rules are one of the form which is simple and easily understandable by humans. So, they are popular in various areas of data mining. Induction of decision rules can be performed from the point of view of (i) knowledge representation or (ii) classification. Since the aims are different, algorithms for construction of rules and quality measures for evaluating of such rules are also different [12].

In the literature, there are many approaches to the construction of decision rules, for instance: greedy algorithms [7], genetic algorithms [10], algorithms based on decision tree construction [1], algorithms based on a sequential covering procedure [4,6], and many others. There are also different rule quality measures that are used for induction or classification tasks [11,13].

We are interested in the construction of short rules. In the paper, a modification of dynamic programming approach for optimization of decision rules relative to length is presented. Optimization with respect to length follows the Minimum Description Length principle [9] stating that: the best hypothesis for a given set of data is the one that leads to the largest compression of this data.

Unfortunately, the problem of minimization of length of decision rules is NP-hard. Using results of Feige [5] it is possible to show that under reasonable assumptions on the class NP there are no approximate algorithms with high accuracy and polynomial complexity for minimization of decision rule length. The most part of approaches mentioned above (with the exception of dynamic

J. Świątek et al. (Eds.): ISAT 2018, AISC 853, pp. 351–360, 2019.
https://doi.org/10.1007/978-3-319-99996-8_32

programming [2] and Boolean reasoning [8]) cannot guarantee the construction of the shortest rules.

In terms of the scope of the paper, a modification of the dynamic programming approach for rule optimization was searched for. Such a heuristic is to provide rules close to optimal ones taking into account their length. In order to assess the proposed modification, the work consists of comparisons of two factors with respect to the classical dynamic programming approach [2]: lengths of rules and sizes of directed acyclic graphs constructed in both scenarios. The size of a directed acyclic graph can be understood as the number of nodes and edges in the graph. Experimental results connected with classification accuracy, for both scenarios, are also presented.

It is worth introducing the principles of the dynamic programming approach for optimization of decision rules with respect to their length. A given decision table T is partitioned into subtables. A directed acyclic graph $\Delta^*_\gamma(T)$ is constructed with nodes being the subtables. The subtables are once again partitioned up to the level when the uncertainty is at most equal to γ. The difference between the dynamic programming approach without modification [2] (called as classical) and the modified one is that in the classical algorithm the partitioning is done for each attribute and its value from the given table T, whilst for the modified one, it is done for all values of one attribute with minimum number of values in T, and for the remaining attributes it is done only for their most common value. As a result, the graph constructed by means of the modified algorithm is smaller than the one obtained with the classical algorithm due to fewer number of nodes and edges.

In order to construct approximate decision rules, it is necessary to introduce an uncertainty measure $G(T)$ of a decision table T. It is defined as a difference between number of rows in a given decision table T and the number of rows labeled with the most common decision from this table, divided by the number of rows in table T. A fixed threshold value γ, $0 \leq \gamma < 1$ is introduced and so-called γ-decision rules are studied that localize rows in subtables of T which uncertainty is at most γ.

Basing on the graph $\Delta^*_\gamma(T)$, sets of γ-decision rules attached to rows of table T are described. Then, using a procedure of optimization of the graph $\Delta^*_\gamma(T)$ relative to length, it is possible to find, for each row r of T, the shortest γ-decision rule. It allows one to study how far the obtained values of length are from the optimal ones, i.e., the minimum length of rules obtained by the classical dynamic programming approach.

In [15], a modified dynamic programming approach for exact decision rules optimization relative to length was studied, in [14], an optimization of approximate rules relative to coverage was presented.

The paper consists of six sections. Section 2 contains main notions. In Sect. 3, a modified algorithm for construction of a directed acyclic graph is presented. Section 4 presents a description of a procedure of optimization relative to length. Section 5 contains experimental results, and Sect. 6 – conclusions.

2 Main Notions

Main notions referring to decision tables and decision rules are depicted in this section.

Decision table [8] can be expressed as $T = (U, A \cup \{d\})$, where $U = \{r_1, \dots, r_k\}$ is a nonempty, finite set of objects (rows) and $A = \{f_1, \dots, f_n\}$ is a nonempty, finite set of attributes called as conditional, $d \notin A$ is a decision attribute. An assumption on consistency of decision tables is made (there are no rows of the same values of conditional attributes, but with different decisions).

The most common decision for T is a minimum value of a decision attribute existing in the maximum number of rows of the table T.

Each decision table T contains a finite number of rows which is denoted as $N(T)$. As for *the most common decision for* T, it is expressed as $N_{mcd}(T)$. The *uncertainty* of a decision table T is denoted by $G(T)$,

$$G(T) = \frac{N(T) - N_{mcd}(T)}{N(T)}.$$

As for a *subtable*, it is defined as a table derived from the table T by removing selected rows from T. Supposing that T is nonempty and $f_{i_1}, \dots, f_{i_m} \in \{f_1, \dots, f_n\}$ and $a_1, \dots, a_m$ are values of conditional attributes, then subtable $T(f_{i_1}, a_1) \dots (f_{i_m}, a_m)$ of the table T has just rows with values $a_1, \dots, a_m$ at the intersection with columns $f_{i_1}, \dots, f_{i_m}$. The subtables constructed in such a way, as well as a the initial table T, can be named *separable subtables.*

A *non-constant* attribute of T is denoted as $f_i \in \{f_1, \dots, f_n\}$ and it has at least two different values. *The most frequent value of an attribute* f_i is the value attached to the maximum number of rows of the table T.

A set of non-constant attributes of the table T is expressed as $E(T)$ whilst a subset of attributes from $E(T)$ associated with a given row r is expressed as $E(T, r)$.

$E^*(T, f_i)$ denotes a set of values of a given attribute $f_i \in E(T)$. *The minimum attribute for* T is the one with the minimum number of values and the minimum index i of all such attributes f_i from the set $E(T)$. If a given attribute is minimum, then $E^*(T, f_i)$ contains all values of f_i on T. Contrarily, $E^*(T, f_i)$ includes merely the most frequent value of f_i on T.

A *decision rule over* T is the following expression:

$$f_{i_1} = a_1 \wedge \dots \wedge f_{i_m} = a_m \rightarrow d \tag{1}$$

if $f_{i_1}, \dots, f_{i_m} \in \{f_1, \dots, f_n\}$ are conditional attributes from T and $a_1, \dots a_m, d$ are these attributes' and a decision attribute's values, respectively. The value of m can be also equal to zero and in such a case (1) is of a special form: $\rightarrow d$.

Supposing that a row r of T is denoted as $r = (b_1, \dots, b_n)$, the rule (1) is *realizable for* r under a condition that $a_1 = b_{i_1}, \dots, a_m = b_{i_m}$. For the special case when $m = 0$, the rule $\rightarrow d$ is realizable for any row of T.

For γ being a nonnegative real number, $0 \leq \gamma < 1$, the rule (1) is *γ-true for* T provided that a decision d is the most common decision for $T' = T(f_{i_1}, a_1) \dots (f_{i_m}, a_m)$ and $G(T') \leq \gamma$.

A given rule can be called a γ-*decision rule for* T *and* r when such a rule is γ-true for T and realizable for r.

The *length* of a decision rule (1) is the number of descriptors (pairs attribute-value) on the left-hand side of this rule. It is expressed as $l(\tau)$.

3 Modified Algorithm for Directed Acyclic Graph Construction $\Delta^*_\gamma(T)$

This section contains presentation of the modified algorithm (see Algorithm 3.1) used to create a directed acyclic graph (denoted by $\Delta^*_\gamma(T)$) for the decision table T. The graph is further used to derive decision rules for each row r of the table T.

Algorithm 3.1. Algorithm for construction of a graph $\Delta^*_\gamma(T)$

```
Input  : Decision table T with attributes f_1, ..., f_n, nonnegative real γ,
         0 ≤ γ < 1.
Output: Graph Δ*_γ(T)
A graph contains a single node T which is not marked as processed;
while all nodes of the graph are not marked as processed do
    Select a node (table) Θ, which is not marked as processed;
    if G(Θ) ≤ γ then
        The node is marked as processed;
    end
    if G(Θ) > γ then
        For each f_i ∈ E(Θ), draw edges from the node Θ.
        Mark the node Θ as processed;
    end
end
return Graph Δ*_γ(T);
```

Nodes of the graph are separable subtables of the table T. During each step, the algorithm processes one node and marks it with the symbol *. At the first step, the algorithm constructs a graph containing a single node T which is not marked with the symbol *.

Let the algorithm have already performed p steps. Let us describe the step $(p+1)$. If all nodes with the symbol * are marked as processed, the algorithm finishes its work and presents the resulting graph as $\Delta^*_\gamma(T)$. Otherwise, choose a node (table) Θ, which has not been processed yet. If $G(\Theta) \le \gamma$ mark the considered node with symbol * and proceed to the step $(p+2)$. If $G(\Theta) > \gamma$, for each $f_i \in E(\Theta)$, draw a bundle of edges from the node Θ. Let $E^*(\Theta, f_i) = \{b_1, \ldots, b_t\}$. Then draw t edges from Θ and label these edges with pairs $(f_i, b_1), \ldots, (f_i, b_t)$ respectively. These edges enter to nodes $\Theta(f_i, b_1), \ldots, \Theta(f_i, b_t)$. If some of nodes $\Theta(f_i, b_1), \ldots, \Theta(f_i, b_t)$ are absent in the graph then add these nodes to the graph. Each row r of Θ is labeled with the set of attributes $E^*_{\Delta_\gamma(T)}(\Theta, r) \subseteq E(\Theta)$. Mark the node Θ with the symbol * and proceed to the step $(p+2)$.

The graph $\Delta^*_\gamma(T)$ is a directed acyclic graph. A node of this graph will be called *terminal* if there are no edges leaving this node. Note that a node Θ of $\Delta^*(T)$ is terminal if and only if $G(\Theta) \leq \gamma$.

The presented algorithm in comparison with the classical one [2] does not construct a complete directed acyclic graph but only its part. Instead of using all the attributes with all their values, the attribute with minimum number of values and all its values are taken. As for the rest of attributes, only their most frequent value is taken into consideration. In such a way only a part of the graph is constructed and the computation time used to generate the graph is saved.

The graph constructed in such a way can be then optimized (as described further in this work). The optimization taken by us into account is the one with respect to the length of rules. The procedure results in a graph denoted as G. The nodes and edges are equal as in $\Delta^*_\gamma(T)$. The difference is in labeling the rows of nonterminal nodes. These sets of attributes follows the following relation: $E_G(\Theta, r) \subseteq E_{\Delta^*_\gamma(T)}(\Theta, r)$ (however, it is also possible that $G = \Delta^*_\gamma(T)$).

Having the graph constructed and optimized, it is possible to move into description of the decision rules. A set of decision rules needs to be generated for each node Θ of G and for each row r of Θ. The dynamic programming nature of the algorithm can be seen in the direction of rule description. It is done starting from terminal nodes up to the root node T.

The procedure is as follows. Supposing that Θ is a terminal node of G labeled with the most common decision d for Θ, it is possible to induce a rule

$$Rul_G(\Theta, r) = \{\to d\}.$$

Next, supposing that Θ is a nonterminal node of G such that for each child Θ' of Θ and for each row r' of Θ', the rule set $Rul_G(\Theta', r')$ has already been determined. Let $r = (b_1, \ldots, b_n)$ be a row of Θ. For any $f_i \in E_G(\Theta, r)$, the set of rules $Rul_G(\Theta, r, f_i)$ is defined as follows:

$$Rul_G(\Theta, r, f_i) = \{f_i = b_i \wedge \sigma \to k : \sigma \to k \in Rul_G(\Theta(f_i, b_i), r)\}.$$

Then $Rul_G(\Theta, r) = \bigcup_{f_i \in E_G(\Theta, r)} Rul_G(\Theta, r, f_i)$.

Example 1. To illustrate the presented algorithm, a simple decision table T_0 depicted on the top of Fig. 1, is considered. In the example, $\gamma = 0.5$, so during the construction of the graph $\Delta^*_{0.5}(T_0)$ the partitioning of a subtable Θ is stopped when $G(\Theta) \leq 0.5$. We denote $G = \Delta_{0.5}(T)$.

Now, for each node Θ of the graph G and for each row r of Θ the set $Rul_G(\Theta, r)$ is described. Let us move from terminal nodes of G to the node T. Terminal nodes of the graph G are Θ_1, Θ_2, Θ_3 and Θ_4. For these nodes,

$$\begin{aligned}
&Rul_G(\Theta_1, r_2) = Rul_G(\Theta_1, r_3) = \{\to 1\};\\
&Rul_G(\Theta_2, r_1) = \{\to 1\};\\
&Rul_G(\Theta_3, r_1) = Rul_G(\Theta_3, r_2) = \{\to 1\};\\
&Rul_G(\Theta_4, r_1) = Rul_G(\Theta_4, r_3) = \{\to 1\}.
\end{aligned}$$

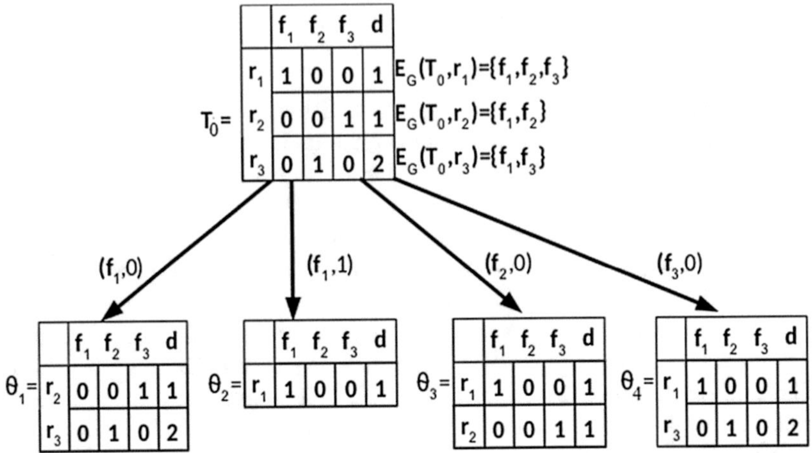

Fig. 1. Directed acyclic graph $G = \Delta^*_{0.5}(T_0)$

Now, the sets of rules attached to rows of T_0 are described:

$Rul_G(T_0, r_1) = \{f_1 = 1 \rightarrow 1, f_2 = 0 \rightarrow 1, f_3 = 0 \rightarrow 1\}$;
$Rul_G(T_0, r_2) = \{f_1 = 0 \rightarrow 1, f_2 = 0 \rightarrow 1\}$;
$Rul_G(T_0, r_3) = \{f_1 = 0 \rightarrow 1, f_3 = 0 \rightarrow 1\}$.

4 Procedure of Optimization Relative to Length

In this section, a procedure of optimization of the graph G relative to length l is presented. Let $G = \Delta^*_\gamma(T)$. Every node Θ of the graph G is taken into consideration and for each of the nodes, the procedure ascribes to each row r of Θ the set $Rul^l_G(\Theta, r)$ of decision rules with the minimum length from $Rul_G(\Theta, r)$. This minimum length of a decision rule from $Rul_G(\Theta, r)$ is denoted as $Opt^l_G(\Theta, r)$.

The number $Opt^l_G(\Theta, r)$ is assigned to every row r of every table Θ. As a result, sets $E_G(\Theta, r)$ associated with rows r of nonterminal nodes can be different. The resulting graph is denoted as G^l.

Let us start from terminal nodes. Supposing that Θ is a terminal node of G and d is the most common decision for Θ. Each row r of a terminal node Θ has the number $Opt^l_G(\Theta, r) = 0$ assigned to it.

Let us now move up to nonterminal nodes. Supposing that Θ is a nonterminal node of G and all children of Θ have already been considered. Supposing that $r = (b_1, \ldots, b_n)$ is a row of Θ, the number

$$Opt^l_G(\Theta, r) = \min\{Opt^l_G(\Theta(f_i, b_i), r) + 1 : f_i \in E_G(\Theta, r)\}$$

is assigned to the row r in the table Θ and:

$$E_{G^l}(\Theta, r) = \{f_i : f_i \in E_G(\Theta, r), Opt^l_G(\Theta(f_i, b_i), r) + 1 = Opt^l_G(\Theta, r)\}.$$

Example 2. Below you can find sets $Rul^l_G(T_0, r_i)$, $i = 1, \ldots, 3$, of γ-decision rules for T_0 (depicted on the top of Fig. 1) and r_i, with the minimum length, and the value $Opt^l_G(T, r_i)$. It is equal to the minimum length of γ-decision rule for T_0 and r_i, and it was obtained during the procedure of optimization of the graph G relative to the length. In the case of presented example, sets of γ-decision rules obtained in Example 1 are the shortest rules for T_0 and r_i, $i = 1, \ldots, 3$.

$$Rul^l_G(T_0, r_1) = \{f_1 = 1 \rightarrow 1, f_2 = 0 \rightarrow 1, f_3 = 0 \rightarrow 1\},\ Opt^l_G(T_0, r_1) = 1;$$
$$Rul^l_G(T_0, r_2) = \{f_1 = 0 \rightarrow 1, f_2 = 0 \rightarrow 1\},\ Opt^l_G(T_0, r_2) = 1;$$
$$Rul^l_G(T_0, r_3) = \{f_1 = 0 \rightarrow 1, f_3 = 0 \rightarrow 1\},\ Opt^l_G(T_0, r_3) = 1.$$

5 Experimental Results

Experiments have been performed on sets from UCI Machine Learning Repository [3]. There was a need for preprocessing of some of these sets. The preprocessing process consisted of 2 stages: removal of attributes with only one value for the whole decision table and filling each missing attribute value with the most common value of the attribute considered. Moreover, the decision tables have been disposed of inconsistencies by replacing a group of inconsistent rows with one row containing the most common decision of such a group.

Let T be one of these decision tables. Values of γ from the set $\Gamma(T) = \{G(T) \times 0.01, G(T) \times 0.1, G(T) \times 0.2\}$ are considered for the table T.

Two methods were used to optimize the generated decision rules with respect to length: modified and classical dynamic programming approach. Having applied each of the mentioned optimization methods to the given decision table T, a minimum length rule for each row r of the table T has been obtained. Then, it was possible to consider minimum, average, and maximum rule lengths for each of the sets taken into account. The results obtained for the modified dynamic programming approach have been gathered in the Table 1.

Table 1. Length of γ-decision rules, $\gamma \in \{G(T) \times 0.01, G(T) \times 0.1, G(T) \times 0.2\}$

Decision table	Rows	Attr	$\gamma = G(T) \times 0.G(T)$			$\gamma = G(T) \times 0.1$			$\gamma = G(T) \times 0.2$		
			min	avg	max	min	avg	max	min	avg	max
adult-stretch	16	4	1	1.75	4	1	1.75	4	1	1.75	4
balance-scale	625	4	3	3.48	4	2	3.1	4	2	3.04	4
cars	1728	6	1	2.72	6	1	2.72	6	1	2.72	6
house-votes	279	16	2	3.13	8	2	3.13	8	1	1.67	3
lymphography	148	18	2	3.14	7	2	3.14	7	2	2.93	7
shuttle-landing	15	6	1	4.40	6	1	4.40	6	1	4.40	6
soybean-small	47	35	1	1.64	2	1	1.64	2	1	1.64	2
teeth	23	8	2	3.35	4	2	3.35	4	2	3.35	4

Table 2 depicts comparison of minimum (column *min*), average (column *avg*) and maximum (column *max*) lengths of rules. The values gathered in this table are quotients of minimum, average, and maximum length of rules obtained by classical dynamic programming approach and the corresponding values constructed by the modified one. Bold values mark equality of lengths of rules for classical and modified dynamic programming approaches. The highest average difference of length of γ-decision rules can be observed for shuttle-landing data set and $\gamma \in \Gamma(T)$.

Table 2. Comparison of length of γ-decision rules

Decision table	$\gamma = \lfloor G(T) \times 0.01 \rfloor$			$\gamma = \lfloor G(T) \times 0.1 \rfloor$			$\gamma = \lfloor G(T) \times 0.2 \rfloor$		
	min	avg	max	min	avg	max	min	avg	max
adult-stretch	**1.00**	0.71	0.50	**1.00**	0.71	0.50	**1.00**	0.71	0.50
balance-scale	**1.00**	0.92	**1.00**	**1.00**	0.92	**1.00**	**1.00**	0.88	**1.00**
cars	**1.00**	0.89	**1.00**	**1.00**	0.89	**1.00**	**1.00**	0.89	**1.00**
house-votes	**1.00**	0.81	0.63	0.50	0.53	0.50	**1.00**	0.80	**1.00**
lymphography	0.50	0.63	0.57	0.50	0.63	0.57	0.50	0.61	0.43
shuttle-landing	**1.00**	0.32	0.67	**1.00**	0.32	0.67	**1.00**	0.32	0.67
soybean-small	**1.00**	0.61	0.50	**1.00**	0.61	0.50	**1.00**	0.61	0.50
teeth	0.50	0.67	**1.00**	0.50	0.67	**1.00**	0.50	0.67	**1.00**

Table 3 compares sizes of directed acyclic graphs constructed by the modified algorithm and the classical one. Columns *nd-diff* compare numbers of nodes whilst columns *edg-diff* compare numbers of edges. The values gathered in this table are quotients of numbers of nodes or edges in the directed acyclic graph constructed by the classical algorithm and the modified one. Taking into account the obtained results, it turns out that the size of the directed acyclic graph obtained by the modified approach is smaller than the size of the directed acyclic graph obtained by the classical one. Bold values mark highest differences (more than three times). It is worth mentioning that for "cars" dataset the difference in nodes number is octuple and in edges number is seventeen-fold. Nevertheless, the length results are comparable for this set (see Table 2). Furthermore, when considering number of edges it turns out that for each dataset and each value of γ, the difference is at least double.

Table 4 collates accuracy of classifiers based on approximate decision rules optimized with respect to length and generated by modified and classical dynamic programming approaches. It comprises average test errors for two-fold cross validation. The experiments have been performed 30 times for each decision table. The procedure was to randomly divide each dataset into three parts: train (30% of rows), validation (20% of rows), and test (50% of rows). Then, exact decision rules (having γ equal to 0) have been derived from the train part of a given data set. Having the rules constructed, they have been pruned and as a

Table 3. Comparision of size of a directed acyclic graph

Decision table	$\gamma = G(T) \times 0.01$		$\gamma = G(T) \times 0.1$		$\gamma = G(T) \times 0.2$	
	nd-diff	edg-diff	nd-diff	edg-diff	nd-diff	edg-diff
adult-stretch	2.00	2.92	2.00	2.92	2.00	2.92
balance-scale	1.85	**4.23**	1.93	**4.54**	1.93	**4.54**
cars	**8.77**	**17.55**	**8.77**	**17.55**	**8.82**	**17.69**
house-votes	1.43	2.66	1.45	2.69	1.50	2.78
lymphography	1.52	**3.89**	1.53	**3.90**	1.55	**3.92**
shuttle-landing	1.09	2.00	1.09	2.00	1.09	2.00
soybean-small	1.19	2.69	1.19	2.69	1.19	2.70
teeth	1.14	2.41	1.14	2.41	1.14	2.41

result γ-decision rules have been obtained. The chosen value of γ was the one for which the rules minimize a validation error. The constructed classifier has then been verified on a test part of the dataset. Classification result is understood as a test error. It is the ratio of false positives and the number of all rows in the test part of the dataset. Columns *test error* and *std* contain average test error and standard deviation, respectively. Table's 4 last row depicts the average test error for all datasets. It proves that the accuracies of classifiers obtained by the modified dynamic programming approach and the classical one, are comparable.

Table 4. Average test error

Decision table	Modified approach		Classical approach	
	test error	std	test error	std
balance-scale	0.32	0.03	0.29	0.02
cars	0.20	0.02	0.30	0.01
house-votes	0.07	0.07	0.09	0.05
lymphography	0.24	0.04	0.28	0.04
soybean-small	0.17	0.08	0.13	0.26
average	0.20		0.22	

6 Conclusions

The paper introduces a modified dynamic programming approach for optimization of approximate decision rules with respect to their length. It is worth mentioning that short rules are important from the viewpoint of knowledge representation.

The experiments prove that the size of the directed acyclic graph built by the modified approach is smaller than the size of the graph constructed by the classical one. It is especially visible in case of numbers of edges. The difference, for each decision table and $\gamma \in \Gamma(T)$, is at least double. As for the lengths of rules, they are comparable and close to the optimal ones. Considering accuracy of classifiers constructed by both of the approaches introduced on the pages of this work, it is comparable.

In our future works, we will investigate other heuristics and we will compare it with the one proposed in this paper.

References

1. Alkhalid, A., Amin, T., Chikalov, I., Hussain, S., Moshkov, M., Zielosko, B.: Optimization and analysis of decision trees and rules: dynamic programming approach. Int. J. Gen. Syst. **42**(6), 614–634 (2013)
2. Amin, T., Chikalov, I., Moshkov, M., Zielosko, B.: Dynamic programming approach for partial decision rule optimization. Fundam. Inf. **119**(3–4), 233–248 (2012)
3. Asuncion, A., Newman, D.J.: UCI Machine Learning Repository (2007). http://www.ics.uci.edu/~mlearn/
4. Błaszczyński, J., Słowiński, R., Szeląg, M.: Sequential covering rule induction algorithm for variable consistency rough set approaches. Inf. Sci. **181**(5), 987–1002 (2011)
5. Feige, U.: A threshold of ln n for approximating set cover. J. ACM **45**(4), 634–652 (1998)
6. Fürnkranz, J.: Separate-and-conquer rule learning. Artif. Intell. Rev. **13**(1), 3–54 (1999)
7. Moshkov, M.J., Piliszczuk, M., Zielosko, B.: Greedy algorithms with weights for construction of partial association rules. Fundam. Inf. **94**(1), 101–120 (2009)
8. Pawlak, Z., Skowron, A.: Rough sets and boolean reasoning. Inf. Sci. **177**(1), 41–73 (2007)
9. Rissanen, J.: Modeling by shortest data description. Automatica **14**(5), 465–471 (1978)
10. Ślęzak, D., Wróblewski, J.: Order based genetic algorithms for the search of approximate entropy reducts. In: Wang, G., Liu, Q., Yao, Y., Skowron, A. (eds.) RSFDGRC 2003. LNCS, vol. 2639, pp. 308–311. Springer (2003)
11. Stańczyk, U., Zielosko, B.: On combining discretisation parameters and attribute ranking for selection of decision rules. In: Polkowski, L., Yao, Y., Artiemjew, P., Ciucci, D., Liu, D., Ślęzak, D., Zielosko, B. (eds.) IJCRS 2017, Part I. LNCS, vol. 10313, pp. 329–349. Springer (2017)
12. Stefanowski, J., Vanderpooten, D.: Induction of decision rules in classification and discovery-oriented perspectives. Int. J. Intell. Syst. **16**(1), 13–27 (2001)
13. Wróbel, Ł., Sikora, M., Michalak, M.: Rule quality measures settings in classification, regression and survival rule induction - an empirical approach. Fundam. Inf. **149**(4), 419–449 (2016)
14. Zielosko, B.: Optimization of approximate decision rules relative to coverage. In: Kozielski, S., Mrozek, D., Kasprowski, P., Małysiak-Mrozek, B., Kostrzewa, D. (eds.) BDAS 2014, CCIS, vol. 424, pp. 170–179. Springer (2014)
15. Zielosko, B.: Optimization of exact decision rules relative to length. In: Czarnowski, I., Howlett, R.J., Jain, L.C. (eds.) KES-IDT 2017, Part I, pp. 149–158. Springer (2018)

Comparison of Fuzzy Multi Criteria Decision Making Approaches in an Intelligent Multi-agent System for Refugee Siting

Maria Drakaki[1(✉)], Hacer Güner Gören[2], and Panagiotis Tzionas[1]

[1] Department of Automation Engineering, Alexander Technological Educational Institute of Thessaloniki, P.O. Box 141, 574 00 Thessaloniki, Hellas
{drakaki,ptzionas}@autom.teithe.gr

[2] Department of Industrial Engineering, Pamukkale University, Kinikli Campus, Denizli, Turkey
hgoren@pau.edu.tr

Abstract. Refugee crisis has escalated into a leading crisis in recent years, including Europe since 2015 after the massive refugee and migrant sea arrivals in the Mediterranean. Therefore, its socio-economic and environmental impact requires complex decision making for the delivery of effective humanitarian aid operations. Refugee settlement and shelter is an operations sector where the application of multi-criteria decision making (MCDM) methods seems appropriate. Additionally, the range of involved decision makers as well as their relationships can be addressed using a multi-agent system (MAS). Different decision making fuzzy methods have been proposed in the literature which can be used by the agents in order to address refugee siting. The purpose of this paper is to perform a comparative analysis of two such methods, namely, hierarchical fuzzy TOPSIS and fuzzy axiomatic design approach used in a MAS for refugee siting. The comparative study has been done by evaluating operating temporary sites in Greece and the obtained results reflect the current situation.

Keywords: Hierarchical fuzzy TOPSIS · Fuzzy Axiomatic Design
Intelligent multi-agent system · Refugee siting · Refugee crisis

1 Introduction

In an evolving geopolitical environment, refugee and migrant arrivals in Europe continue at a non-constant rate. In 2015, 1.015.078 people crossed the Mediterranean to arrive in Europe, whereas those arriving in Greece reached 851.319 people in the same year [1]. In 2016, 173.450 people arrived in Greece, whereas sea arrivals were recorded as 29.718 in 2017. Overall, currently, an estimated 51.000 refugees and migrants reside in Greece, located in both mainland (39.500 people) and the islands (11.500 people) [1]. In the first quarter of 2018, 16.640 refugees and migrants entered Europe by sea, a 106% decrease compared to the same period last year. However, at a country level, in the first quarter of 2018, 5.318 sea arrivals were recorded in Greece, a 33% increase compared to the same period in 2017. Moreover, the increase was even larger for land arrivals. 1.480 land arrivals were recorded in March 2018 at the Greece-Turkey border

J. Świątek et al. (Eds.): ISAT 2018, AISC 853, pp. 361–370, 2019.
https://doi.org/10.1007/978-3-319-99996-8_33

in Evros, an increase by seven times in 2018 compared to the same period in 2017. The situation has led to an aggravation of living conditions of already overcrowded border reception centers and increased protection risks for the refugees and migrants. Meanwhile, currently operating accommodation sites have reached full capacity [1].

The minimum standards for settlement and shelter when disasters occur include both strategic planning and settlement planning according to the SPHERE Handbook [2]. The right to adequate housing is a basic human right, while the SPHERE project [3] serves as a guide for settlement and shelter design. Strategic planning should contribute to the "security, safety, health and well-being" of displaced populations. Key actions include development of response plans in coordination with the relevant authorities, agencies and the affected population; accommodation in temporary communal settlements; ensuring safe distance from potential threats and minimization of hazards' risks; access to water, sanitation, health and education services, among others. Moreover, risks, vulnerability and hazards assessments should be undertaken. Furthermore, security issues, risks and vulnerabilities related to age, gender, as well as relationships between the affected population and host communities should be investigated. Settlement planning should aim to safe and secure use of accommodation and basic services of the affected population, However, the type of crisis or disaster affects the planning processes. Refugee siting should take into account the long-term implications of planning decisions.

A range of refugee settlement options are available such as planned camps, public buildings, reception centers, houses and apartments. Reception and Identification Centers (RICs), as well as Temporary Accommodation Sites are used in Greece. In addition, apartments and hotels are also used [1]. Minimum standards should ensure security for refugees in a healthy environment which improves their quality of life. Overcrowding results in increased morbidity and stress, therefore the camp size and average area per person are considered as basic indicators for camp settlement design. Relationships between refugees and migrants affect their satisfaction, therefore should be accounted for [4]. Refugee settlement and shelter involves many stakeholders, such as refugees and migrants, the government, host communities, civic and private sector, EU, United Nations High Commissioner for Refugees (UNHCR), national and international Non-Governmental Organizations (NGOs) and donors [5].

The uniqueness of crises and disasters, both of which are dynamic in nature and complex events with diverse impacts on populations and infrastructure has shown the applicability of Operations Research (OR) research for real-time and effective solutions [6]. Refugee siting has to consider a range of criteria and risks [3, 4]. Settlement and shelter research has focused on design, sustainability, long-term planning [2, 8] and reports on existing shelter status [9]. Multi-criteria decision making (MCDM) methods are appropriate to address refugee settlement and planning. MCDM methods are a growing research area in humanitarian aid and logistics [10]. This is mainly due to the multiple objectives of humanitarian logistics in order to achieve beneficiaries' satisfaction. Cetinkaya et al. [7] proposed a geographic information system (GIS) based multi-criteria decision analysis method for refugee camp siting. The authors identified relevant criteria including risk related criteria for ten cities in Turkey. Then, they entered criteria into a GIS. FAHP was then used to obtain weights of GIS layers and indicate alternatives. The alternative sites were finally ranked using the technique for order preference by similarity to ideal solution (TOPSIS).

Complex real-life problems in many application areas have shown that centralized solutions are inappropriate. Distributed problem solving using a MAS has been identified as a suitable approach to address these problems. Agents in the MAS are individual intelligent, autonomous, goal-oriented problem solvers which interact using communication and cooperation, in order to achieve the global system goal [11]. The individual knowledge of each agent is not enough to achieve the global objective, however each agent is capable of decision making in its local environment. Agent properties include autonomous decision making and social behavior expressed with interactions such as coordination, cooperation and negotiation. Therefore, the multiple objectives of different decision makers as well as the complexity and dynamic nature of the refugee settlement siting problem refer to an MAS approach, in which individual agents respect each other's' beliefs and goals and collaboratively decide on and optimize the solution.

Refugee settlement siting using MCDM methods in an intelligent MAS has been proposed in [12]. The agents in the MAS represented different decision makers which used two MCDM methods to rank alternative sites for refugee siting. The criteria as well as potential risks were identified and categorized by a site planner agent. In particular, the Fuzzy Analytic Hierarch Process (FAHP) was used by an agent to determine the weights of criteria, whereas the Fuzzy Axiomatic Design extended with risk factors (RFAD) was used by the respective agent to rank alternative sites for refugee settlement. However, hierarchical fuzzy TOPSIS can also be used in the MAS for the final ranking.

Therefore, in this paper, a comparative study of Fuzzy Axiomatic Design (FAD) and hierarchical fuzzy TOPSIS approaches used by intelligent agents in an MAS is done in order to identify the most effective approach for refugee settlement siting. The comparative study is applied to rank refugee sites operating in Greece. The MAS consists of five agents, namely a site planner agent, a FAHP with extent analysis agent, a supervisor agent, and an agent using the hierarchical fuzzy TOPSIS method. The MAS evaluates and ranks four available alternative refugee sites on the greenfield in Greece [1], by distributing the decision making initially to the site planner agent in order to decide on the criteria and risk factors. Then, the FAHP agent prioritizes the criteria whereas the RFAD agent ranks the alternative sites. The site planner agent after receiving the ranking of alternative sites made by the RFAD agent requests from the supervisor agent to initiate the conversation with the hierarchical fuzzy TOPSIS agent. Finally, the site planner agent decides on the final ranking of alternative sites based on a comparison of results between the two agents performing site ranking. The MAS is implemented with JADE (Java Agent Development Framework) [13].

The organization of the paper follows. A literature review on fuzzy methods used for decision making in disaster management and MAS is given in the next section. Then the proposed method is presented. Finally, the findings are presented and a comparative analysis of the MAS performance for refugee siting using the different fuzzy methods is done.

2 Literature Review

Decisions for humanitarian aid operations involve many stakeholders and are taken in an uncertain environment. Moreover, the ultimate goal of humanitarian operations is the effective aid delivery to the affected population. Refugee settlement siting shows similarities to the facility location selection problems. Whether commercial or humanitarian logistics, MCDM approaches have been widely used for this set of problems of strategic importance. Fuzzy MCDM methods have been applied to address the uncertain nature in criteria in both commercial [13–15] and humanitarian type of facility location problems [10, 16, 17]. Alternative locations are ranked against a set of quantitative and qualitative criteria, in many cases arranged in a hierarchy of criteria and sub-criteria, whereas linguistic variables are represented as triangular fuzzy numbers. Although cost minimization is the main objective in commercial facility location selection, the interests of city residents, as well as sustainability factors have been considered for urban distribution center location determination [13], and the quality of life was considered as a main criterion in [18]. In [17] socioeconomic features were introduced for warehouse location determination in humanitarian relief operations in Nepal. Roh [16] identified cooperation as the most important attribute in warehouse location determination for pre-positioning relief items, followed by national stability. Fuzzy AHP and fuzzy TOPSIS have both been used for facility location selection. A comparison of the two methods for facility selection determination has been done in [18]. AHP [19] and fuzzy AHP are used when a hierarchy of criteria exists, however the techniques are limited to a relatively low number of criteria. TOPSIS ranks alternatives using the similarity to the positive ideal solution and the negative ideal solution, whereas hierarchical fuzzy TOPSIS is used when a hierarchy of criteria is considered [20]. Fuzzy AHP and fuzzy TOPSIS have been used to prioritize and rank alternative locations, respectively, for refugee siting in [7]. The authors proposed a hierarchy of main criteria and sub-criteria, where main criteria included geographical, risk-related, infrastructural and social ones. Drakaki et al. [12] presented a MAS based approach for refugee siting, in which the agents used fuzzy MCDM to acquire intelligence acquisition. Different agents used FAHP and RFAD. In MCDM, RFAD integrates risk factors in the methodology [21], therefore risk factors were not considered as separate criteria. Final decision was made by a site planner agent.

Agents in the MAS represent different decision makers who need to interact in order to make the final decision. Different stakeholders are involved in refugee siting, such as refugees and migrants, host communities, government, UNHCR, NGOs, yet they have different and possibly conflicting goals. Under these circumstances, cooperation, coordination, as well as negotiation are necessary in order to reach a mutual decision, therefore a MAS based approach seems an appropriate solution. Human-like social behavior is a characteristic of MAS, whereas the final solution is a result of interaction between agents [22]. Moreover, MAS can operate in real-time, thus removing part of the uncertainty present in the refugee siting problem.

MAS are developed in software platforms, with dedicated agent architecture, message protocols and communication languages. JADE (Java Application Development Environment) is an established FIPA (Foundation for Intelligent Physical Agents) - compliant

MAS development software platform [22]. It supports reactive agent architectures, as well as (Belief-Desire-Intention) BDI architectures. Reactive agents acquire intelligence through interaction with their environment, whereas each agent has local knowledge of the environment. The global system goal is achieved through interaction between agents.

Applications of MAS in many areas, including manufacturing [23], diagnostics [24], energy management [25], as well as humanitarian logistics [26] can be found in the literature.

Next section explains the proposed decision support system.

3 Methodology

The Elements of the Proposed Decision Support System

Similar to the work presented in Drakaki et al. [29], the proposed decision support system in this study is also based on an intelligent MAS implemented with JADE reactive architecture. The agents use two MCDM methods such as FAHP and hierarchical TOPSIS. The elements are described in detail as follows.

The MAS

Four agents are created in the proposed MAS namely a site planner agent (SPA), a TOPSIS agent (TOPA), a fuzzy AHP agent (FAHPA) and a supervisor agent (SA). Agents interact using the FIPA request IP. Table 1 shows the agents and their goals in the proposed MAS.

Table 1. MAS agents and their respective goals.

Agent	Goal
Site planner agent (SA)	To decide the set of criteria and risk factors for site selection. To approve or disapprove the ranking of the alternatives and site selection
Supervisor agent (SA)	To coordinate and control agent interactions
FAHP agent (FAHPA)	To calculate priority weights of all criteria and sub-criteria, using FAHP
TOPSIS agent (TOPA)	To make the ranking of alternatives and site selection, using TOPSIS

The Site Planner Agent (SPA)

The SPA consults a database that contains site selection criteria collected from a range of data sources including UNHCR guidelines, the SPHERE project and literature. After discussions with all stakeholders, including government and host communities, as well as engineers, decides on the set of site selection criteria, i.e. main criteria and sub-criteria. The main and sub-criteria used in this study are given in Table 2.

The site planner receives the final ranking of alternatives from the coordinator agent and depending on the results, it approves it or initiates a new decision making cycle.

Table 2. The main and sub-criteria for the refugee siting [29]

Basic characteristics of the land (C1)
Camp settlement size (C1_1)
Drainage (C1_2)
Water availability (C1_3)
Sanitation (C1_4)
Location (C2)
Distance from major towns (C2_1)
Distance from protected areas (C2_2)
Distance from tourism attractiveness (C2_3)
Supportive factors (C3)
Accessibility to national services: health (C3_1)
Accessibility to roadway (C3_2)
Availability of electricity (C3_3)
Accessibility to national services: education (C3_4)

The FAHP Agent (FAHPA)

The FAHPA uses the FAHP method with extent analysis to prioritize the weights of criteria. A fuzzy comparison matrix contains the pairwise comparisons related to determining the weights of criteria.

AHP [19] is a process for developing a numerical score to rank each decision alternative based on how well each alternative meets the decision maker's criteria. It allows users to assess the relative weight of multiple criteria or multiple options against given criteria in an intuitive manner. In some decision problems, all data are not available so in order to deal with such a decision problem, fuzziness should be added to the solution approaches. Chang's Extent Analysis method [30] on Fuzzy AHP is used in this paper.

The FAHPA utilizes linguistic terms in building the decision matrices in this study. Similar to Gumus [31], linguistic terms have been converted to fuzzy numbers using the scale shown in Table 3.

Table 3. Comparison scale.

Linguistic terms (Abbreviation)	Triangular fuzzy numbers
Very Good (VG)	(7,9,9)
Good (G)	(5,7,9)
Preferable (P)	(3,5,7)
Weak Advantage (WA)	(1,3,5)
Equal (EQ)	(1,1,1)

Using this scale, the final weights of the main criteria obtained are shown in Table 4. Based on the results, the most important criterion in refugee siting is the basic characteristics of the land followed by supportive factors. The least important is the location. The weights of sub-criteria are calculated in the same manner.

Table 4. The final priority weights for the main criteria [29].

Main criteria	Final priority weights
Basic characteristics of the land	0.549
Location	0.113
Supportive factors	0.339

The TOPSIS Agent (TOPA)

The TOPA uses hierarchical fuzzy TOPSIS for evaluating and ranking the camp sites. Hwang and Yoon [27] have proposed TOPSIS which has been the most widely used MCDM approach. The main idea of TOPSIS is that the best or chosen alternative should be very close to the Positive Ideal Solution and far away from the Negative Ideal Solution. Therefore, this solution minimizes the cost criteria and maximizes the benefit criteria. Since the problem on hand has a hierarchy, using conventional TOPSIS might lead a wrong decision. Different from the conventional TOPSIS, hierarchical TOPSIS considers the hierarchical structure in the decision problem. More information can be found in Wang and Chan [20]. In defining the decision matrix, the linguistic scale in [20] have been utilized.

The same camp sites evaluated in Drakaki et al. [29] have also been used in this study. Four accommodation sites operating in June 2017, namely Trikala (Atlantic), Pieria (Ktima Iraklis), Kara Tepe and Souda, which are settlements on the greenfield, based on data provided by UNHCR, have been evaluated using TOPA. Using the weights of all criteria, TOPA evaluates and ranks the camp sites. The relative closeness index showing the distances from the positive and negative ideal solutions for each alternative is given in Table 5. Based on the results, Trikala is the most appropriate camp site for refuges followed by Pieria.

Table 5. The relative closeness index of alternatives along with the final ranking

	d^+	d^-	C_k	Ranking
Trikala (Atlantic)	0.030	0.296	0.910	1
Pieria (Ktima Iraklis)	0.203	0.244	0.546	2
Kara Tepe	0.183	0.122	0.399	3
Souda	0.261	0.105	0.287	4

The Agent Interaction Protocol

SPA sends the selected criteria to the SA with a request communication act. SA sends the selected criteria to the FAHPA with a request communication act for evaluation of the criteria. FAHPA sends the priority weights to the SA with an inform-result communication act. SA sends the priority weights to the TOPA with a request communication act. The TOPA sends the final site location decision and ranking to the SA with an inform-result communication act. The SA sends the final decision to the SPA with an inform-result communication act. If the SPA agrees on the site selection, the process has completed. Otherwise, it can restart the whole process.

3.1 Comparison

This section compares the results obtained using Fuzzy Axiomatic Design (FAD) and Fuzzy Hierarchical TOPSIS in evaluating the camp sites. FAD [28] is one of the MCDM approaches widely used in decision making field in recent years.

As seen in Table 6, none of the alternatives has satisfied the functional requirements when the MAS employed FAHP and FAD. Results obtained by the MAS using FAHP and hierarchical fuzzy TOPSIS present a ranking, according to which Trikala is the most appropriate refugee site location on the greenfield. However, the currently operating sites are the ones at Kara Tepe and Souda, which have been reported to operate overcrowded, at a potentially unsafe environment.

Table 6. Comparison of fuzzy MDCM approaches

Alternatives	FAHP + FAD	FAHP + Hierarchical Fuzzy TOPSIS
Trikala (Atlantic)	NA	1
Pieria (Ktima Iraklis)	NA	2
Kara Tepe	NA	3
Souda	NA	4

4 Conclusions

In this paper, an intelligent MAS is presented in decision making of refugee camp siting. Refugee camp siting involves a complex decision-making process involving different decision makers. The multi-agent system models and solves the problem by distributing related tasks to different agents representing the decision makers, until an optimal or near optimal solution is obtained. Agents in the MAS use a hybrid MCDM method based on FAHP and TOPSIS. They are implemented with JADE. A coordinator agent supervises and controls agent interaction and communicates with the site planner agent, initially to receive the list of criteria, and finally to deliver the ranking of alternatives for approval. The learning agents, i.e. FAHPA and TOPA, acquire knowledge by employing FAHP and TOPSIS respectively. The final normalized weights of criteria are calculated by the FAHP agent and transferred to the coordinator agent. The coordinator agent forwards the data to the TOPSIS agent that calculates the ranking of the alternatives. The procedure can be repeated until the site planner agent approves the results. The proposed method has been applied to evaluate four alternative used accommodation sites in Greece and the results obtained have been reasonable.

References

1. UNHCR data portal (2017). https://data2.unhcr.org. Accessed 17 Nov 2017
2. Moore, B.: Refugee settlements and sustainable planning. Forced Migr. Rev. **55**, 5–7 (2017)
3. SPHERE (2011): Sphere Project, Sphere Handbook: Humanitarian Charter and Minimum Standards in Disaster Response (2011). http://www.ifrc.org/docs/idrl/I1027EN.pdf. Accessed 20 Nov 2017

4. UNHCR emergency handbook: UNHCR Handbook for Emergencies. https://www.unicef.org/emerg/files/UNHCR_handbook.pdf. Accessed 20 Nov 2017
5. UNHCR settlement shelter, 2017, Global Strategy for Settlement and Shelter, A UNHCR strategy 2014-2018. http://www.unhcr.org/530f13aa9.pdf. Accessed 20 Nov 2017
6. Altay, N., Green III, W.G.: OR/MS research in disaster operations management. Eur. J. Oper. Res. **175**, 475–493 (2006)
7. Çetinkaya, C., Özceylan, E., Erbaş, M., Kabak, M.: GIS-based fuzzy MCDA approach for siting refugee camp: a case study for southeastern Turkey. Int. J. Disaster Risk Reduct. **18**, 218–231 (2016)
8. Terne, M., Karlsson, J., Gustafsson, C.: The diversity of data needed to drive design. Forced Migr. Rev. **55**, 25–26 (2017)
9. Wain, J.F.: Shelter for refugees arriving in Greece, 2015-17. Forced Migr. Rev. **55**, 20–22 (2017)
10. Gutjahr, W.J., Nolz, P.C.: Multicriteria optimization in humanitarian aid. Eur. J. Oper. Res. **252**, 351–366 (2016)
11. Wooldridge, M., Jennings, N.R.: Intelligent agents: theory and practice. Knowl. Eng. Rev. **10**, 115–152 (1995)
12. Drakaki, M., Goren, H.G., Tzionas, P.: An intelligent multi-agent system using fuzzy analytic hierarchy process and axiomatic design as a decision support method for refugee settlement siting. In: Proceedings of ICDSST-PROMETHEE 2018. Lecture Notes in Business Information Processing, LNBIP. Springer (2018)
13. Awasthi, A., Chauhan, S.S., Goyal, S.K.: A multi-criteria decision making approach for location planning for urban distribution centers under uncertainty. Math. Comput. Modell. **53**, 98–109 (2011)
14. Özcan, T., Çelebi, N., Sakir, E.: Comparative analysis of multi-criteria decision making methodologies and implementation of a warehouse location selection problem. Expert Syst. Appl. **38**, 9773–9779 (2011)
15. Kahraman, C., Ruan, D., Dogan, I.: Fuzzy group decision-making for facility location selection. Inf. Sci. **157**, 135–153 (2003)
16. Roh, S.-Y., Jang, H.-M., Han, C.-H.: Warehouse location decision factors in humanitarian relief logistics. Asian J. Shipp. Logist. **29**(1), 103–120 (2013)
17. Maharjana, R., Hanaoka, S.: Warehouse location determination for humanitarian relief distribution in Nepal. Transp. Res. Procedia **25**, 1151–1163 (2017)
18. Ertugrul, I., Karakasoglu, N.: Comparison of fuzzy AHP and fuzzy TOPSIS methods for facility location selection. Int. J. Adv. Manuf. Technol. **39**, 783–795 (2008)
19. Saaty, T.L.: The Analytic Hierarchy Process. McGraw-Hill, New York (1980)
20. Wang, X., Chan, H.K.: A hierarchical fuzzy TOPSIS approach to assess improvement areas when implementing green supply chain initiatives. Int. J. Prod. Res. **51**(10), 3117–3130 (2013)
21. Gören, H.G., Kulak, O.: A new fuzzy multi-criteria decision making approach: extended hierarchical fuzzy axiomatic design approach with risk factors. Lecture Notes in Business Information Processing, LNBIP, vol. 184, pp. 141–156. Springer (2014)
22. Bellifemine, F.L., Caire, G., Greenwood, D.: Developing Multi-Agent Systems with JADE. Wiley, Chichester (2007)
23. Leitao, P.: Agent-based distributed manufacturing control: a state-of-the-art survey. Eng. Appl. Artif. Intell. **22**, 979–991 (2009)

24. Drakaki, M., Karnavas, Y.L., Chasiotis, I.D., Tzionas, P.: An intelligent multi-agent system framework for fault diagnosis of squirrel-cage induction motor broken bars. In: Świątek J., Borzemski L., Wilimowska Z. (eds.) Proceedings of 38th International Conference on Information Systems Architecture and Technology – ISAT 2017. Advances in Intelligent Systems and Computing, vol. 656. Springer, Cham (2018)
25. Dou, C.X., Wang, W.Q., Hao, D.W., Li, X.B.: MAS-based solution to energy management strategy of distributed generation system. Electr. Power Energy Syst. **69**, 354–366 (2015)
26. Edrissi, A., Poorzahedy, H., Nassiri, H., Nourinejad, M.: A multi-agent optimization formulation of earthquake disaster prevention and management. Eur. J. Oper. Res. **229**, 261–275 (2013)
27. Hwang, C.L., Yoon, K.P.: Multiple Attribute Decision Making: Methods and Applications, A State-of-the-Art Survey. Springer-Verlang, Berlin (1981)
28. Kulak, O., Kahraman, C.: Multi-attribute comparison of advanced manufacturing systems using fuzzy vs. crisp axiomatic design approach. Int. J. Prod. Econ. **95**(3), 415–424 (2005)
29. Drakaki, M., Goren, H.G., Tzionas, P.: An intelligent multi-agent based decision support system for refugee settlement siting. Int. J. Disaster Risk Reduct. **31**, 576–588 (2018)
30. Chang, D.Y.: Applications of the extent analysis method on fuzzy AHP. Eur. J. Oper. Res. **95**(3), 649–655 (1996)
31. Gumus, A.T.: Evaluation of hazardous waste transportation firms by using a two-step fuzzy-AHP and TOPSIS methodology. Expert Syst. Appl. **36**(2), 4067–4074 (2009)

Selected Aspects of Crossover and Mutation of Binary Rules in the Context of Machine Learning

Bartosz Skobiej and Andrzej Jardzioch(✉)

West Pomeranian University of Technology, Szczecin,
17 Piastow Avenue, 70-310 Szczecin, Poland
andrzej.jardzioch@zut.edu.pl

Abstract. The study focuses on two operators of a genetic algorithm (GA): a crossover and a mutation in the context of machine learning of fuzzy logic rules. A decision support system (DSS) is placed in a simulation environment created in accordance with the complex adaptive system (CAS) concept. In a multi-agent CAS system, the learning classifier system (LCS) paradigm is used to develop a learning system. The aim of the learning system is to discover binary rules that allow an agent to perform efficient actions in a simulation environment. The agent's objective is to make an effective decision on which order, from the set of the awaiting orders, should be transferred into a production zone next. The decision is based on the fuzzy logic system response. In the conducted study, two input signals and one output signal of the fuzzy logic system are considered. The concept of the presented fuzzy logic system affects the construction of rules of a specific agent. The paper focuses on the problem of coding the agent's rules and modification of the coding by the GA.

Keywords: Crossover · Mutation · Binary rules · Machine learning

1 Introduction

1.1 Complex Adaptive System (CAS)

The CAS term has been present in the modern science since early 1990s. As Holland describes it in his book "Hidden Order: How Adaptation Builds Complexity" [1], the city of New York is a system that exists in a steady state of operation, made up of "buyers, sellers, administrations, streets, bridges, and buildings [that] are always changing. Like the standing wave in front of a rock in a fast-moving stream, a city is a pattern in time" and can be regarded as a CAS itself. In [1], Holland also proposed a use of agents that are controlled by rules and work simultaneously, but sometimes against each other. According to that concept, every agent is able to make independent or partially independent decisions that result in a macroscopic behavior of a complex system. A dynamic network of interactions of agents causes an inability to predict the behavior of a CAS [2]. In the light of the above description, an agent is regarded as a singular component of an examined system. A practical implementation of Holland's CAS concept can be found in [3], where the authors use a genetic programming (GP) to

J. Świątek et al. (Eds.): ISAT 2018, AISC 853, pp. 371–381, 2019.
https://doi.org/10.1007/978-3-319-99996-8_34

optimize agent's behavior in a dynamic logistic system. Figure 1 shows a schematic diagram of the examined CAS of a production line built on the basis of Holland's concept and employing a multi-agent approach.

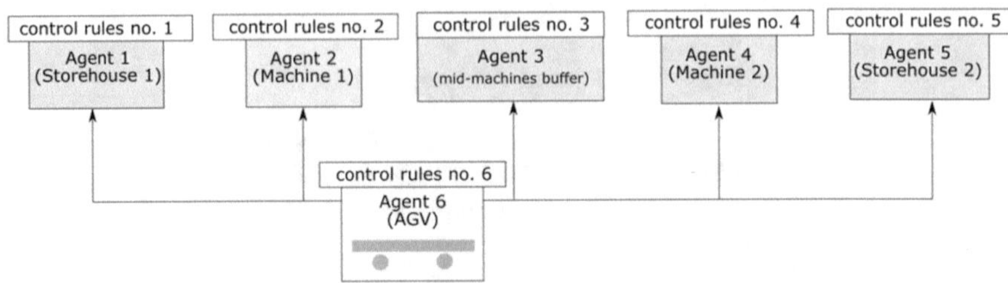

Fig. 1. A schematic diagram of the examined CAS of a production line, built on basis of John Holland's concept, employing a multi-agent approach. (source: authors)

In the study conducted, the examined agent is a decision support sub-system that can be placed in a Storehouse 1 – Agent 1 in Fig. 1, or can be implemented as a part of an automated guided vehicle (AGV) – Agent 6 in Fig. 1. The primary objective of the examined agent is to select an order from awaiting orders so as to transfer it into a production zone.

The investigated CAS represents a two-machine flow-shop system with an unconstrained buffer between the machines and an AGV of capacity of one order. A similar model of a production system was used by the authors in [4] for investigating a job scheduling problem using other methods.

1.2 Learning Classifier System (LCS) Paradigm

According to the Merriam-Webster English Dictionary and the Oxford English Dictionary, a "paradigm" can be described as a pattern, an example and a model. However, in the field of theory of science, it is assumed that a modern definition of a paradigm word was introduced by Kuhn in his book "The Structure of Scientific Revolutions" [5]. Kuhn characterizes a scientific paradigm as "universally recognized scientific achievements that, for a time, provide model problems and solutions for a community of practitioners". It seems that the paradigm definition cited above is not an algorithm or a method – rather, it is a group of definitions, methods and solutions that are regarded as a foundation of selected branch of science.

One of the most important features of a paradigm is its ability to change. Such a change usually occurs as a result of experimentum crucis, then a new theory emerges which contradicts the old theory, or complements the old one. Considering the above statement and the dynamic development of the field of science discussed in this work, the theories and concepts contained in the paradigm should be treated with some precaution despite their positive verification as of today.

The origins of the LCS concept can be found in the book of Holland [1] as well as in many modern papers discussing LCS [6, 7] or CAS [8, 9]. As follows from the works cited above, the concept of the LCS paradigm emerged at the beginning of the 21st century. Of the basic literature available, two reference papers deserve attention, i.e. "Learning Classifier Systems: Then and Now" by Piera Luci Lanzi [10] and "Learning Classifier Systems: A Complete Introduction, Review, and Roadmap" by Urbanowicz and Moore [11].

On the basis of the definition from [11], the LCS paradigm defines machine learning methods of rule systems that combine the discovery component – usually a genetic algorithm (GA) – and a learning component represented by supervised learning, reinforcement learning (RL) or unsupervised machine learning. Such a broad definition of the LCS paradigm results from the fact that a paradigm is not a single algorithm or method by nature. The very application of the genetic algorithm paradoxically prevents the use of the word algorithm in this case, which comply with that alternative methods of constructing this algorithm may be used. A more precise term would be a set of methods and techniques, some of which are optional, and some may even be omitted when constructing a learning system according to the LCS paradigm. Therefore, it is the researcher's responsibility to determine which components should be present in the structure of the learning system. On the one hand, it gives the researcher a huge set of possibilities; on the other hand, it requires some experience and knowledge in the field of operation and application of selected components of the learning system.

The general LCS scheme was presented, for example, by Urbanowicz and Moor in [11]. However, it is interpreted differently depending on the practical implementation of the LCS model [12–14]. One of the simplified interpretations of the LCS paradigm was introduced in [15] and is presented in Fig. 2.

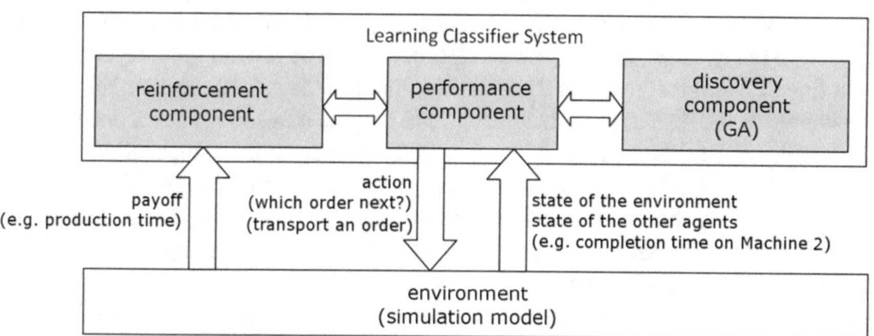

Fig. 2. Simplified schematic diagram of LCS paradigm. (source: adapted from [15])

When analyzing the diagram presented in Fig. 2, one can specify the subsequent stages of LCS operations, which are implemented in the machine learning system:

a. The genetic algorithm creates a population of sets of rules controlling an agent;
b. Before making a decision, the examined agent sends a query to the simulation model and receives selected model parameters (e.g. completion time of the current task on Machine 2);

c. On the basis of the examined control rules and with the use of fuzzy logic, the agent decides which order should go into the production zone;
d. Depending on the decision made, the agent may be rewarded (high quality of the assessed value) or may be punished (low quality of the assessed value);
e. The learning cycle (points b to d) lasts until the number of pending orders reaches the value of 1 (the last order does not need to be evaluated);
f. The final rating for the agent is calculated (in fact, it is the assessment of the agent's rules set);
g. Unless all agents have been tested, the genetic algorithm starts the procedure of testing the next set of rules (the next instant of the agent) - return to point b;
h. The ratings of all agents (all rules sets) are the starting point for the genetic algorithm to initiate the selection procedure.

2 Fuzzy Logic Decision Support System

The primary objective of the decision support agent is to select an order from the awaiting orders and to introduce it into the production zone. The key performance index (KPI) in the simulation model of production zone is defined as minimization of production time (makespan). Decisions made by the agent, will eventually build a production schedule of orders. The quality of obtained schedule, according to the defined KPI, is measured by the overall production time.

There are two input signals identified in presented fuzzy logic system (see Fig. 3.). The first one is named Input Signal 1 and it is a completion time of an awaiting order on Machine 1. For the purpose of building a ranking of orders, every awaiting order has to be analyzed by the agent. As a consequence, the examination of the orders is initiated as many times, as many awaiting orders are. The second input signal is named Input Signal 2 and it consists of dynamically imported data from the simulation model. The import of the data is performed each time there is a need to make a decision as to which order should be delivered into the production zone next. Input Signal 2 is the sum of current completion time on Machine 2 and the sum of completion times of all orders awaiting in the mid-machine buffer on Machine 2 (see Fig. 1.). A visualization of exemplary orders, selected states of the model and corresponding input signals are shown in Table 1. Each order has two attributes: completion time on Machine 1 and completion time on Machine 2. The number of input signals and the list of the attributes for each order are considered as the minimal but efficient number of attributes. The above mentioned hypothesis is assumed on basis of the input data range for the Johnson's algorithm [16], which is known to produce optimal solutions for two machine flow-shop systems. It is also worth mentioning that the agent's decision is made only in the situation when Machine 1 is empty in the simulation model.

The behavior of the agent is determined by its chromosome. The chromosome activates selected genes and determines the decision making process. If the genes of poor quality activate selected rules, the overall production time will not fulfill the expectations and an evaluation of the agent will produce poor results. The machine

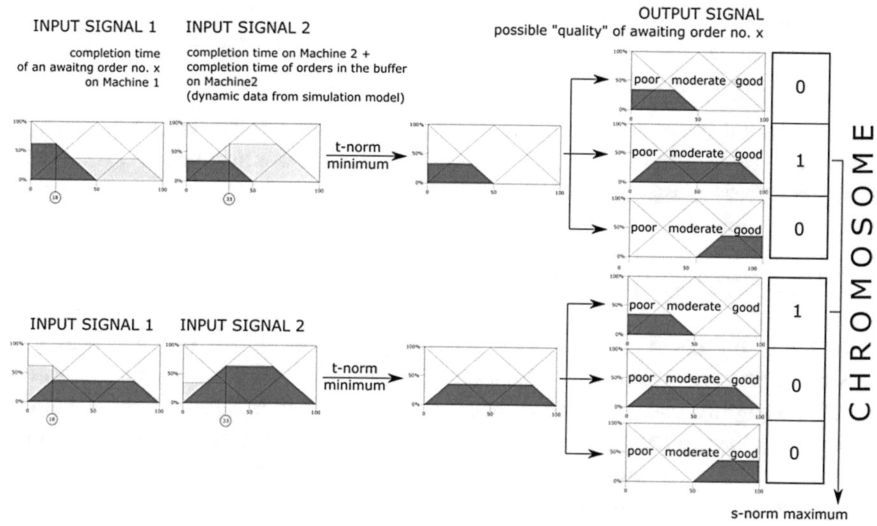

Fig. 3. Schematic diagram of exemplary fuzzy logic decision support system with chromosome coding. (source: authors)

Table 1. Exemplary orders, selected states of the model and the corresponding input signals.

Order examined (completion time on Machine 1)	Data from the simulation model		Input signal 1	Input signal 2
	Completion time of orders in the buffer on Machine 2	Current completion time on Machine 2		
60	0	23	60	23
80	20	15	80	35
25	10	5	25	15

learning process is thereby used to investigate various chromosomes in order to understand general rules controlling the best agents.

In the study presented, the unparalleled construction of a chromosome (see Fig. 3) is based on binary coding, where one pair of input signals results in three genes constituting the output. Behind the concept of chromosome construction, there is a basic truth that one combination of input signals shall result in one response signal of the system. The above mentioned response can be observed in one of the three fuzzy sets: poor, moderate or good. At first, the agent is unable to determine which pair of input signals is "good" and which is "poor". However, by means of a simulation process, the agent is able to determine the KPI value needed to evaluate agent's behavior, which is the quality of the decision made with the use of fuzzy logic rules (chromosomes). Every rule in a chromosome represents conflicting hypothesis, for example:

- IF Input signal 1 = “low” and Input signal 2 = “mid” THEN result = “poor”;
- IF Input signal 1 = “low” and Input signal 2 = “mid” THEN result = “moderate”;
- IF Input signal 1 = “low” and Input signal 2 = “mid” THEN result = “good”.

The total length of each chromosome equals 27 bits, grouped in 9 structures of 3 alleles of possible system responses. Such make-up of a chromosome poses a challenge for the discovery component (GA).

3 Discovery Component – GA

In the study presented, the GA as a part of LCS is developed with the use of the classical approach. There is a selection process based on rank selection, an elite function, a single crossover point on both parents described in Sect. 3.1 and a mutation process described in Sect. 3.2. All the GA parameters, e.g. number of elite individuals, crossover technique or mutation probability, are set by a user via a website interface of the machine learning system.

3.1 Crossover

A single GA crossover point operator employed in the conducted study is widely used and discussed in literature [17, 18]. However, as far as the problem described in this paper is concerned, it is not a ready-to-go solution. The principal rule that one pair of input signals results in one system response signal is a crucial element of the system. As it is explained in Sect. 2, a chromosome consists of 9 triplets of output signals. A random place cut-off procedure of a single point crossover may result in an error in the coding of a child’s chromosome (see Fig. 4).

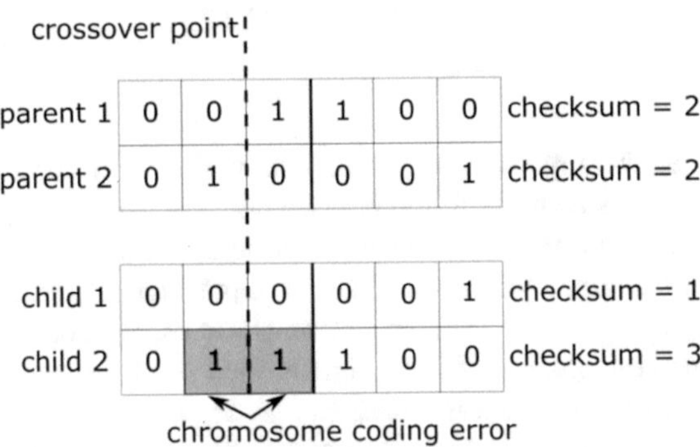

Fig. 4. Example of a single crossover point operation resulting in an error. (source: authors)

The checksum function implemented in the developed software calculates the sum of all non-zero genes of a chromosome to verify a crossover and mutation operations. Since there are 9 pairs of incoming signals, there must be 9 outputs of the system, each coded as a gene value 1. Therefore, the checksum must always equal 9. In order to adapt a single point crossover operator, the cut-off point cannot be set "fully" randomly. As a result, two possibilities of performing a valid crossover emerged.

The first possibility is to leave a random point pick procedure (number between 3 and 25) as it is and to add a modulo condition to check – if the random number of a gene in chromosome modulo 3 equals 0, then the crossover point is correct. If not, check modulo of the random number – 1. If not, check modulo of the random number – 2. The modulo check procedure is performed 3 times for every crossover procedure at the most.

The second possibility to perform a valid crossover is to pick a number between 1 and 8 randomly (0 and 9 are at the ends of a chromosome) and multiply the number by 3. In other words, the additional coding of a chromosome is introduced and then converted to the number in a domain of 27 bit chromosome coding. The above mentioned procedures are not the only ones that can be implemented in such case. Nevertheless, the first solution was chosen for implementation. The selected pre-crossover procedure consists of 3 steps:

a. Pick a random integer number between 3 and 25;
b. If random number mod 3 = 0, perform crossover; if not go to c);
c. Random number – 1, go to b).

3.2 Mutation

Among many techniques of mutating of a binary chromosome, two of them seem to be commonly used – an inverse mutation and a swap mutation [19–21]. Both of them need to be investigated critically before implementation. As regards the inverse mutation, there is a possibility of damaging an individual by means of the improper execution of the mutation technique (see Fig. 5).

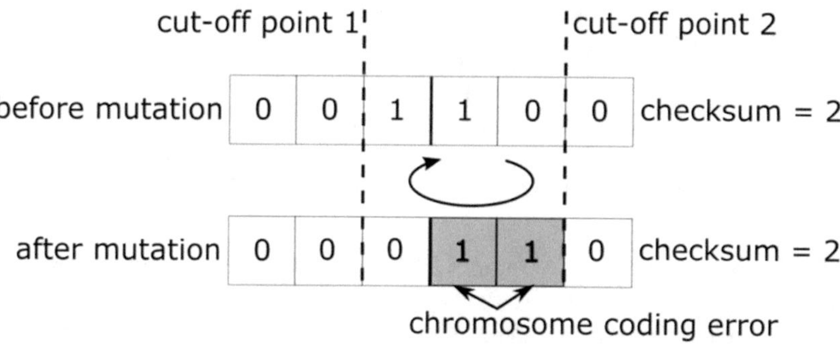

Fig. 5. Example of an inverse mutation resulting in an error. (source: authors)

Despite the possible difficulties in the implementation of the inverse mutation, the presented mutation technique can be regarded as a useful one. The only factor in the successful usage of the inverse mutation is to determine correct cut-off points. The above mentioned problem is solved and described in Sect. 3.1. However, one should keep in mind that even a successful implementation of the inverse mutation may in some cases bring about unexpected effects, as shown in Fig. 6.

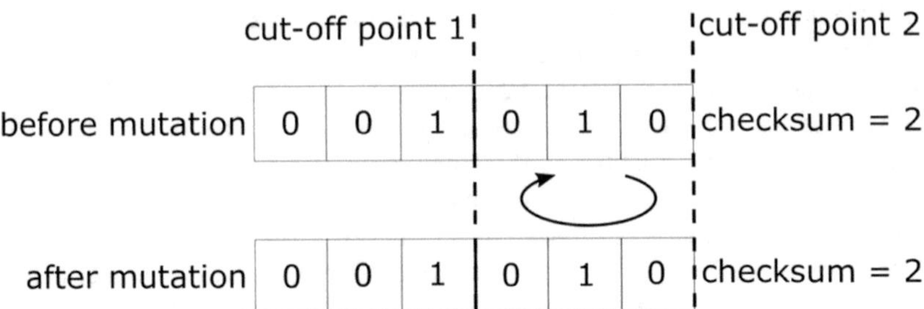

Fig. 6. Example of an unexpected effect of the inverse mutation. (source: authors)

As far as the swap mutation is concerned, there are two possible options: swapping two genes, or swapping two or more 3-allele sections. Both approaches seem problematic to apply. When swapping two genes, one expects to change a chromosome. Unfortunately, it is impossible to swap 0 with 1, or 1 with 0 not causing an error in chromosome coding (see Fig. 7.), and obviously, it is the only way to change a chromosome. All possible swaps of 0 with 0 and 1 with 1 cause no difference in a chromosome.

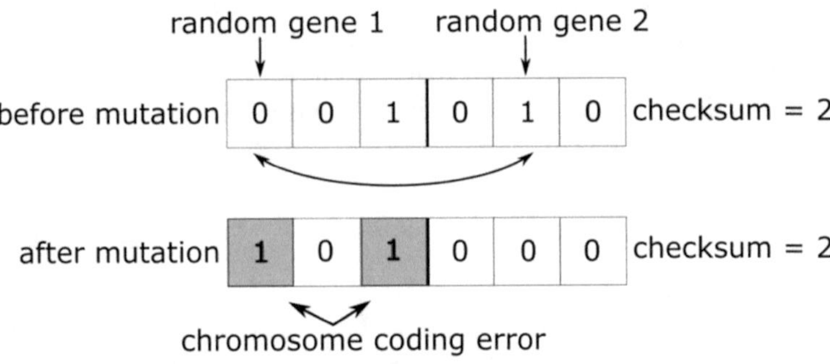

Fig. 7. Example of a swap mutation resulting in an error. (source: authors)

The idea of swapping two or more 3-allele sections appears tempting. The possible advantage of such a mutation is the lack of chromosome coding errors, since no changes to 3-allele sections are made. Still, there are also two possible disadvantages. First of all, when two 3-allele sections have the same bit sequence, no change is observed in a chromosome. Theoretically, the probability that such a situation will take place is rather high and equals ca. 33%. The second disadvantage is an extensive usage of the computational technique described in Sect. 3.1 to identify a beginning of 3-allele section in a chromosome. The example of a correct swap mutation procedure performed on 3-allele sections is shown in Fig. 8.

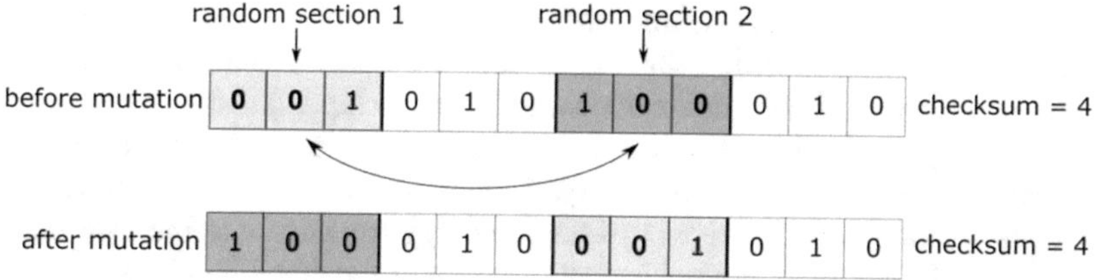

Fig. 8. Example of a swap mutation successfully performed on two 3-allele sections. (source: authors)

Since the mutation operators are heavily dependent on the computational technique to identify the beginning of 3-allele sections, the authors decided to investigate another approach to the mutation problem. The identification of a 3-allele section is indispensible, but in the authors' concept, it is limited to one run. While the 3-allele section is read and cut-off from a chromosome, two random operations are possible: shifting all genes one locus left or shifting all genes one locus right (see Fig. 9).

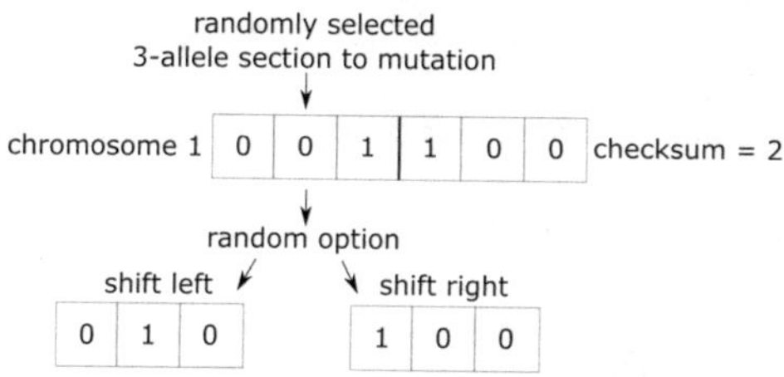

Fig. 9. Example of the shift mutation used in the study. (source: authors)

In programming terms, a gene from the beginning of an array is taken and pushed to the end of an array, moving all genes one locus left, or a gene at the end of an array is popped out and placed to the front of an array, moving all genes one locus right. From the authors' perspective, the selected mutation method constitutes an interesting alternative for the demanding computational process of machine learning.

4 Conclusion

The study presented focuses on two GA operators – a crossover and a mutation. The operators act in a demanding computational environment of a machine learning system. Therefore, one of the major constraints considered by the authors is the use of techniques of low computational costs. Another challenge is posed by a specific chromosome construction, which makes it difficult or even impossible to implement the well-known crossover and mutation techniques. From many available methods, the authors decided to use a modified, single point crossover operator and a shift mutation which is based on the frameshift mutation in genetics. In order to present the scope of the study conducted, the CAS and the machine learning system based on the LCS paradigm are introduced. Within the machine learning system, the decision support system and its rules are described, and their non-classical features are highlighted.

As for today, the authors confirm that GA techniques presented in this paper and selected for implementation fulfill the expectations. In the course of machine learning system examination, it was noticed that the discovery component of the LCS performs relatively fast. Ipso facto, the need for optimization of the other components emerged.

References

1. Holland, J.: Hidden Order: How Adaptation Builds Complexity. Basic Books, New York (1995)
2. Miller, J., Page, S.: Complex Adaptive Systems: An Introduction to Computational Models of Social Life. Princeton University Press, Princeton (2007)
3. Lon, Rv, Branke, J., Holvoet, T.: Optimizing agents with genetic programming: an evaluation of hyper-heuristics in dynamic real-time logistics. Genet. Program Evolvable Mach. **19**, 93–120 (2018)
4. Jardzioch, A., Skobiej, B.: Job scheduling problem in a flow shop system with simulated hardening algorithm. In: Advances in Manufacturing. Lecture Notes in Mechanical Engineering, pp. 101–109 (2018)
5. Kuhn, T.S.: The Structure of Scientific Revolutions. University of Chicago Press, Chicago (1962)
6. Booker, L., Goldberg, J.H.D.: Classifier systems and genetic algorithms. Artif. Intell. **40**(1–3), 235–282 (1989)
7. Lanzi, P.: Learning classifier systems from a reinforcement learning perspective. Soft. Comput. **6**(3–4), 162–170 (2002)
8. McCarthy, I., Tsinopoulos, C., Allen, P., Rose-Anderssen, C.: New product development as a complex adaptive system of decisions. J. Prod. Innov. Manag. **23**(5), 437–456 (2006)
9. Holland, J.: Complex Adaptive Systems. Daedalus **121**(1), 17–30 (1992)
10. Lanzi, P.: Learning classifier systems: then and now. Evol. Intell. **1**(1), 63–82 (2008)
11. Urbanowicz, R., Moore, J.: Learning classifier systems: a complete introduction, review, and roadmap. J. Artif. Evol. Appl. **2009**, 1–25 (2009)
12. Zhong, Y., Wyns, B., Keyser, R., Pinte, G.: An implementation of genetic-based learning classifier system on a wet clutch system. In: 14th Applied Stochastic Models and Data Analysis Conference, Rome (2011)
13. Holmes, J., Sager, J.: Rule discovery in epidemiologic surveillance data using EpiXCS: an evolutionary computation approach. In: Artificial Intelligence in Medicine. Lecture Notes in Computer Science, vol. 3581, pp. 444–452 (2005)
14. Bull, A., Sha'Aban, J., Tomlinson, A., Addison, J., Heydecker, B.: Towards distributed adaptive control for road traffic junction signals using learning classifier systems. In: Applications of Learning Classifier Systems. Studies in Fuzziness and Soft Computing, vol. 150, pp. 279–299 (2004)
15. Wasilewska, K., Seredyński, F.: Learning classifier systems: a way of reinforcement learning based on evolutionary techniques. In: Algorytmy Ewolucyjne i Optymalizacja Globalna, Warszawa (2006)
16. Johnson, D.B.: Efficient algorithms for shortest paths in sparse networks. J. ACM **24**(1), 1–13 (1977)

17. Kellegoz, T., Toklu, B., Wilson, J.: Comparing efficiencies of genetic crossover operators for one machine total weighted tardiness problem. Appl. Math. Comput. **199**, 590–598 (2008)
18. Reeves, C.R., Rome, J.E.: Genetic Algorithms Principles and Perspectives. Kluwer Academic Publishers, Dordrecht (2003)
19. Chieng, H.H., Wahid, N.: A performance comparison of genetic algorithm's mutation operators in n-cities open loop travelling salesman problem. In: Recent Advances on Soft Computing and Data Mining. Advances in Intelligent Systems and Computing, vol. 287 (2014)
20. Ryan, E., Azad, R., Ryan, C.: On the performance of genetic operators and the random key representation. In: Genetic Programming. EuroGP 2004. Lecture Notes in Computer Science, vol. 3003 (2004)
21. Maheswaran, R., Ponnambalam, S.: An intensive search evolutionary algorithm for single-machine total-weighted-tardiness scheduling problems. Int. J. Adv. Manuf. Technol. **26**(9–10), 1150–1156 (2005)

Author Index

J. Świątek et al. (Eds.): ISAT 2018, AISC 853, pp. 383–384, 2019.
https://doi.org/10.1007/978-3-319-99996-8

Printed in the United States
By Bookmasters